Biomedical Materials for Multi-functional Applications

Series Editors

Arnab Chanda, Centre for Biomedical Engineering, Indian Institute of Technology Delhi, New Delhi, Delhi, India

Sarabjeet Sidhu, Department of Mechanical Engineering, Sardar Beant Singh State University, Gurdaspur, Punjab, India

The series 'Biomedical Materials for Multi-functional Applications' is intended for the audience with an interest in biomaterials for various applications, including bio-implants, bio-instrumentation, biomechanics, and tissue engineering. The book titles under this series will bring different areas of biomedical engineering, bioengineering, materials science, mechanical engineering, and biosciences under one umbrella of biomedical materials. The books in this series will provide deep insights into the state-of-art of biomedical materials and deal with diverse and interdisciplinary aspects of biomaterials, ranging from basic science to host responses, as well as real-time applications. Some of the topics include but are not limited to: functional materials, smart materials, biomimetic materials, polymers, composites, ceramics, metals, alloys, and nanomaterials. This series will provide the most recent advances in biomedical materials, including design, synthesis, processing, and characterization. This series will accept both authored and edited volumes, including monographs, textbooks, reference works, and handbooks. This series aims to publish new advances in biomedical materials that have been made in the fabrication and development of biomaterials for different applications and the advancement of principles, theories, and designs. Each book title in the series will give a comprehensive overview of pioneering topics, and the readers will gain new knowledge and insights about the topic. As a collection, the series on 'Biomedical Materials for Multi-functional Applications' is anticipated to benefit a wide audience in academia, the research community, and industry.

Deepa Suhag

Extracellular Matrix Biomimicking for Theranostics

Springer

Deepa Suhag
Amity University
Gurgaon, Haryana, India

ISSN 2731-9695 ISSN 2731-9709 (electronic)
Biomedical Materials for Multi-functional Applications
ISBN 978-981-95-7899-3 ISBN 978-981-95-7900-6 (eBook)
https://doi.org/10.1007/978-981-95-7900-6

This Springer imprint is published by the registered company Springer Nature Singapore Pte Ltd.
The registered company address is: 152 Beach Road, #21-01/04 Gateway East, Singapore 189721, Singapore

Preface

The extracellular matrix (ECM) is no longer seen as a mere structural scaffold but is now widely recognized as a dynamic and bioactive component of cellular microenvironments, influencing a vast range of physiological and pathological processes. In recent decades, a profound shift has occurred in how scientists, clinicians, and biomedical engineers perceive and interact with the ECM—ushering in an era of biomimetic innovations that bridge biology, materials science, and medicine.

"Extracellular Matrix Biomimicking for Theranostics: Cellular Microenvironment Dynamics in Biomedicine" is a comprehensive exploration of the ECM's evolving role in health and disease and its transformative potential in diagnostics and therapeutics. This book is born out of a desire to unify the rapidly growing demand of knowledge on ECM biology with cutting-edge technological advancements in biomaterials and regenerative medicine. It offers an integrated view of ECM dynamics—from molecular interactions and remodeling mechanisms to real-world medical applications.

The organization of the book follows a logical progression, starting with foundational aspects of ECM structure and function, progressing through cellular interactions, pathological remodeling, and culminating in advanced applications such as tissue engineering, biofabrication, and ECM-inspired diagnostics. Special attention is given to the translational potential of ECM biomimicry, with dedicated chapters on clinical case studies, regulatory perspectives, and future innovations.

This text is designed for researchers, clinicians, graduate students, and biomedical innovators seeking a deep understanding of the ECM's role in theranostics. Whether you are exploring cell-ECM communication pathways or engineering ECM-like scaffolds for clinical use, this book serves as both a reference and an inspiration.

It is my hope that *Extracellular Matrix Biomimicking for Theranostics* serves as both a knowledge repository and a source of inspiration for those passionate about the cellular microenvironment. As our ability to replicate and manipulate ECM components grows, so does our potential to revolutionize health care—offering smarter diagnostics, more effective treatments, and ultimately, a deeper understanding of life itself.

Gurgaon, India Dr. Deepa Suhag

Acknowledgements This book would not have been possible without the invaluable contributions of those who dedicated their expertise and time. I am particularly grateful to my Project Research Scientist Ms. Raksha Rathore, whose tireless efforts have enriched the content of this monograph.

I would also like to acknowledge Dr. Arnab Chanda for his deep insightful input throughout the development of this work.

Additionally, I owe a special note of gratitude to Dr. Swati Meherishi for her unwavering support and expert advice. Her contributions not only enhanced the scholarly rigor of this work but also made the publication process remarkably smooth.

Thank you all for your dedication and support, which have been vital in bringing this project to fruition.

Dr. Deepa Suhag

Competing Interests The author has no competing interests to declare that are relevant to the content of this manuscript.

Contents

1 Introduction to Extracellular Matrix (ECM) Dynamics 1
1.1 Introduction 1
1.2 Importance of ECM Dynamics in Biological Systems 2
1.3 Structure of ECM 3
1.3.1 Composition of ECM Components 4
1.3.2 Organization and Functionality 6
1.4 Cell-ECM Interactions 8
1.4.1 Receptors and Signaling Pathways 8
1.4.2 Role in Cellular Adhesion and Migration 9
1.5 ECM Remodeling 10
1.5.1 Mechanisms and Regulation 11
1.5.2 Physiological and Pathological Implications 13
1.6 Techniques for Studying ECM Dynamics 14
1.7 Clinical Relevance 15
1.7.1 ECM Dynamics in Disease Pathogenesis 16
1.7.2 Therapeutic Targeting of ECM 18
1.8 Future Directions and Challenges 20
1.9 Summary 21
References 22

2 Cellular Interactions with the ECM 25
2.1 Introduction 25
2.2 Structural Components of ECM 26
2.2.1 Collagen 27
2.2.2 Elastin 28
2.2.3 Proteoglycans 29
2.2.4 Glycoproteins 30
2.3 Cellular Receptors and ECM 32
2.3.1 Integrins 32
2.3.2 Discoidin Domain Receptors (DDRs) 33

2.3.3 Syndecans ... 34
2.4 Mechanisms of Cellular Adhesion ... 36
2.4.1 Focal Adhesions ... 36
2.4.2 Hemidesmosomes ... 37
2.4.3 Cadherin-Mediated Adhesion ... 38
2.5 ECM Remodeling ... 40
2.5.1 Matrix Metalloproteinases (MMPs) ... 40
2.5.2 Tissue Inhibitors of Metalloproteinases (TIMPs) ... 41
2.5.3 Role in Tissue Repair and Pathology ... 42
2.6 Cellular Signaling Through ECM ... 44
2.6.1 Growth Factors and Cytokines ... 44
2.6.2 Signal Transduction Pathways ... 45
2.7 Cell Migration in ECM ... 46
2.7.1 Mesenchymal Migration ... 47
2.7.2 Amoeboid Migration ... 48
2.7.3 Role in Development and Disease ... 49
2.8 ECM and Stem Cell Niche ... 50
2.8.1 Role of ECM in Stem Cell Differentiation ... 51
2.8.2 Engineering ECM for Stem Cell Therapies ... 52
2.9 Future Perspectives ... 54
2.10 Summary ... 54
References ... 57

3 Mechanisms of ECM Remodeling ... 61
3.1 Introduction ... 61
3.2 Importance of ECM Remodeling in Physiology ... 62
3.3 Cellular Components Involved in ECM Remodeling ... 63
3.3.1 Fibroblasts and Myofibroblasts ... 64
3.3.2 Macrophages and Immune Cells ... 65
3.4 Biochemical Processes in ECM Remodeling ... 67
3.4.1 Matrix Metalloproteinases (MMPs) ... 67
3.4.2 Tissue Inhibitors of Metalloproteinases (TIMPs) ... 69
3.5 Biophysical Mechanisms of ECM Remodeling ... 70
3.5.1 Mechanical Forces and ECM Adaptation ... 71
3.5.2 Role of Cytoskeletal Dynamics ... 72
3.6 Role of Growth Factors and Signaling Pathways ... 73
3.6.1 Transforming Growth Factor-Beta (TGF-B) ... 74
3.6.2 Integrin Signaling ... 75
3.7 Pathological ECM Remodeling ... 76
3.7.1 Fibrosis and Scar Formation ... 77
3.7.2 Cancer Metastasis and ECM Alterations ... 78
3.8 Therapeutic Approaches Targeting ECM Remodeling ... 79
3.8.1 Anti-fibrotic Therapies ... 79

3.8.2 ECM Modulation in Regenerative Medicine 80
3.9 Future Directions and Emerging Research 83
3.10 Summary 83
References 85

4 ECM in Health and Disease 87
4.1 Introduction 87
4.2 ECM Alterations in Cancer 89
4.2.1 Role of ECM in Tumor Microenvironment 89
4.2.2 ECM Remodeling and Cancer Progression 91
4.2.3 Therapeutic Implications Targeting ECM in Cancer Treatment 93
4.3 ECM and Cardiovascular Diseases 94
4.3.1 ECM Composition in Cardiovascular Tissues 95
4.3.2 ECM Remodeling in Heart Disease 97
4.3.3 Impact of ECM on Vascular Health 98
4.4 Role of ECM in Fibrosis 99
4.4.1 Mechanisms of ECM Deposition in Fibrotic Diseases 100
4.4.2 Consequences of ECM Dysregulation in Organ Fibrosis 102
4.4.3 Therapeutic Strategies Targeting ECM in Fibrotic Conditions 103
4.5 ECM in Chronic Inflammatory Conditions 105
4.5.1 ECM Involvement in Chronic Inflammation 105
4.5.2 Crosstalk Between Immune Cells and ECM 107
4.5.3 ECM-Based Therapies for Chronic Inflammatory Diseases 108
4.6 ECM in Neurological Disorders 110
4.7 Novel Techniques for Studying ECM Dynamics 112
4.7.1 Potential Biomarkers Related to ECM Alterations 113
4.7.2 Translational Opportunities for ECM-Based Therapies 114
4.8 Summary 115
References 117

5 Biomimicry of ECM in Medical Applications 121
5.1 Introduction 121
5.2 Importance of ECM in Tissue Engineering and Medical Applications 122
5.3 Structural and Functional Properties of ECM 124
5.3.1 Composition and Functionality 125
5.3.2 Functional Roles of ECM 127
5.4 Biomimetic Approaches in ECM Design 130

5.4.1 Principles of Biomimicry ... 130
5.5 Applications in Regenerative Medicine and Medical Devices ... 134
5.5.1 Tissue Engineering with ECM Mimics ... 135
5.5.2 ECM Based Therapies ... 136
5.6 Biophysical and Biochemical Properties of Biomimetic ECM ... 139
5.6.1 Mechanical Properties ... 140
5.6.2 Biochemical Signaling ... 142
5.7 Current Limitations and Challenges ... 146
5.8 Summary ... 147
References ... 148

6 ECM in Diagnostic Innovations ... 151
6.1 Introduction ... 151
6.2 Role of ECM in Disease Pathophysiology ... 152
6.3 ECM-Based Biomarkers for Disease Detection ... 154
6.4 Identification of ECM Components as Biomarkers ... 155
6.4.1 ECM Remodeling as a Diagnostic Tool ... 157
6.4.2 Biomarkers for Cancer, Fibrosis, and Cardiovascular Diseases ... 159
6.5 Imaging Techniques for ECM Visualization ... 161
6.5.1 Traditional Imaging Techniques (MRI, CT, Ultrasound) ... 162
6.5.2 Advanced Imaging Techniques: Optical Coherence Tomography, Multiphoton Microscopy ... 164
6.5.3 Molecular Imaging of ECM Components ... 166
6.5.4 Application of Imaging in Disease Diagnosis and Monitoring ... 167
6.6 Microfluidic Devices for ECM Analysis ... 167
6.6.1 ECM-on-a-Chip Models for Disease Detection ... 168
6.6.2 Integration of Microfluidics with Biomarker Detection ... 169
6.7 Technological Advances in ECM Analysis ... 171
6.7.1 Nanotechnology in ECM Diagnostics ... 171
6.7.2 Computational Models and AI in ECM Data Analysis ... 173
6.7.3 Integrating Multi-omics Data for ECM Analysis ... 174
6.8 Clinical Applications and Future Directions ... 176
6.8.1 Translational Potential of ECM-Based Diagnostics ... 177
6.9 Summary ... 178
References ... 179

7 Therapeutic Applications of ECM Mimicry 183
7.1 Introduction 183
7.2 Principles of ECM Mimicry in Therapeutic Applications 185
7.2.1 Understanding ECM Structure and Function 185
7.2.2 Biochemical Cues in ECM 187
7.2.3 Mechanical Properties of ECM 187
7.2.4 Cellular Interactions with ECM 189
7.2.5 ECM Mimicry in Tissue Engineering 190
7.3 Design and Fabrication of ECM-Mimicking Scaffolds 191
7.3.1 Material Selection for ECM Mimicry 192
7.3.2 Techniques for Scaffold Fabrication 193
7.3.3 Surface Modification Strategies 194
7.3.4 Integration of Growth Factors and Bioactive Molecules 195
7.3.5 Characterization of ECM-Mimicking Scaffolds 196
7.4 ECM Mimicry in Wound Healing 197
7.4.1 ECM-Mimicking Scaffolds in Acute Wound Healing and Skin Regeneration 199
7.4.2 Chronic Wound Healing and ECM Mimicry 200
7.5 ECM Mimicry in Cancer Therapy 202
7.5.1 Tumor Microenvironment and ECM 202
7.5.2 ECM Mimicry for Targeted Drug Delivery 204
7.5.3 Impact on Tumor Progression and Metastasis 205
7.5.4 ECM-Based Therapies in Clinical Trials 206
7.6 ECM Mimicry in Bone and Cartilage Regeneration 207
7.6.1 Scaffolds for Bone Regeneration 207
7.6.2 ECM-Mimicking Strategies for Cartilage Repair 208
7.7 ECM Mimicry in Cardiovascular Applications 209
7.7.1 Role of ECM in Cardiovascular Health 210
7.7.2 Scaffold Design for Vascular Tissue Engineering 211
7.7.3 ECM-Mimicking Approaches in Heart Tissue Repair 213
7.8 ECM Mimicry in Nervous System and Neural Tissue Regeneration 214
7.9 ECM Mimicry in Organ Regeneration 215
7.10 Smart ECM Mimicking Scaffolds 217
7.10.1 Customizable ECM Mimicry Through AI and Machine Learning 218
7.11 Summary 220
References 221

8 Advanced Technologies in ECM Mimicry 225
8.1 Introduction 225
8.2 Biomaterials for ECM Mimicry 226

8.2.1 Natural Versus Synthetic Biomaterials for ECM Mimicry 227
8.2.2 Polymers in ECM Scaffold Development 227
8.2.3 Bioactive Molecules for Functional Mimicry 228
8.3 3D Printing and Bioprinting Techniques 229
8.3.1 3D Printing Technologies for ECM Scaffolds 230
8.3.2 Bioprinting of Cells and ECM Components 231
8.4 Hydrogels and Bioinks for ECM Mimicry 232
8.4.1 Functional Hydrogels as ECM Mimetics 233
8.4.2 Bioinks for Tissue-Specific ECM Mimicry 234
8.5 Decellularization and Recellularization Strategies 236
8.5.1 Principles of Decellularization 237
8.5.2 Recellularization with Native Cells and Stem Cells 238
8.6 Microfluidics and Organ-on-a-Chip Models 240
8.6.1 Microfluidic Platforms for ECM Analysis 241
8.6.2 Organ-on-a-Chip for Disease Modeling 242
8.6.3 ECM Role in Microfluidic Systems for Drug Testing 243
8.7 Integration of AI and Machine Learning in ECM Scaffold Design 244
8.7.1 Personalized ECM Mimicry for Precision Medicine 245
8.8 Summary 246
References 247

9 Clinical Case Studies and Trials 251
9.1 Introduction 251
9.2 Success Stories in ECM-Based Therapies 252
9.2.1 Breakthrough Cases in Wound Healing 253
9.2.2 ECM-Based Approaches in Cardiovascular Regeneration 254
9.2.3 Cancer Treatments Using ECM Mimics 256
9.2.4 Case Studies of Regenerative Medicine Using ECM Biomaterials 257
9.2.5 Lessons Learned from Successful Clinical Implementations 259
9.3 Ongoing Clinical Trials Involving ECM-Based Therapies 260
9.3.1 Overview of Current ECM-Related Trials 261
9.3.2 ECM and Tissue Engineering in Orthopedics 261
9.3.3 ECM in Dermatology: Skin Grafts and Burn Wounds 263
9.3.4 ECM-Based Therapies in Neurology and Neurosurgery 264
9.3.5 Biomimetic ECM in Organ Transplants 265

9.3.6 Success Rates and Challenges in Ongoing Trials 267
9.4 Future Directions and Challenges ... 268
9.5 Ethical Considerations in ECM-Based Clinical Trials ... 269
9.6 Summary ... 272
References ... 273

10 Future Perspectives and Innovations ... 275
10.1 Introduction ... 275
10.2 Emerging Trends in ECM Research ... 277
10.2.1 Advances in ECM-Based Therapeutics ... 277
10.2.2 Innovations in ECM Engineering for Regenerative Medicine ... 278
10.2.3 Integration of ECM with Nanotechnology and Smart Materials ... 279
10.2.4 Role of ECM in Immunomodulation and Cancer Therapy ... 280
10.3 Potential Breakthroughs in ECM Biomimicry ... 281
10.3.1 Bio Fabrication Techniques for ECM-Like Scaffolds ... 281
10.3.2 Synthetic ECM Models for Precision Medicine ... 284
10.3.3 Biomimetic Hydrogels for Controlled Drug Delivery ... 285
10.3.4 Application of AI in ECM-Related Diagnostics ... 285
10.4 The Road Ahead: Opportunities and Challenges ... 287
10.4.1 Clinical Translation of ECM Innovations ... 287
10.4.2 Regulatory and Ethical Considerations in ECM Applications ... 288
10.4.3 Overcoming Barriers in ECM-Integrated Therapeutics ... 289
10.4.4 Future Directions in ECM-Driven Tissue Engineering ... 290
10.5 Summary ... 291
References ... 294

11 Conclusion ... 295
11.1 Summary of Key Findings ... 295
11.2 Impact of ECM Mimicry on Healthcare ... 296
11.2.1 Contributions to Tissue Engineering and Regenerative Medicine ... 297
11.2.2 Improvements in Diagnostic Tools and Imaging Techniques ... 298
11.2.3 Innovations in Drug Delivery and Personalized Medicine ... 299
11.3 Future Directions in ECM Research ... 299
11.3.1 Exploring Novel Biomaterials for ECM Mimicry ... 300

11.3.2 Integration of Advanced Technologies Like AI and Machine Learning in ECM Studies 301
11.3.3 Ethical and Regulatory Challenges in Clinical Translation 301
11.3.4 Potential for Addressing Unmet Medical Needs and Global Health Challenges 302
11.4 Summary 303
References 305

Resources and Further Reading 307

Answers 331

Glossary 339

Abbreviations

3D	Three-Dimensional
AFM	Atomic Force Microscopy
ALS	Amyotrophic Lateral Sclerosis
BMPs	Bone Morphogenetic Proteins
CAFs	Cancer-Associated Fibroblasts
CAMs	Cell Adhesion Molecules
COPD	Chronic Obstructive Pulmonary Disease
CT	Computed Tomography
CTGF	Connective Tissue Growth Ffactor
DDRs	Discoidin Domain Receptors
DTI	Diffusion Tensor Imaging
DWI	Diffusion-Weighted Imaging
ECM	Extracellular Matrix
EGF	Epidermal Growth Factor
ELISA	Enzyme-linked immunosorbent assays
EMT	Epithelial-Mesenchymal Transition
FAK	Focal Adhesion Kinase
FEFs	Fibroblast Growth Factors
FTIR	Fourier-Transform Infrared Spectroscopy
GAGs	Glycosaminoglycans
GPCRs	G Protein-Coupled Receptors
HA	Hyaluronic Acid
HSPGs	Heparan Sulfate Proteoglycans
IFNs	Interferons
IGFs	Insulin-like Growth Factors
ILs	Interleukins
IPF	Idiopathic Pulmonary Fibrosis
JAKs	Janus Kinases
LOX	Lysyl Oxidase
MHC	Major Histocompatibility Complex
Micro CT	Micro-Computed Tomography

MMPs	Matrix Metalloproteinases
MPM	Multiphoton Microscopy
MRE	Magnetic Resonance Elastography
MRI	Magnetic Resonance Imaging
MS	Multiple Sclerosis
MSC	Mesenchymal Stem Cell
NGF	Nerve Growth Factor
OCT	Optical Coherence Tomography
PCL	Polycaprolactone
PCR	Polymerase Chain Reaction
PDGF	Platelet-Derived Growth Factor
PEG	Polyethylene Glycol
PKC	Protein Kinase C
PLGA	Polylactic-co-glycolic Acid
PNS	Peripheral Nervous System
RTK	Receptor Tyrosine Kinase
SEM	Scanning Electron Microscopy
STATs	Signal Transducer and Activator of Transcription Proteins
TGF-β	Transforming Growth Factor-Beta
TIMPs	Tissue Inhibitor of Metalloproteinases
TME	Tumor Microenvironment
TNFs	Tumor Necrosis Factors
TNF-α	Tumor Necrosis Factor-Alpha
VEGF	Vascular Endothelial Growth Factor
XPS	X-ray Photoelectron Spectroscopy
α-SMA	α-Smooth Muscle Actin

Chapter 1
Introduction to Extracellular Matrix (ECM) Dynamics

1.1 Introduction

The extracellular matrix (ECM) is a complex network of macromolecules that provides structural and biochemical support to cells within tissues and organs. Composed primarily of proteins such as collagen, elastin, fibronectin, and laminin, as well as glycosaminoglycans (GAGs) and proteoglycans, the ECM plays a crucial role in maintaining tissue integrity and function [1]. This dynamic scaffold not only provides physical support but also regulates various cellular processes including adhesion, migration, proliferation, and differentiation through interactions with cell surface receptors. Structurally, the ECM is organized into a three-dimensional meshwork that varies in composition depending on the tissue type and physiological conditions. For instance, in connective tissues like bone and cartilage, collagen fibers predominate, imparting tensile strength and resilience [2]. In contrast, in basement membranes that underlie epithelial and endothelial layers, laminins and collagen IV form a specialized scaffold that regulates permeability and cell polarity. This structural diversity reflects the ECM's adaptive nature, tailored to meet the specific functional requirements of different tissues and organs. Functionally, the ECM serves as a reservoir for growth factors, cytokines, and other signaling molecules that influence cellular behavior and tissue homeostasis. It acts as a molecular sieve, selectively regulating the passage of ions, nutrients, and other molecules between cells and their environment [3]. The ECM participates actively in tissue repair and regeneration processes by providing cues for cell migration, proliferation, and differentiation during wound healing or in response to injury. The ECM is also integral to cellular communication and signaling. Cell surface receptors such as integrins and proteoglycans interact with ECM components, triggering intracellular signaling cascades that modulate gene expression, cytoskeletal dynamics, and cell survival. These interactions are crucial not only for maintaining tissue architecture but also for orchestrating complex biological processes such as embryonic development, immune responses, and tissue morphogenesis.

D. Suhag, *Extracellular Matrix biomimicking for Theranostics*, Biomedical Materials for Multi-functional Applications, https://doi.org/10.1007/978-981-95-7900-6_1

The ECM represents a dynamic and multifunctional network that not only provides structural support to tissues but also regulates cellular behavior and tissue function through intricate biochemical and biomechanical interactions [4]. Understanding the complexity and versatility of the ECM is essential for unraveling its role in health and disease, and for developing targeted therapeutic strategies aimed at manipulating ECM dynamics for clinical benefit.

1.2 Importance of ECM Dynamics in Biological Systems

The dynamics of the extracellular matrix (ECM) play a pivotal role in biological systems, influencing various physiological processes essential for tissue development, homeostasis, and repair (Fig. 1.1). One of the key functions of ECM dynamics is its role in providing structural support to cells and tissues. Through its composition of collagen, elastin, proteoglycans, and glycoproteins, the ECM forms a scaffold that not only maintains tissue architecture but also regulates mechanical properties such as stiffness and elasticity. This structural support is crucial for maintaining tissue integrity and ensuring proper function of organs throughout the body [5]. Beyond its structural role, ECM dynamics are integral to regulating cellular behavior and function. The ECM serves as a reservoir for bioactive molecules such as growth factors, cytokines, and enzymes, which are crucial for cell signaling and communication. These molecules bind to ECM components and are released in response to physiological cues, thereby modulating cellular processes such as proliferation, differentiation, and migration. By controlling the availability and spatial distribution of these signaling molecules, ECM dynamics influence the behavior of neighboring cells and orchestrate complex tissue-level responses.

ECM dynamics are essential for tissue morphogenesis and organ development during embryogenesis. Spatial and temporal regulation of ECM deposition, organization, and remodeling are critical for guiding cell fate determination, tissue patterning, and morphogenetic movements [6]. For instance, in embryonic development, gradients of ECM components and their interactions with growth factors provide positional information that directs cell migration and differentiation, contributing to the formation of distinct tissues and organs.

In adult tissues, ECM dynamics continue to play a crucial role in tissue homeostasis and adaptive remodeling in response to physiological changes or injury. During wound healing and tissue repair processes, the ECM undergoes dynamic remodeling to create a provisional matrix that supports cell migration, proliferation, and tissue regeneration. This remodeling process involves the coordinated action of proteases, such as matrix metalloproteinases (MMPs), and their inhibitors, which regulate ECM turnover and degradation, thereby facilitating tissue remodeling and scar formation [7]. Dysregulation of ECM dynamics is implicated in various pathological conditions. Aberrant ECM remodeling and deposition contribute to the pathogenesis of fibrotic diseases, where excessive ECM accumulation leads to tissue stiffness and loss of function. In cancer, altered ECM composition and stiffness promote

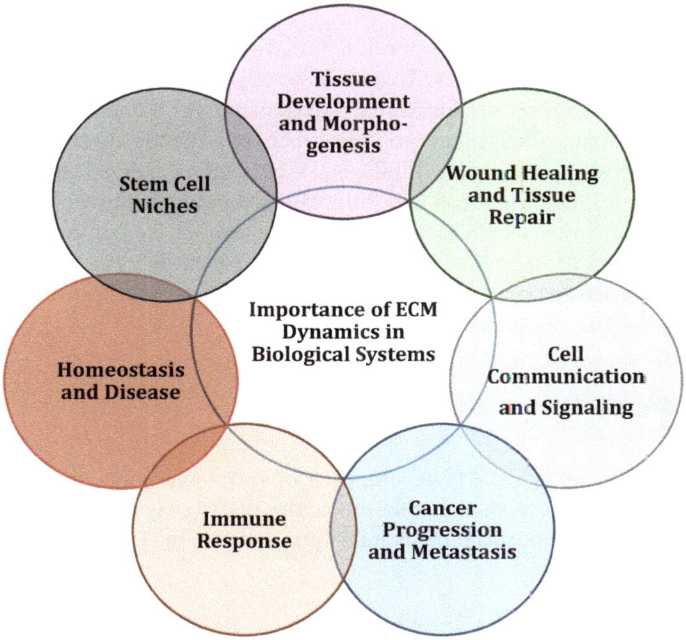

Fig. 1.1 Importance of ECM dynamics in biological systems

tumor progression by enhancing cell survival, proliferation, and metastatic potential. Understanding the mechanisms underlying ECM dynamics in health and disease is therefore critical for developing novel therapeutic strategies targeting ECM components or their associated signaling pathways. ECM dynamics are essential for maintaining tissue structure, regulating cellular behavior, guiding tissue development, and facilitating adaptive responses to physiological changes and injury. By influencing a wide range of biological processes, from embryogenesis to wound healing and disease progression, the ECM represents a dynamic and multifunctional component of biological systems that warrants further investigation and therapeutic targeting in clinical contexts.

1.3 Structure of ECM

The extracellular matrix (ECM) is a complex and dynamic network of macromolecules that surrounds cells within tissues and organs, providing essential structural and biochemical support. Composed primarily of proteins such as collagen, elastin, fibronectin, and laminin, along with glycosaminoglycans (GAGs) and proteoglycans, the ECM forms a three-dimensional scaffold that contributes to tissue architecture and mechanical properties. This intricate organization varies across different tissue types, reflecting specific functional requirements [8]. For example, in connective tissues like bone and cartilage, collagen fibers provide tensile strength, while in basement membranes, laminins and collagen IV create a specialized barrier that regulates molecular exchange and cell adhesion. Understanding the diverse composition and spatial arrangement of ECM components is essential for comprehending their roles in supporting cellular functions, maintaining tissue integrity, and influencing physiological processes throughout the body.

1.3.1 Composition of ECM Components

The extracellular matrix (ECM) is composed of various proteins, glycosaminoglycans (GAGs), proteoglycans, and other molecules. ECM plays crucial roles in tissue architecture, cell signaling, and mechanical properties (Fig. 1.2).

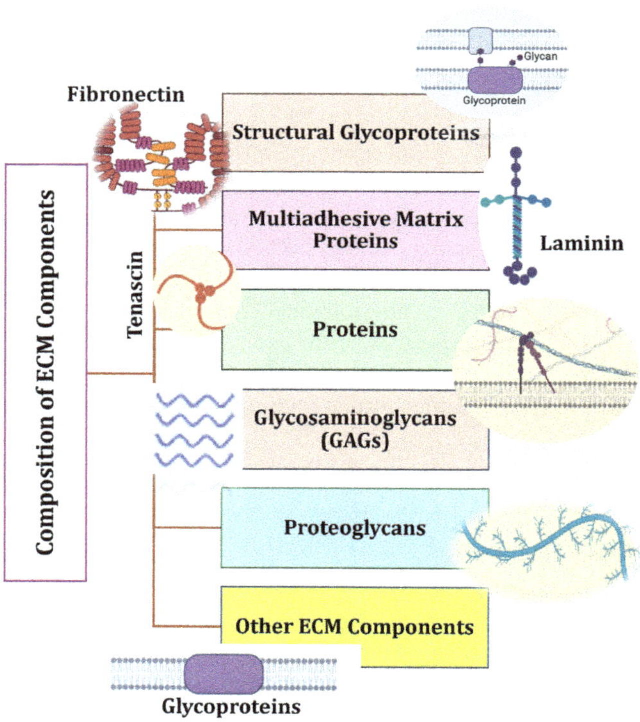

Fig. 1.2 Composition of ECM components

- **Proteins**: Collagens are the predominant proteins in ECM, accounting for up to 30% of total body protein. These fibrous proteins form long, triple-helical structures that provide tensile strength to tissues. Collagen types I, II, and III are abundant in connective tissues such as skin, tendons, and cartilage, respectively. Elastin, another key protein, confers elasticity and recoil properties to tissues like blood vessels and lungs, enabling them to stretch and return to their original shape [9]. Fibronectin and laminins are adhesive glycoproteins that facilitate cell adhesion to the ECM, mediating processes such as cell migration, differentiation, and wound healing.
- **Glycosaminoglycans (GAGs)**: GAGs are long, linear polysaccharides composed of repeating disaccharide units that are highly negatively charged due to sulfate or carboxyl groups. They are covalently attached to core proteins to form proteoglycans. Examples of GAGs include hyaluronic acid, chondroitin sulfate, dermatan sulfate, heparan sulfate, and keratan sulfate [10]. These molecules play critical roles in regulating ECM hydration, viscosity, and tissue compressibility. Hyaluronic acid, for instance, forms a viscous gel-like substance that contributes to the lubrication and shock absorption properties of synovial fluid in joints.
- **Proteoglycans**: Proteoglycans are proteins with covalently attached GAG chains. They are abundant in the ECM and play diverse roles in cell–matrix interactions, signal transduction, and ECM organization. Aggrecan, found in cartilage, is a large proteoglycan that provides resilience and compressive strength to the tissue due to its high content of chondroitin sulfate and keratan sulfate chains [11]. Other examples include decorin, which regulates collagen fibrillogenesis and matrix assembly, and syndecans, which serve as co-receptors for growth factors and cytokines.
- **Multiadhesive Matrix Proteins**: These proteins, such as fibronectin, laminin, and thrombospondin, mediate interactions between cells and ECM components. Fibronectin binds to integrins on cell surfaces and ECM molecules like collagen and proteoglycans, facilitating cell adhesion, migration, and wound healing processes. Laminins are large, cross-shaped glycoproteins that contribute to the structural integrity of basement membranes and regulate cell differentiation and tissue organization. Thrombospondins modulate cell–matrix interactions, influence cell adhesion and migration, and participate in tissue remodeling processes [12].
- **Structural Glycoproteins**: ECM structural glycoproteins include fibronectin, elastin, and laminin, which play crucial roles in maintaining ECM integrity and function. Fibronectin, in addition to its adhesive functions, regulates cell behavior and ECM assembly. Elastin provides elasticity and resilience to tissues such as arteries and skin, ensuring their ability to withstand stretching and recoil. Laminins contribute to the structural organization of basement membranes and participate in tissue-specific signaling events that regulate cell behavior and tissue development.
- **Other ECM Components**: Besides proteins and GAGs, the ECM also contains glycoproteins, growth factors, and enzymes that regulate ECM turnover and remodeling. These components influence cell behavior, tissue morphogenesis, and

responses to injury or disease. Growth factors such as transforming growth factor-beta (TGF-β) and vascular endothelial growth factor (VEGF) regulate cell proliferation, differentiation, and ECM synthesis, contributing to tissue repair and angiogenesis. Enzymes like matrix metalloproteinases (MMPs) and tissue inhibitors of metalloproteinases (TIMPs) control ECM degradation and remodeling processes, maintaining ECM homeostasis and tissue integrity.

Understanding the composition and interactions of ECM components is essential for elucidating their roles in physiological processes and disease pathogenesis. Advances in ECM research continues to uncover new therapeutic targets and strategies for treating diseases characterized by ECM dysregulation, such as fibrosis, cancer, and musculoskeletal disorders.

Exercise 1

Fill Ups
(Cytoskeleton, Transmembrane receptors, Matrix metalloproteinases, Fibronectin, Extracellular matrix)

1. The .... is a complex network of proteins and polysaccharides that surrounds cells, providing structural and biochemical support.

2. Integrins are .... that mediate the attachment between a cell and the extracellular matrix (ECM).

3. The .... protein in the ECM binds to integrins and plays a crucial role in cell adhesion, migration, and wound healing.

4. During cell migration, the interaction between the actin .... and the extracellular matrix (ECM) is crucial for generating the necessary forces.

5. .... are enzymes that degrade components of the extracellular matrix to allow for tissue remodeling and cell migration.

1.3.2 Organization and Functionality

The organization and functionality of the extracellular matrix (ECM) are critical for supporting tissue structure, regulating cellular behavior, and maintaining physiological homeostasis throughout the body. The ECM is organized into intricate networks that vary in composition and architecture across different tissue types, reflecting their specialized functions and mechanical requirements.

- **Organization**: The ECM is organized into distinct structural entities depending on the tissue context. In connective tissues such as bone and tendon, collagen fibers are arranged in parallel bundles, providing tensile strength and resilience to withstand mechanical stress. In contrast, in epithelial tissues and blood vessels, the ECM forms thin, sheet-like structures known as basement membranes, composed primarily of laminins, collagen IV, and other glycoproteins. These basement membranes serve as barriers that regulate the movement of molecules between adjacent tissue compartments while providing structural support to epithelial and endothelial cells [13].

- **Fibrous Networks**: Collagen, the most abundant protein in the ECM, forms fibrils and networks that contribute to tissue integrity and mechanical properties. Collagen fibrils can assemble into larger fibers, creating a scaffold that supports cells and tissues. Different types of collagens (e.g., Type I, II, III) are present in varying proportions depending on the tissue's functional requirements, such as providing tensile strength in skin or flexibility in cartilage.
- **Basement Membranes**: Basement membranes are specialized ECM structures that underlie epithelial and endothelial cell layers. Composed of laminins, collagen IV, nidogens, and heparan sulfate proteoglycans, basement membranes provide a platform for cell adhesion, migration, and differentiation [14]. They also regulate the passage of molecules and ions between different tissue compartments, contributing to tissue polarity and barrier function.
- **Glycosaminoglycan-rich Matrix**: Glycosaminoglycans (GAGs), such as hyaluronic acid, chondroitin sulfate, and heparan sulfate, are long, unbranched polysaccharides that bind to proteoglycans within the ECM. These GAGs have negative charges due to sulfate or carboxyl groups, contributing to the ECM's ability to retain water and maintain tissue hydration and viscosity [15]. This hydrated gel-like matrix provides a compressive resistance and facilitates diffusion of nutrients and signaling molecules between cells.
- **Cell-ECM Interactions**: The ECM interacts dynamically with cells through cell surface receptors such as integrins, which bind to specific ECM components. These interactions transmit mechanical and biochemical signals that regulate various cellular processes including adhesion, migration, proliferation, and differentiation. Integrins are linked to the cell's cytoskeleton, allowing cells to sense and respond to changes in ECM composition and mechanical properties.
- **Regulation of Biological Processes**: Beyond structural support, the ECM regulates diverse biological processes critical for tissue development, repair, and homeostasis. It serves as a reservoir for growth factors, cytokines, and proteases that are involved in cell signaling, ECM remodeling, and tissue remodeling in response to physiological cues or injury. ECM components and their interactions with cells and soluble factors influence tissue morphogenesis during embryonic development and maintain tissue function throughout life.
- **Pathophysiological Implications**: Dysregulation of ECM organization and functionality is implicated in various diseases. For example, aberrant ECM remodeling contributes to fibrosis in organs such as the liver, lungs, and kidneys, leading to tissue stiffness and loss of function. In cancer, altered ECM composition and stiffness promote tumor progression and metastasis by enhancing cell survival, migration, and angiogenesis. Understanding the organization and functionality of the ECM is therefore crucial for elucidating disease mechanisms and developing targeted therapies aimed at restoring ECM homeostasis and tissue function.

1.4 Cell-ECM Interactions

Cell-ECM interactions constitute a fundamental aspect of cellular biology, crucial for orchestrating various physiological processes within tissues and organs. The extracellular matrix (ECM) serves not only as a structural scaffold but also as a dynamic interface that mediates communication between cells and their surrounding microenvironment. This interaction is primarily facilitated by cell surface receptors, such as integrins, which bind specifically to ECM components like collagen, fibronectin, and laminin. Through these interactions, cells are able to sense mechanical cues, receive biochemical signals, and regulate essential processes such as adhesion, migration, proliferation, and differentiation [16]. Cell-ECM interactions play pivotal roles in tissue development, homeostasis, and repair, highlighting their significance in maintaining cellular function and tissue integrity throughout physiological and pathological contexts.

1.4.1 Receptors and Signaling Pathways

Receptors and signaling pathways involved in cell-ECM interactions are crucial for mediating cellular responses to the extracellular matrix (ECM) environment, thereby influencing various physiological processes and tissue behaviors. Central to these interactions are cell surface receptors such as integrins, which are transmembrane heterodimeric proteins composed of α and β subunits. Integrins bind specifically to ECM components like collagen, fibronectin, and laminin, forming focal adhesions that link the ECM to the cell's cytoskeleton [17]. This mechanical coupling allows cells to sense and respond to changes in ECM composition and stiffness, facilitating processes such as cell adhesion, migration, and mechano transduction.

Beyond integrins, other receptor families such as discoidin domain receptors (DDRs) and syndecans also participate in cell-ECM interactions. DDRs are receptor tyrosine kinases that bind to collagen and are involved in regulating cell adhesion, migration, and matrix remodeling. Syndecans are transmembrane proteoglycans that bind to both ECM proteins and growth factors, modulating signaling pathways that control cell proliferation, differentiation, and tissue morphogenesis. Signaling pathways activated by these receptors play critical roles in translating ECM signals into cellular responses. For instance, integrin engagement activates intracellular signaling cascades such as focal adhesion kinase (FAK) and Src family kinases, which regulate cytoskeletal dynamics, gene expression, and cell survival. These pathways integrate mechanical and biochemical signals from the ECM to coordinate cellular behaviors and maintain tissue homeostasis [18]. The ECM also serves as a reservoir for bioactive molecules, including growth factors and cytokines, which bind to ECM components and are released in response to cell-mediated cues or tissue injury. Binding of these molecules to their cognate receptors on cell surfaces initiates signaling cascades such as the MAPK/ERK pathway or the PI3K/Akt pathway, which regulate cell proliferation, differentiation, and survival. Thus, ECM-bound growth factors act as spatial and temporal regulators of cellular responses, influencing tissue development, repair, and disease progression.

Importantly, dysregulation of ECM receptor signaling pathways is implicated in various pathological conditions. For example, aberrant integrin signaling contributes to cancer progression by enhancing tumor cell survival, invasion, and metastasis. Similarly, defects in ECM receptor signaling are associated with fibrotic diseases, where excessive ECM deposition and altered matrix stiffness disrupt tissue architecture and function [19]. Understanding the intricacies of ECM receptor interactions and signaling pathways is therefore crucial for developing targeted therapies aimed at modulating these pathways to treat diseases and promote tissue repair.

Receptors and signaling pathways involved in cell-ECM interactions are integral to maintaining tissue homeostasis and regulating cellular behaviors such as adhesion, migration, proliferation, and differentiation. These interactions not only facilitate mechanical coupling between cells and their microenvironment but also enable cells to respond dynamically to changes in ECM composition and biochemical signals. Through these mechanisms, ECM receptors and signaling pathways play pivotal roles in orchestrating tissue development, repair, and disease progression, underscoring their importance in biological and biomedical research.

1.4.2 Role in Cellular Adhesion and Migration

Cellular adhesion and migration are critical processes mediated by interactions between cells and the extracellular matrix (ECM), which play essential roles in tissue development, homeostasis, and repair. Adhesion involves the attachment of cells to specific ECM components through transmembrane receptors such as integrins, which bind to ECM proteins like collagen, fibronectin, and laminin. This interaction forms focal adhesions, anchoring the cell to the ECM and transmitting mechanical signals that regulate cell morphology, cytoskeletal dynamics, and gene expression [20]. These adhesive contacts are dynamic, allowing cells to detach and migrate in response to environmental cues. Cell migration, facilitated by ECM interactions, is essential for various physiological processes including embryogenesis, immune response, and wound healing. During migration, cells extend protrusions called lamellipodia and filopodia, which interact with the ECM to propel the cell forward. Integrins and other adhesion receptors cluster at the leading edge of migrating cells, forming focal complexes that link the actin cytoskeleton to the ECM and generate traction forces necessary for movement [21]. Integrins not only mediate adhesion but also serve as signaling hubs that coordinate cellular responses during migration. Engagement of integrins with ECM ligands activates signaling pathways such as focal adhesion kinase (FAK), Src family kinases, and Rho GTPases, which regulate cytoskeletal dynamics and adhesion turnover. These pathways control the balance between cell adhesion and detachment, allowing cells to efficiently navigate through complex ECM environments. The ECM composition and organization also influence cellular adhesion and migration. Different ECM components provide varying degrees of adhesive cues and mechanical support that guide cell behavior. For instance, fibronectin-rich matrices promote cell spreading and migration through integrin-mediated interactions, whereas collagen matrices provide structural support and cues that direct cell movement along specific pathways [22]. In addition to integrins, other ECM receptors such as syndecans and discoidin domain receptors (DDRs) contribute to cellular adhesion and migration. Syndecans interact with both ECM proteins and growth factors, modulating signaling pathways that regulate cell adhesion, migration, and tissue morphogenesis. DDRs, on the other hand, are collagen-binding receptor tyrosine kinases involved in ECM remodeling and cell-ECM interactions during development and tissue repair. Dysregulation of ECM-mediated adhesion and migration processes is implicated in various pathological conditions, including cancer metastasis and fibrosis. Aberrant integrin signaling promotes tumor cell invasion and dissemination by enhancing adhesion to ECM components in the tumor microenvironment. Similarly, altered ECM composition and stiffness contribute to fibrotic diseases, where excessive ECM deposition impairs tissue function and promotes aberrant cell migration.

Understanding the roles of ECM in cellular adhesion and migration is crucial for elucidating mechanisms of tissue development, repair, and disease progression. By studying how cells interact with and respond to their ECM environment, researchers can identify potential therapeutic targets for modulating cell behavior in pathological conditions and enhancing tissue regeneration strategies. Thus, ECM-mediated adhesion and migration represent fundamental processes that integrate mechanical and biochemical signals to regulate cellular behaviors and tissue dynamics in health and disease.

1.5 ECM Remodeling

Extracellular matrix (ECM) remodeling refers to the dynamic process by which cells continuously modify and reorganize the ECM's composition and structure in response to physiological cues or pathological stimuli. This process involves a coordinated interplay between ECM-degrading enzymes, such as matrix metalloproteinases (MMPs) and tissue inhibitors of metalloproteinases (TIMPs), and ECM-producing cells, such as fibroblasts and smooth muscle cells [23]. ECM remodeling is essential for maintaining tissue homeostasis, regulating mechanical properties, and facilitating adaptive responses to changes in the cellular environment (Fig. 1.3). It plays critical roles in tissue development, wound healing, and disease progression, highlighting its significance in both physiological and pathological contexts.

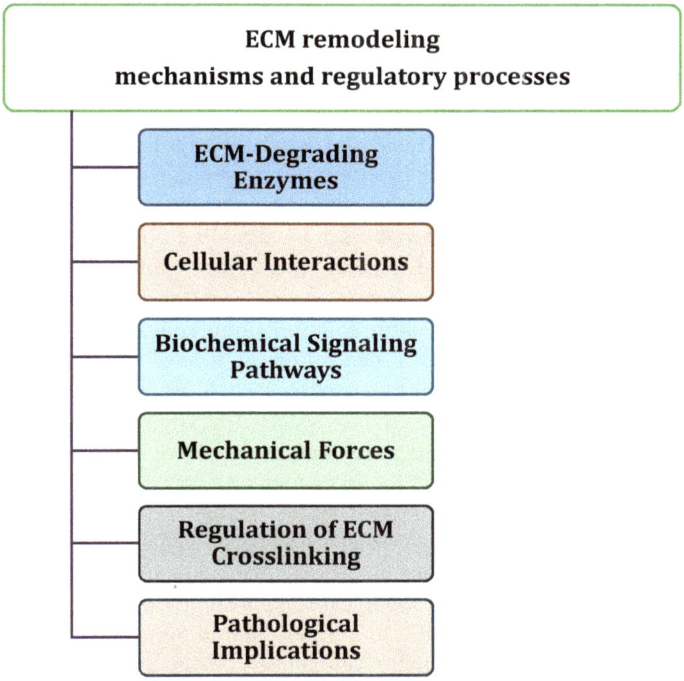

Fig. 1.3 Regulatory processes involved in remodeling of ECM

1.5.1 Mechanisms and Regulation

ECM remodeling involves a complex interplay of mechanisms and regulatory processes that govern the turnover, composition, and structural organization of the extracellular matrix (ECM) in tissues and organs. This dynamic process is essential for maintaining tissue homeostasis, adapting to physiological changes, and responding to injury or disease. Several key mechanisms contribute to ECM remodeling, primarily mediated by ECM-degrading enzymes, cellular interactions, and biochemical signaling pathways.

- **ECM-Degrading Enzymes**: Matrix metalloproteinases (MMPs) are pivotal enzymes involved in ECM remodeling by cleaving and degrading ECM proteins such as collagen, elastin, and fibronectin. MMPs are produced by various cell types including fibroblasts, macrophages, and endothelial cells in response to inflammatory cytokines, growth factors, and mechanical stimuli [24]. The activity of MMPs is tightly regulated at multiple levels, including transcriptional control, activation of zymogens, and inhibition by tissue inhibitors of metalloproteinases (TIMPs). The balance between MMPs and TIMPs is critical for maintaining ECM integrity and preventing excessive ECM degradation.

- **Cellular Interactions**: Cells within the ECM, such as fibroblasts and smooth muscle cells, actively participate in ECM remodeling through their synthesis and secretion of ECM components. Fibroblasts, for example, produce collagen and fibronectin, contributing to ECM deposition and organization. These cells also sense changes in ECM stiffness and composition through integrin-mediated signaling, which regulates their proliferation, migration, and ECM synthesis or degradation activities in response to mechanical cues.
- **Biochemical Signaling Pathways**: Various signaling pathways regulate ECM remodeling by modulating the activity of ECM-degrading enzymes and the synthesis of ECM components. For instance, transforming growth factor-beta (TGF-β) signaling plays a central role in stimulating ECM production by fibroblasts and promoting ECM deposition in tissues undergoing repair or fibrosis. Conversely, inflammatory cytokines such as interleukins and tumor necrosis factor-alpha (TNF-α) can induce MMP expression and ECM degradation, contributing to tissue remodeling in response to injury or inflammation [25].
- **Mechanical Forces**: Mechanical forces exerted on tissues influence ECM remodeling by altering cell-ECM interactions and activating mechano transduction pathways. Cells sense and respond to mechanical cues such as stretch, compression, and shear stress through integrins and focal adhesion complexes, which activate signaling cascades that regulate MMP expression and ECM turnover. Mechanical forces also induce changes in ECM stiffness and alignment, influencing tissue architecture and function in organs like the heart, lungs, and blood vessels.
- **Regulation of ECM Crosslinking**: Crosslinking of ECM proteins such as collagen and elastin by enzymes like lysyl oxidase (LOX) contributes to ECM stability and mechanical properties. Crosslinking enzymes modify ECM structure by forming covalent bonds between adjacent protein molecules, which enhances tissue resilience and resistance to degradation. Dysregulation of ECM crosslinking is implicated in diseases characterized by altered tissue stiffness and fibrosis, highlighting the importance of maintaining proper enzymatic balance in ECM remodeling processes.
- **Pathological Implications**: Dysregulated ECM remodeling is associated with various pathological conditions, including fibrosis, cardiovascular diseases, and cancer metastasis. Excessive ECM degradation or deposition can disrupt tissue architecture, impair organ function, and promote disease progression. Understanding the mechanisms and regulatory pathways involved in ECM remodeling is therefore crucial for developing therapeutic strategies aimed at modulating ECM dynamics in disease states, promoting tissue repair, and improving clinical outcomes.

ECM remodeling is a highly orchestrated process involving ECM-degrading enzymes, cellular interactions, biochemical signaling pathways, mechanical forces, and ECM crosslinking mechanisms. These processes collectively regulate the turnover, composition, and structural organization of the ECM in response to physiological cues and pathological stimuli. Dysregulation of ECM remodeling contributes to various diseases, underscoring the importance of elucidating these mechanisms for developing targeted therapies and interventions aimed at restoring tissue homeostasis and function.

Exercise 2

Fill Ups
(Extracellular matrix, Fibroblasts, Tissue inhibitors of metalloproteinases, Matrix metalloproteinases, Collagen)

1. .... are enzymes that play a crucial role in ECM remodeling by degrading various components of the extracellular matrix.

2. ECM remodeling is essential during processes like wound healing, where .... must be broken down and rebuilt to facilitate tissue repair.

3. The balance between ECM synthesis and degradation is regulated by a group of proteins known as ...., which inhibit the activity of matrix metalloproteinases (MMPs).

4. During fibrosis, excessive ECM deposition occurs due to increased activity of .... and decreased ECM degradation.

5. .... is a structural protein in the ECM that must be remodeled to allow for changes in tissue architecture during development and disease.

1.5.2 Physiological and Pathological Implications

ECM remodeling has profound physiological implications in maintaining tissue integrity, function, and adaptation to dynamic environmental changes. Physiologically, ECM remodeling is essential during tissue development and growth, ensuring proper tissue architecture and organ function. For example, in embryogenesis, ECM dynamics guide cell migration, differentiation, and tissue patterning, crucial for the formation of complex structures and organ systems. Throughout life, ECM remodeling continues to support tissue homeostasis by regulating cell behavior, maintaining mechanical properties, and facilitating tissue repair and regeneration in response to physiological demands such as exercise or hormonal changes. ECM remodeling plays a critical role in wound healing and tissue repair processes. Following injury, ECM components undergo controlled degradation and synthesis to create a provisional matrix that supports cell migration, proliferation, and tissue re-organization [26]. This dynamic remodeling process is orchestrated by inflammatory cells, fibroblasts, and endothelial cells, which coordinate ECM deposition and remodeling to restore tissue architecture and function. Disruption of ECM remodeling can impair wound healing, leading to chronic wounds or fibrotic tissue formation. In pathological contexts, dysregulated ECM remodeling contributes to the pathogenesis of various diseases. Excessive ECM deposition, as seen in fibrotic diseases of the liver, lungs, and kidneys, results in tissue stiffness and dysfunction. In fibrosis, persistent activation of fibroblasts and aberrant ECM synthesis disrupts tissue architecture and impairs organ function, highlighting the detrimental consequences of uncontrolled ECM remodeling. Conversely, insufficient ECM remodeling, characterized by ECM degradation imbalance, can lead to tissue degeneration and loss of structural integrity, as observed in degenerative diseases such as osteoarthritis.

ECM remodeling also plays a role in cardiovascular diseases, where alterations in ECM composition and stiffness contribute to hypertension, atherosclerosis, and myocardial infarction. Changes in ECM organization and crosslinking affect vessel compliance and endothelial function, influencing blood flow regulation and vascular remodeling. In cancer progression, dysregulated ECM remodeling promotes tumor invasion, metastasis, and resistance to therapy [27]. ECM components create a permissive microenvironment that facilitates tumor cell survival, migration through tissues, and interaction with stromal cells and immune cells, shaping the aggressive behavior of malignant tumors.

Understanding the physiological and pathological implications of ECM remodeling is crucial for developing therapeutic strategies to target these processes in disease settings. Approaches aimed at modulating ECM turnover, targeting ECM-degrading enzymes or crosslinking mechanisms, and promoting ECM homeostasis represent promising avenues for treating fibrotic diseases, cancer metastasis, and other conditions characterized by ECM dysregulation.

1.6 Techniques for Studying ECM Dynamics

Studying the dynamics of the extracellular matrix (ECM) involves employing a variety of techniques that allow researchers to investigate its composition, organization, and functional interactions with cells in both physiological and pathological contexts (Fig. 1.4). These techniques encompass a broad range of experimental approaches, from advanced imaging methods to biochemical assays and molecular biology tools. High-resolution imaging techniques such as confocal microscopy, electron microscopy, and atomic force microscopy enable visualization of ECM structures at different scales, providing insights into ECM organization, fiber morphology, and interactions with cells. Biochemical assays, including immunohistochemistry, enzyme-linked immunosorbent assays (ELISA), and western blotting, allow quantification and characterization of ECM components such as collagen types, proteoglycans, and ECM-bound growth factors. Molecular biology techniques such as polymerase chain reaction (PCR) and gene expression analysis provide information on ECM gene expression profiles and regulatory mechanisms underlying ECM synthesis and remodeling. Proteomics approaches, such as mass spectrometry, enable comprehensive profiling of ECM proteins and post-translational modifications, offering insights into ECM composition and changes associated with disease states. Advanced biophysical techniques, including rheology and traction force microscopy, assess ECM mechanical properties and cellular forces exerted on the ECM, crucial for understanding mechano transduction pathways and cell-ECM interactions [21].

Three-dimensional (3D) tissue engineering models and organoids mimic in vivo ECM environments, allowing researchers to study ECM dynamics in physiologically relevant contexts. These models facilitate investigations into ECM remodeling processes, cell behavior, and therapeutic interventions for tissue regeneration and

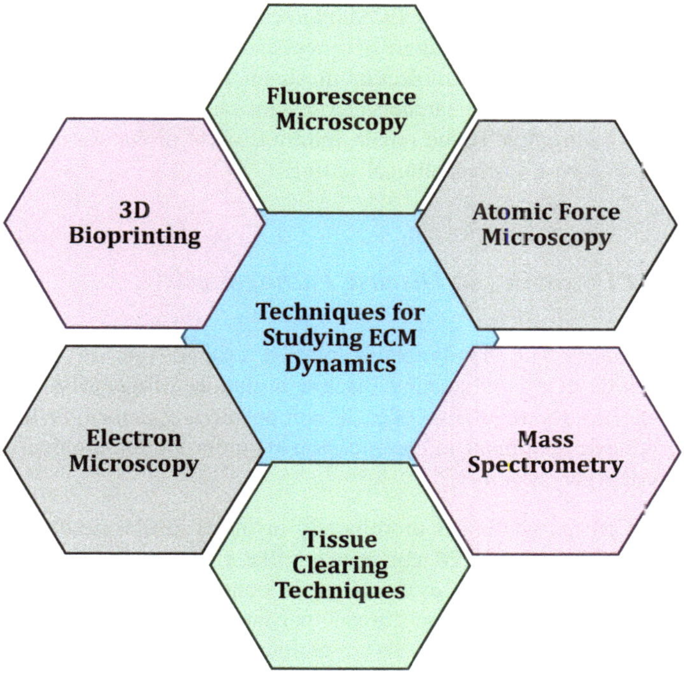

Fig. 1.4 Techniques for studying ECM dynamics

disease modeling. Collectively, these diverse techniques provide multidimensional insights into ECM dynamics, advancing our understanding of its roles in tissue development, homeostasis, and disease progression.

1.7 Clinical Relevance

The clinical relevance of studying the extracellular matrix (ECM) lies in its pivotal role in maintaining tissue structure, function, and homeostasis, thereby influencing various aspects of health and disease. Understanding ECM dynamics is crucial for diagnosing, treating, and managing a wide range of medical conditions, from chronic diseases to tissue injuries and degenerative disorders. For instance, in cardiovascular diseases, ECM remodeling contributes to arterial stiffening and plaque formation, affecting vascular integrity and blood flow regulation. In orthopedics, ECM research informs strategies for treating musculoskeletal injuries and disorders like osteoarthritis, where ECM degradation and altered composition impact joint function and mobility [28]. In cancer biology, ECM interactions influence tumor growth, invasion, and metastasis, shaping the aggressiveness and therapeutic responses of malignant tumors. By elucidating the molecular mechanisms underlying ECM changes in disease states, clinicians and researchers can develop targeted therapies that modulate ECM dynamics to promote tissue repair, inhibit disease progression, and improve patient outcomes across various clinical settings.

1.7.1 ECM Dynamics in Disease Pathogenesis

Extracellular matrix (ECM) dynamics play a critical role in the pathogenesis of various diseases, influencing disease initiation, progression, and clinical outcomes (Fig. 1.5). Dysregulation of ECM composition, organization, and turnover contributes to the development and progression of numerous pathological conditions across different organ systems.

- **Fibrosis**: ECM remodeling is prominently involved in fibrotic diseases, where excessive deposition of ECM components like collagen and fibronectin leads to tissue scarring and organ dysfunction. In conditions such as liver cirrhosis, pulmonary fibrosis, and cardiac fibrosis, persistent activation of fibroblasts and aberrant ECM synthesis disrupt tissue architecture and impair organ function. Dysregulated ECM remodeling alters tissue stiffness and elasticity, creating a microenvironment that promotes fibroblast activation and ECM deposition, perpetuating the fibrotic process.

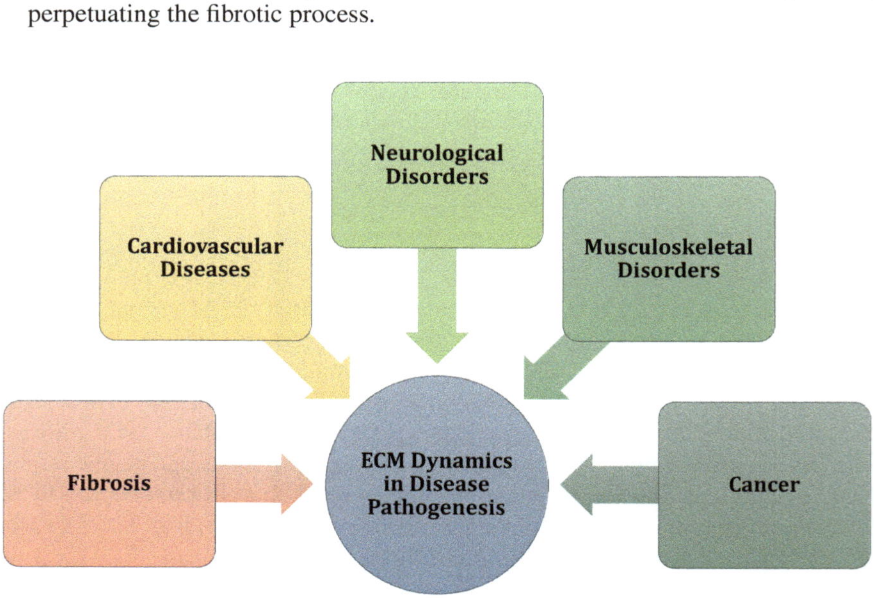

Fig. 1.5 ECM dynamics involved in disease pathogenesis

- **Cancer**: The ECM plays a crucial role in cancer progression and metastasis. Tumor cells interact dynamically with the ECM, remodeling their microenvironment to facilitate invasion into surrounding tissues and dissemination to distant sites. ECM components like collagen and hyaluronic acid promote tumor cell migration and invasion through physical and biochemical cues. ECM-bound growth factors and cytokines regulate tumor cell proliferation, survival, and angiogenesis, supporting tumor growth and metastatic spread. Changes in ECM stiffness and composition also influence therapeutic resistance, impacting the efficacy of anticancer treatments.
- **Cardiovascular Diseases**: ECM remodeling contributes to the pathogenesis of cardiovascular diseases, including atherosclerosis, hypertension, and myocardial infarction. Alterations in ECM composition, particularly increased collagen deposition and crosslinking, contribute to arterial stiffening and impaired vascular compliance. These changes affect blood vessel elasticity, blood flow dynamics, and endothelial function, predisposing individuals to hypertension and cardiovascular events. ECM remodeling in cardiac tissues influences myocardial remodeling and contractile function, contributing to heart failure progression.
- **Neurological Disorders**: In neurological diseases such as Alzheimer's disease, Parkinson's disease, and stroke, ECM dynamics play a role in disease pathogenesis and tissue repair mechanisms. ECM components in the brain's extracellular space influence neuronal connectivity, synaptic plasticity, and neuroinflammatory responses. Alterations in ECM composition and proteolytic processing contribute to neurodegenerative processes and impair tissue repair mechanisms following injury or ischemic events. Understanding ECM dynamics in neurological disorders is essential for developing therapies aimed at modulating neuronal ECM interactions and promoting neuro regeneration.
- **Musculoskeletal Disorders**: ECM dynamics are critical in musculoskeletal disorders such as osteoarthritis, rheumatoid arthritis, and muscular dystrophies. Degenerative changes in ECM components, including collagen, proteoglycans, and elastin, contribute to joint inflammation, cartilage degradation, and impaired tissue biomechanics. ECM remodeling influences the balance between tissue repair and degeneration in skeletal muscles, affecting muscle function and mobility. Therapeutic strategies targeting ECM dynamics aim to restore tissue integrity, reduce inflammation, and improve functional outcomes in musculoskeletal diseases.

ECM dynamics play diverse and pivotal roles in the pathogenesis of diseases across multiple organ systems [29]. Dysregulated ECM remodeling alters tissue structure, mechanical properties, and cellular interactions, contributing to disease progression and clinical manifestations.

1.7.2 Therapeutic Targeting of ECM

Therapeutic targeting of the extracellular matrix (ECM) represents a promising approach for treating various diseases characterized by ECM dysregulation, tissue remodeling, and structural abnormalities. Advances in understanding ECM biology have revealed multiple strategies to modulate ECM dynamics and restore tissue homeostasis, offering new avenues for therapeutic intervention.

- **Inhibition of ECM-Degrading Enzymes**: Targeting ECM-degrading enzymes such as matrix metalloproteinases (MMPs) represents a strategy to mitigate excessive ECM degradation observed in conditions such as arthritis and cardiovascular diseases. Small molecule inhibitors of MMPs have been developed to selectively inhibit their activity and reduce ECM breakdown, thereby preserving tissue integrity and function. By controlling ECM turnover, these inhibitors aim to alleviate disease progression and symptoms associated with tissue degradation [30].
- **ECM Crosslinking Inhibition**: Alterations in ECM crosslinking contributes to tissue stiffness and fibrosis in diseases like liver cirrhosis and pulmonary fibrosis. Therapeutic approaches targeting enzymes involved in ECM crosslinking, such as lysyl oxidases (LOX), aim to disrupt excessive crosslink formation and restore tissue flexibility and function. Inhibitors of LOX enzymes have shown potential in preclinical studies to attenuate ECM stiffening and reduce fibrotic tissue deposition, offering a novel strategy for treating fibrotic diseases.
- **Modulation of ECM-Cell Interactions**: ECM components interact with cell surface receptors such as integrins and syndecans, influencing cellular behavior and disease progression. Therapeutic strategies aim to modulate these interactions to restore normal cell-ECM signaling and function [31]. For instance, integrin-blocking antibodies or peptide mimetics can disrupt aberrant ECM-integrin interactions in cancer cells, inhibiting tumor cell migration, invasion, and metastasis. Similarly, targeting syndecan interactions with ECM components and growth factors may regulate cell proliferation, differentiation, and tissue repair processes in diseases involving ECM dysregulation.
- **Biomaterial-Based Therapies**: Biomaterials engineered to mimic ECM properties offer innovative approaches for tissue repair and regeneration. Scaffold materials composed of ECM-derived components or synthetic polymers can provide structural support, promote cell adhesion, and deliver therapeutic agents to targeted tissues. These biomaterials integrate into the host ECM, facilitating tissue remodeling and regeneration in applications ranging from wound healing to tissue engineering. By harnessing the bioactive properties of ECM-derived materials, these therapies aim to enhance tissue repair mechanisms and improve clinical outcomes in patients with chronic wounds, bone defects, and other tissue injuries.

Therapeutic targeting of the ECM represents a multifaceted approach to treat diseases characterized by ECM dysregulation and tissue remodeling. By modulating

ECM composition, turnover, and interactions with cells, these strategies aim to restore tissue homeostasis, alleviate disease symptoms, and improve patient outcomes across diverse clinical settings. Ongoing research efforts continue to advance our understanding of ECM biology and develop novel therapeutic interventions aimed at harnessing the therapeutic potential of ECM modulation in treating complex diseases.

Exercise 3

Multiple Choice Questions

1. What is the primary function of the extracellular matrix (ECM)?
A) Energy production
B) Genetic information storage
C) Structural support and regulation of cell behavior
D) Protein synthesis

2. Which of the following is NOT a major component of the extracellular matrix?
A) Collagen
B) Elastin
C) Cytoplasm
D) Proteoglycans

3. How does the extracellular matrix influence cell differentiation?
A) By providing nutrients directly to cells
B) Through signaling pathways initiated by ECM components
C) By altering the cell's DNA
D) By generating electrical impulses

4. Which enzyme is primarily responsible for the degradation of the extracellular matrix?
A) Kinase
B) Ligase
C) Matrix metalloproteinases (MMPs)
D) Amylase

5. Which of the following ECM components provides tensile strength?
A) Hyaluronan
B) Collagen
C) Laminin
D) Fibronectin

6. What role does fibronectin play in the extracellular matrix?
A) It provides compressive strength to tissues.
B) It serves as a scaffold for tissue repair.
C) It initiates DNA replication.
D) It generates ATP for cells.

7. Which ECM component is known for its ability to stretch and recoil, giving tissues their elasticity?
A) Collagen
B) Elastin
C) Fibronectin
D) Integrin

1.8 Future Directions and Challenges

Future directions in ECM dynamics research are increasingly focusing on emerging areas that aim to deepen our understanding of ECM biology and address current challenges in therapeutic targeting and clinical translation. These directions encompass several key research areas and challenges:

- **Mechanisms of ECM Regulation**: Future research aims to elucidate the complex regulatory mechanisms that govern ECM synthesis, degradation, and remodeling in health and disease. This includes investigating novel signaling pathways, epigenetic regulation, and post-translational modifications that influence ECM dynamics. Understanding these mechanisms will provide insights into how ECM alterations contribute to disease pathogenesis and identify new targets for therapeutic intervention.
- **ECM in Regenerative Medicine**: There is growing interest in leveraging ECM-based approaches for tissue regeneration and repair. Future research will explore the use of ECM-derived biomaterials, such as decellularized matrices and ECM-mimicking scaffolds, to promote tissue-specific regeneration. Challenges include optimizing scaffold design, enhancing bioactivity, and ensuring integration with host tissues to support functional recovery in diverse clinical applications.
- **Role of ECM in Immune Regulation**: The ECM is increasingly recognized as a modulator of immune responses in health and disease. Future studies will investigate how ECM components and their interactions with immune cells regulate inflammation, tissue repair, and immune surveillance. Understanding these interactions may uncover new therapeutic strategies for autoimmune disorders, chronic inflammation, and cancer immunotherapy.
- **ECM in Precision Medicine**: Advances in genomic and proteomic technologies are paving the way for personalized medicine approaches that consider individual variations in ECM composition and dynamics. Future research will focus on identifying ECM biomarkers associated with disease susceptibility, progression, and treatment response. Integrating ECM data with patient-specific profiles may enable tailored therapeutic strategies that optimize outcomes and minimize adverse effects.
- **Engineering Complex ECM Models**: Future research will advance the development of sophisticated in vitro and in vivo models that recapitulate the complexity of native ECM environments. This includes engineering 3D organotypic cultures, microfluidic devices, and organ-on-a-chip platforms that mimic tissue-specific ECM architecture, mechanical properties, and cell-ECM interactions. Challenges include recreating dynamic ECM remodeling processes and integrating multiple cell types to model disease states accurately.
- **Therapeutic Challenges**: Despite significant progress, translating ECM research into clinical therapies poses several challenges. These include optimizing delivery strategies for ECM-targeted therapies, overcoming barriers to ECM penetration in dense tissues, and ensuring safety and efficacy in clinical trials. Future research

will focus on addressing these challenges to accelerate the development and implementation of ECM-based therapies for a wide range of diseases.

Future directions in ECM dynamics research are poised to advance our understanding of ECM biology, uncover new therapeutic targets, and translate scientific discoveries into clinical applications [5]. Addressing emerging research areas and overcoming current challenges will enhance our ability to harness the therapeutic potential of ECM modulation in treating complex diseases and improving patient outcomes.

1.9 Summary

This chapter on ECM dynamics provides a comprehensive overview of the extracellular matrix's pivotal roles in physiological processes and disease pathogenesis. The ECM serves not only as a structural scaffold but also as a dynamic entity that regulates cellular behavior through biochemical and mechanical signals. Key components such as collagen, elastin, proteoglycans, and glycoproteins interact with cells via specific receptors like integrins, influencing processes such as cell adhesion, migration, proliferation, and differentiation. These interactions are essential for tissue development, maintenance, and repair across various organ systems.

In disease contexts, dysregulation of ECM dynamics contributes significantly to pathogenesis. Excessive ECM deposition and crosslinking lead to tissue fibrosis in conditions like liver cirrhosis and pulmonary fibrosis, impairing organ function. Conversely, altered ECM composition and stiffness promote cancer progression by enhancing tumor cell invasion and metastasis. Cardiovascular diseases also involve ECM remodeling, affecting arterial stiffness and vascular function. Understanding these mechanisms is crucial for developing targeted therapies that modulate ECM turnover, cell-ECM interactions, and biochemical signaling pathways to restore tissue homeostasis and improve clinical outcomes. Looking forward, future research in ECM dynamics is poised to explore emerging areas such as the role of ECM in immune regulation, regenerative medicine applications using ECM-derived biomaterials, and the development of personalized therapies based on ECM biomarkers. Advances in these areas will not only deepen our understanding of ECM biology but also pave the way for innovative treatments that harness the therapeutic potential of ECM modulation in diverse disease contexts. Addressing challenges in ECM research, such as engineering complex ECM models and translating findings into clinical practice, will be essential for realizing the full therapeutic promise of targeting ECM dynamics in medicine.

Points to Remember

- **The ECM is highly diverse across different tissues and organs, varying in composition, organization, and mechanical properties. For instance, bone ECM is rich in mineralized collagen, providing strength and rigidity, while brain ECM is specialized for supporting neuronal connections and synaptic transmission.**

- **Recent studies suggest that the ECM plays a role in memory formation and synaptic plasticity in the brain. Changes in ECM composition around synapses affect neuronal communication and learning processes, highlighting its dynamic role beyond structural support.**

- **The ECM microenvironment influences stem cell behavior, guiding differentiation into specific cell types. Different ECM compositions can direct stem cells towards bone, cartilage, muscle, or nerve cell lineages, demonstrating its regulatory role in tissue regeneration and engineering.**

- **Drugs that inhibit ECM-degrading enzymes like MMPs are being developed to treat conditions involving excessive ECM degradation, such as arthritis and cardiovascular diseases.**

- **Changes in ECM stiffness and composition facilitate tumor cell escape from the primary site and colonization of distant organs, influencing cancer prognosis and treatment outcomes.**

- **Aging leads to alterations in ECM composition and organization, contributing to tissue stiffness and loss of elasticity seen in skin aging, joint degeneration, and cardiovascular diseases.**

- **Advances in tissue engineering involve the development of ECM-mimicking biomaterials and scaffolds that promote tissue regeneration. These biomaterials, derived from ECM components or synthetic polymers, support cell adhesion, proliferation, and differentiation, offering potential solutions for repairing damaged tissues and organs.**

References

1. Yue B (2014) Biology of the extracellular matrix: an overview. J Glaucoma 23(8 Suppl 1):S20–S23. https://doi.org/10.1097/IJG.0000000000000108
2. Karamanos NK, Theocharis AD, Piperigkou Z, Manou D, Passi A, Skandalis SS, Vynios DH, Orian-Rousseau V, Ricard-Blum S, Schmelzer CEH, Duca L, Durbeej M, Afratis NA, Troeberg L, Franchi M, Masola V, Onisto M (2021) A guide to the composition and functions of the extracellular matrix. FEBS J 288(24):6850–6912. https://doi.org/10.1111/febs.15776
3. Kular JK, Basu S, Sharma RI (2014) The extracellular matrix: Structure, composition, age-related differences, tools for analysis and applications for tissue engineering. J Tissue Eng 5:2041731414557112. https://doi.org/10.1177/2041731414557112
4. Mecham RP (2012) Overview of extracellular matrix. Curr Protoc Cell Biol 10.1.1–10.1.16. https://doi.org/10.1002/0471143030.cb1001s57
5. Hu M, Ling Z, Ren X (2022) Extracellular matrix dynamics: tracking in biological systems and their implications. J Biol Eng 16(1):13. https://doi.org/10.1186/s13036-022-00292-x
6. Muncie JM, Weaver VM (2018) The physical and biochemical properties of the extracellular matrix regulate cell fate. Curr Top Dev Biol 130:1–37. https://doi.org/10.1016/bs.ctdb.2018.02.002

7. Walma DAC, Yamada KM (2020) The extracellular matrix in development. Dev (Cambridge, England) 147(10):dev175596. https://doi.org/10.1242/dev.175596
8. Theocharis AD, Skandalis SS, Gialeli C, Karamanos NK (2016) Extracellular matrix structure. Adv Drug Deliv Rev 97:4–27. https://doi.org/10.1016/j.addr.2015.11.001
9. McKee TJ, Perlman G, Morris M et al (2019) Extracellular matrix composition of connective tissues: a systematic review and meta-analysis. Sci Rep 9:10542. https://doi.org/10.1038/s41598-019-46896-0
10. Zimmermann R, Werner C, Sterling J (2018) Exploring structure–property relationships of GAGs to tailor ECM-mimicking hydrogels. Polymers 10(12):1376. https://doi.org/10.3390/polym10121376
11. Yanagishita M (1993) Function of proteoglycans in the extracellular matrix. Acta Pathol Jpn 43(6):283–293. https://doi.org/10.1111/j.1440-1827.1993.tb02569.x
12. Ahn S, Lee KY, Parker KK et al (2018) Formation of multi-component extracellular matrix protein fibers. Sci Rep 8:1913. https://doi.org/10.1038/s41598-018-20371-8
13. Huang J, Zhang L, Wan D et al (2021) Extracellular matrix and its therapeutic potential for cancer treatment. Sig Transduct Target Ther 6:153. https://doi.org/10.1038/s41392-021-00544-0
14. Naba A (2024) Mechanisms of assembly and remodelling of the extracellular matrix. Nat Rev Mol Cell Biol. https://doi.org/10.1038/s41580-024-00767-3
15. Zhang W, Liu Y, Zhang H (2021) Extracellular matrix: an important regulator of cell functions and skeletal muscle development. Cell Biosci 11(1):65. https://doi.org/10.1186/s13578-021-00579-4
16. Rosso F, Giordano A, Barbarisi M, Barbarisi A (2004) From cell-ECM interactions to tissue engineering. J Cell Physiol 199(2):174–180. https://doi.org/10.1002/jcp.10471
17. Sainio A, Järveläinen H (2020) Extracellular matrix-cell interactions: Focus on therapeutic applications. Cell Signal 66:109487. https://doi.org/10.1016/j.cellsig.2019.109487
18. Novoseletskaya ES, Evdokimov PV, Efimenko AY (2023) Extracellular matrix-induced signaling pathways in mesenchymal stem/stromal cells. Cell Commun Signal 21:244. https://doi.org/10.1186/s12964-023-01252-8
19. Bao Y, Wang L, Shi L, Yun F, Liu X, Chen Y, Chen C, Ren Y, Jia Y (2019) Transcriptome profiling revealed multiple genes and ECM-receptor interaction pathways that may be associated with breast cancer. Cell Mol Biol Lett 24:38. https://doi.org/10.1186/s11658-019-0162-0
20. Berrier AL, Yamada KM (2007) Cell-matrix adhesion. J Cell Physiol 213(3):565–573. https://doi.org/10.1002/jcp.21237
21. Dibus M, Joshi O, Ivaska J (2024) Novel tools to study cell-ECM interactions, cell adhesion dynamics and migration. Curr Opin Cell Biol 88:102355. https://doi.org/10.1016/j.ceb.2024.102355
22. Mishra YG, Manavathi B (2021) Focal adhesion dynamics in cellular function and disease. Cell Signal 85:110046. https://doi.org/10.1016/j.cellsig.2021.110046
23. Bonnans C, Chou J, Werb Z (2014) Remodelling the extracellular matrix in development and disease. Nat Rev Mol Cell Biol 15(12):786–801. https://doi.org/10.1038/nrm3904
24. Mohammed Butt A, Rupareliya V, Hariharan A, Kumar H (2023) Building a pathway to recovery: targeting ECM remodeling in CNS injuries. Brain Res 1819:148533. https://doi.org/10.1016/j.brainres.2023.148533
25. Cox TR, Erler JT (2011) Remodeling and homeostasis of the extracellular matrix: implications for fibrotic diseases and cancer. Dis Model Mech 4(2):165–178. https://doi.org/10.1242/dmm.004077
26. Sonbol HS (2018) Extracellular matrix remodeling in human disease. J Microsc Ultrastruct 6(3):123–128. https://doi.org/10.4103/JMAU.JMAU_4_18
27. Zhang Y, Zhao Y, An C, Guo Y, Ma Y, Shao F, Zhang Y, Sun K, Cheng F, Ren C, Zhang L, Sun B, Zhang Y, Wang H (2024) Material-driven immunomodulation and ECM remodeling reverse pulmonary fibrosis by local delivery of stem cell-laden microcapsules. Biomaterials 313:122757. https://doi.org/10.1016/j.biomaterials.2024.122757

28. Prakash J, Shaked Y (2024) The interplay between extracellular matrix remodeling and cancer therapeutics. Cancer Discov 14(8):1375–1388. https://doi.org/10.1158/2159-8290.CD-24-0002
29. Lu P, Weaver VM, Werb Z (2012) The extracellular matrix: a dynamic niche in cancer progression. J Cell Biol 196(4):395–406. https://doi.org/10.1083/jcb.201102147
30. Lu P, Takai K, Weaver VM, Werb Z (2011) Extracellular matrix degradation and remodeling in development and disease. Cold Spring Harb Perspect Biol 3(12):a005058. https://doi.org/10.1101/cshperspect.a005058
31. Sechler JL, Corbett SA, Wenk MB, Schwarzbauer JE (1998) Modulation of cell-extracellular matrix interactions. Ann N Y Acad Sci 857:143–154. https://doi.org/10.1111/j.1749-6632.1998.tb10114.x

Chapter 2
Cellular Interactions with the ECM

2.1 Introduction

The Extracellular Matrix (ECM) is a complex network of macromolecules that surrounds cells within tissues and organs of multicellular organisms. It consists primarily of structural proteins such as collagen, elastin, and fibronectin, embedded within a gel-like substance composed of proteoglycans and glycosaminoglycans. This intricate structure not only provides physical support to cells but also regulates various cellular functions such as adhesion, migration, proliferation, and differentiation [1]. Essentially, the ECM acts as a dynamic scaffold that not only maintains tissue integrity but also plays a crucial role in tissue development, homeostasis, and repair processes. Cellular interactions with the ECM are pivotal for numerous physiological processes. Cells interact with the ECM through specialized receptors known as integrins, which bind to specific ECM components [2]. These interactions are essential for transmitting biochemical and mechanical signals between the ECM and the cell interior, influencing cellular behavior and function. For instance, integrin-mediated adhesion to the ECM regulates cell polarity and shape, as well as activates intracellular signaling pathways that control gene expression and cell cycle progression. The ECM serves as a reservoir for various growth factors and cytokines, which are crucial for cell proliferation, differentiation, and tissue remodeling. The spatial and temporal distribution of these molecules within the ECM environment dictates cellular responses, thereby influencing tissue morphogenesis and function [3]. ECM remodeling, mediated by enzymes such as matrix metalloproteinases (MMPs), facilitates dynamic changes in tissue architecture during processes like wound healing, angiogenesis, and organ morphogenesis.

The ECM plays a critical role in regulating the biomechanical properties of tissues. It confers mechanical strength and elasticity to tissues like skin, cartilage, and blood vessels, enabling them to withstand physical forces and maintain structural integrity. Beyond its structural role, the ECM also acts as a barrier that separates different tissue

D. Suhag, *Extracellular Matrix biomimicking for Theranostics*, Biomedical Materials for Multi-functional Applications, https://doi.org/10.1007/978-981-95-7900-6_2

compartments, thereby compartmentalizing cellular activities and maintaining tissue organization [4].

Importantly, dysregulation of ECM composition or function is associated with various pathological conditions, including fibrosis, cancer metastasis, and cardiovascular diseases. For instance, aberrant ECM remodeling can lead to excessive deposition of collagen and formation of scar tissue, impairing organ function and promoting disease progression. Understanding the intricate interplay between cells and their ECM microenvironment is therefore essential for developing therapeutic strategies aimed at targeting ECM-related pathologies. ECM is a dynamic and multifaceted component of tissues that regulates cellular behavior and tissue function through physical, biochemical, and mechanical cues. Cellular interactions with the ECM are integral to processes ranging from embryonic development to tissue homeostasis and repair. By deciphering the complexities of ECM biology, researchers aim to uncover new insights into disease mechanisms and advance innovative therapies for various ECM-related disorders.

2.2 Structural Components of ECM

The Extracellular Matrix (ECM) is composed of a diverse array of structural components that collectively provide tissues with mechanical support, resilience, and functional versatility (Figs. 2.1 and 2.2). Key structural elements include fibrous proteins such as collagen, which forms strong fibrils and networks capable of withstanding tensile forces, and elastin, which imparts elasticity to tissues like skin and blood vessels, allowing them to stretch and recoil. Glycoproteins like fibronectin and laminin facilitate cell adhesion and migration through their binding sites, thereby anchoring cells within the ECM and mediating interactions critical for tissue organization and integrity [5]. Together, these structural components form a dynamic matrix that not only supports cellular functions but also influences tissue architecture and biomechanical properties essential for physiological processes and organ function.

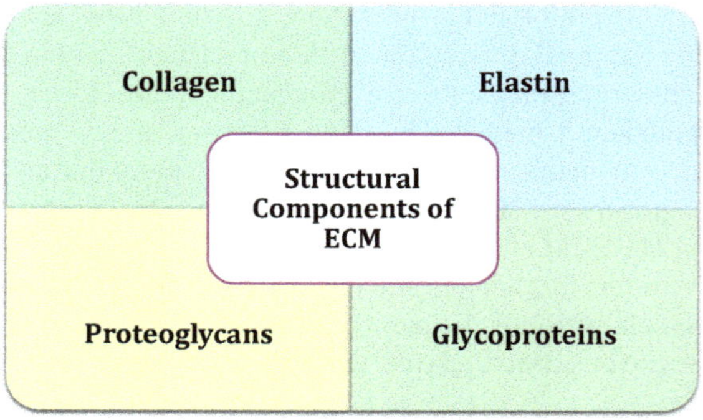

Fig. 2.1 Various structural components of ECM

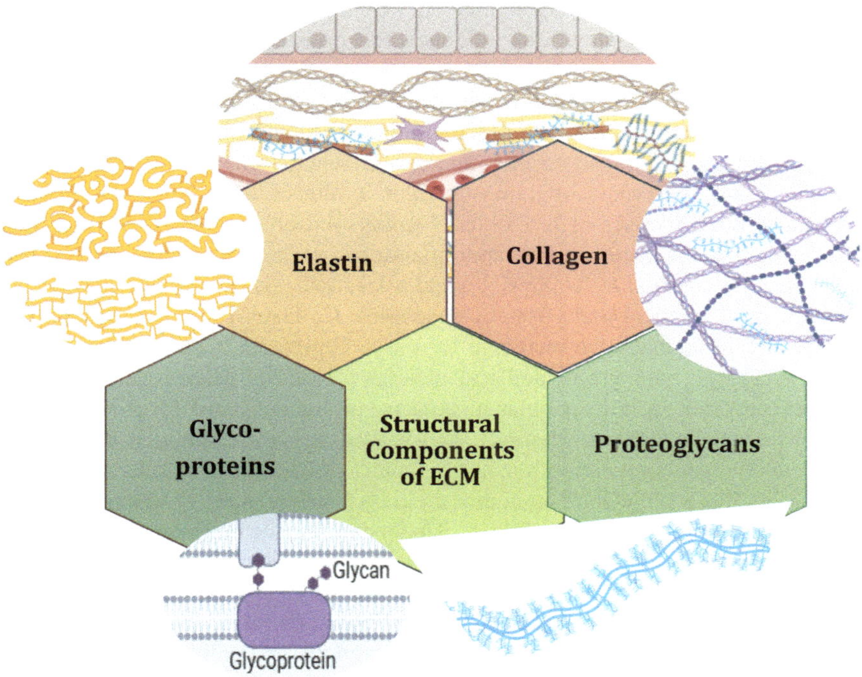

Fig. 2.2 Structural components of ECM

2.2.1 Collagen

Collagen is the most abundant protein in the human body, accounting for about 30% of total protein mass. It plays a critical role in maintaining the structural integrity and mechanical strength of various tissues and organs. Structurally, collagen is a fibrous protein characterized by its triple-helical structure, composed of three polypeptide chains intertwined in a helical conformation. This unique structure provides collagen with high tensile strength, making it well-suited for withstanding mechanical forces such as stretching and tension within tissues. Collagen serves as a major component of the Extracellular Matrix (ECM), where it forms dense networks and fibrils that provide structural support to tissues like skin, bones, tendons, ligaments, and cartilage. In skin, collagen fibers are crucial for maintaining elasticity and firmness, contributing to the skin's resilience and ability to resist wrinkles and sagging. In bones and cartilage, collagen fibers provide a framework that mineralizes to form strong, yet flexible, structures that support and protect internal organs and facilitate movement. Beyond its structural role, collagen is involved in various biological processes [6]. It acts as a scaffold for cell adhesion, migration, and tissue development, providing cues for cellular behavior and regulating processes such as wound healing and tissue repair. Collagen interacts with integrin receptors on cell surfaces, initiating signaling pathways that influence cell proliferation, differentiation, and gene expression.

Collagen synthesis is a tightly regulated process involving multiple enzymes and cofactors. Procollagen molecules are initially synthesized within cells and then secreted into the ECM, where they undergo enzymatic cleavage to form mature collagen fibrils. Disorders affecting collagen synthesis or structure can lead to connective tissue diseases such as Ehlers-Danlos syndrome and osteogenesis imperfecta, which are characterized by abnormalities in skin, joints, and other connective tissues. In addition to its physiological roles, collagen has garnered significant attention in biomedical and cosmetic industries. Collagen-based biomaterials are widely used in regenerative medicine for tissue engineering applications, such as wound dressings, bone grafts, and scaffolds for promoting tissue regeneration [7]. Collagen peptides are also popular in skincare products, touted for their potential to improve skin hydration, elasticity, and reduce wrinkles, although the efficacy of topical collagen application remains a subject of ongoing research. Age-related changes in collagen metabolism contribute to the visible signs of aging, including loss of skin elasticity and joint stiffness. Understanding the mechanisms underlying collagen synthesis, degradation, and turnover is therefore crucial for developing strategies to mitigate age-related degenerative processes and enhance tissue repair.

Overall, collagen's importance extends far beyond its role as a structural protein. It is essential for tissue integrity, mechanical strength, and cellular function across various physiological contexts.

2.2.2 Elastin

Elastin is a vital protein found in the Extracellular Matrix (ECM) of vertebrate tissues, known for its unique ability to provide elasticity and resilience to various organs and structures throughout the body. Unlike collagen, which provides strength and rigidity, elastin imparts elasticity and recoil properties to tissues, allowing them to stretch and return to their original shape. This characteristic elasticity is crucial for tissues subjected to repeated stretching and contraction, such as skin, lungs, arteries, and ligaments. Structurally, elastin is composed of an insoluble matrix of elastin fibers cross-linked with other proteins like fibrillin. These fibers are arranged in a highly organized network within the ECM, forming a dynamic scaffold that supports tissue architecture while enabling elastic recoil. This arrangement allows tissues to withstand mechanical stress and deformation without permanent damage, contributing to their functional integrity and performance [8]. Elastin synthesis and assembly are tightly regulated processes involving specialized cells called fibroblasts and smooth muscle cells. These cells secrete precursor molecules known as tropoelastin into the ECM, where they undergo enzymatic modification and cross-linking to form mature elastin fibers. This process is essential for maintaining tissue elasticity and responding to mechanical cues such as blood pressure changes in arteries or stretching of the skin during movement. In addition to its structural role, elastin plays a critical role in various physiological processes. In the lungs, elastin fibers within alveolar walls allow for efficient gas exchange by facilitating the expansion and contraction of lung tissue during breathing. In arteries, elastic fibers in the walls help maintain blood pressure by expanding and recoiling with each heartbeat, ensuring efficient blood flow throughout the cardiovascular system.

Disruption or degradation of elastin fibers can lead to significant pathological conditions. For example, in diseases like emphysema, chronic obstructive pulmonary disease (COPD), or Marfan syndrome, mutations or deficiencies in elastin or its associated proteins (such as fibrillin) can impair tissue elasticity and function [9]. This can result in reduced lung function, arterial stiffness, and cardiovascular complications, highlighting the critical role of elastin in maintaining tissue health and physiological homeostasis. In biomedical research and clinical applications, elastin's unique properties have inspired the development of biomaterials and therapeutic strategies aimed at restoring tissue elasticity and function. Elastin-based scaffolds and biomimetic materials are used in tissue engineering to mimic the ECM environment and promote regeneration in damaged tissues such as skin, blood vessels, and cardiac muscle. Understanding the mechanisms underlying elastin synthesis, assembly, and degradation is essential for developing treatments for elastin-related disorders and age-related changes in tissue elasticity. Research continues to explore novel approaches to enhance elastin production, prevent elastin degradation, and improve tissue repair and regeneration, underscoring the ongoing importance of elastin in both health and disease contexts.

Elastin is a critical component of the ECM that confers elasticity and resilience to tissues throughout the body. Its unique structural and functional properties support tissue integrity, enable physiological processes like breathing and circulation, and contribute to overall health and well-being. Advances in elastin biology and technology hold promise for future therapies aimed at enhancing tissue elasticity and treating elastin-related disorders, emphasizing its significance in biomedical research and clinical practice.

2.2.3 Proteoglycans

Proteoglycans are essential components of the Extracellular Matrix (ECM), playing crucial roles in maintaining tissue structure, hydration, and regulating cellular interactions. These macromolecules consist of a core protein to which one or more glycosaminoglycan (GAG) chains are covalently attached. Glycosaminoglycans are long, linear polysaccharides characterized by repeated disaccharide units that can be sulfated, providing proteoglycans with negative charges that attract and bind water molecules, thereby contributing to tissue hydration and viscosity. The diversity of proteoglycans allows them to fulfill various functions depending on their tissue and cellular context. For instance, in cartilage, aggrecan is a prominent proteoglycan that forms aggregates with hyaluronic acid, creating a hydrated gel-like matrix that cushions joints and provides resistance to compressive forces. This structure is critical for joint function and mobility, as well as for absorbing shock during physical activities [10]. Beyond their structural roles, proteoglycans interact with growth factors, cytokines, and other ECM components to regulate cellular behavior and tissue homeostasis. They serve as reservoirs for signaling molecules, modulating their availability and activity within the ECM. For example, heparan sulfate proteoglycans (HSPGs) participate in cell signaling pathways by binding growth factors such as fibroblast growth factors (FGFs) and transforming growth factor-beta (TGF-β), thereby regulating processes like cell proliferation, differentiation, and migration. Proteoglycans also play pivotal roles in development and tissue morphogenesis. During embryogenesis, they guide cell migration and tissue patterning through interactions with integrins and other ECM components. In neural tissues, proteoglycans influence axon guidance and synaptic plasticity, contributing to proper nervous system development and function. In wound healing and tissue repair, proteoglycans regulate the inflammatory response, promote cell adhesion and migration, and modulate the deposition and remodeling of ECM components essential for tissue regeneration.

Disruption of proteoglycan function or composition is associated with various pathological conditions. For instance, defects in aggrecan or other cartilage-specific proteoglycans can lead to conditions like osteoarthritis, characterized by joint pain and degeneration [11]. Inflammatory diseases such as atherosclerosis can be exacerbated by alterations in ECM proteoglycans, which influence the retention and modification of lipoproteins within arterial walls, contributing to plaque formation and cardiovascular complications.

In biomedical research, proteoglycans are targets for therapeutic interventions aimed at modulating ECM dynamics and cellular responses. Strategies to enhance proteoglycan synthesis or stability may promote tissue regeneration and repair, particularly in diseases characterized by ECM degradation or dysfunction. Furthermore, advances in proteoglycan biology are paving the way for innovative approaches in drug delivery and tissue engineering, where proteoglycan-based biomaterials are utilized for their ability to mimic native ECM environments and support cell growth and differentiation. Proteoglycans are integral components of the ECM with diverse roles in tissue structure, hydration, cell signaling, and development. Their interactions with growth factors and other ECM molecules regulate essential biological processes, influencing tissue integrity, repair, and disease progression.

2.2.4 Glycoproteins

Glycoproteins are a class of proteins that are covalently linked to carbohydrates, forming glycoconjugates that play critical roles in various biological processes. These molecules are found throughout the body, both on cell surfaces and within the Extracellular Matrix (ECM), where they contribute to structural integrity, cell signaling, and molecular recognition. Structurally, glycoproteins consist of a protein core with one or more carbohydrate chains attached via glycosidic bonds. The types and arrangements of carbohydrates can vary widely, influencing the function and properties of glycoprotein. For example, glycoproteins on cell surfaces often serve as recognition sites for specific molecules, including hormones, antibodies, and other cells, facilitating cell–cell interactions and communication. In the ECM, glycoproteins such as fibronectin and laminin play pivotal roles in cell adhesion and migration. Fibronectin forms fibrillar networks that bind to cell surface receptors called integrins, anchoring cells to the ECM and mediating their interactions with surrounding tissues [12]. This process is crucial for tissue organization, wound healing, and embryonic development, where coordinated cell movements and tissue morphogenesis rely on dynamic interactions with ECM glycoproteins. Beyond their structural functions, glycoproteins are involved in regulating cellular processes through signaling pathways. For instance, growth factors and cytokines often bind to glycoproteins like syndecans and glypicans, which are proteoglycans containing attached glycosaminoglycan chains. These interactions modulate growth factor signaling, influencing cell proliferation, differentiation, and tissue repair mechanisms. Glycoproteins can serve as carriers for lipid molecules or facilitate the transport of ions and nutrients across cellular membranes, contributing to metabolic regulation and cellular homeostasis.

In the immune system, glycoproteins on the surface of cells are involved in antigen recognition and immune response activation. Major histocompatibility complex (MHC) glycoproteins, for example, present antigens to T cells, triggering immune responses against pathogens or abnormal cells. These glycoproteins are essential for adaptive immunity, enabling the immune system to distinguish between self and non-self-molecules and mount targeted responses to infections or cancerous cells. Disruptions in glycoprotein structure or function can lead to various diseases and disorders. For instance, deficiencies in glycoproteins involved in clotting factors can result in bleeding disorders such as hemophilia [13]. Similarly, mutations affecting glycoproteins in ECM can impair tissue integrity and lead to conditions like Ehlers-Danlos syndrome, characterized by hypermobility of joints and fragile skin. Understanding the role of glycoproteins in disease mechanisms is crucial for developing targeted therapies and diagnostic tools to address these conditions effectively. In biomedical research, glycoproteins are targets for therapeutic interventions and drug development. Antibodies and therapeutic proteins often exploit glycoprotein-specific interactions for targeted delivery and modulation of cellular functions. Glycoprotein-based biomaterials are also utilized in tissue engineering and regenerative medicine to mimic ECM environments and support cell growth and differentiation, offering potential solutions for repairing damaged tissues and organs.

Glycoproteins are versatile biomolecules with diverse functions in cellular recognition, adhesion, signaling, and immune response regulation. Their roles extend across physiological processes, influencing tissue development, homeostasis, and disease pathology.

2.3 Cellular Receptors and ECM

Cellular receptors play a pivotal role in mediating interactions between cells and the Extracellular Matrix (ECM), facilitating dynamic communication and signaling essential for tissue development, homeostasis, and response to physiological cues. Integral to this process are transmembrane proteins such as integrins, syndecans, and discoidin domain receptors (DDRs), which bind specific ECM components and transmit biochemical and mechanical signals into the cell interior. These receptors not only anchor cells to the ECM but also regulate cellular behaviors such as adhesion, migration, proliferation, and differentiation through intricate signaling pathways [14]. The diversity and specificity of cellular receptors allow cells to sense and respond to changes in their microenvironment, influencing tissue architecture, function, and adaptive responses during development, wound healing, and disease processes. Understanding the mechanisms by which cellular receptors interact with the ECM is crucial for deciphering fundamental biological processes and developing therapeutic strategies targeting ECM-related disorders and tissue regeneration.

2.3.1 Integrins

Integrins are a family of transmembrane receptors that play pivotal roles in mediating cellular interactions with the Extracellular Matrix (ECM) and with other cells. They are heterodimeric proteins composed of α and β subunits, with different combinations of these subunits yielding diverse integrin types that recognize specific ECM ligands. Integrins are essential for anchoring cells to the ECM and for transmitting bidirectional signals across the plasma membrane, thereby regulating cellular behaviors such as adhesion, migration, proliferation, and differentiation. Integrins have an extracellular domain that binds ECM proteins such as fibronectin, collagen, laminin, and vitronectin, among others [15]. This binding is mediated by specific amino acid sequences within the ECM proteins, which interact with integrin binding sites. On the cytoplasmic side, integrins connect to the cell's internal cytoskeleton through linker proteins such as talin, vinculin, and focal adhesion kinase (FAK). This linkage enables integrins to transmit mechanical forces between the ECM and the cell interior, influencing cell shape, polarity, and motility. Integrins also serve as signaling hubs that coordinate various intracellular signaling pathways in response to ECM cues. Upon ligand binding, integrins undergo conformational changes that activate downstream signaling molecules such as kinases (e.g., FAK and Src family kinases), GTPases (e.g., Rho and Rac), and adaptor proteins (e.g., paxillin). These signaling events regulate processes like cytoskeletal rearrangement, gene expression, and cell cycle progression, thereby influencing cellular responses to changes in the ECM environment. The specificity and diversity of integrin interactions with ECM components and other cellular ligands contribute to their critical roles in tissue development, homeostasis, and repair. During embryogenesis, integrins facilitate cell adhesion, migration, and tissue morphogenesis, helping to shape organs and establish tissue architecture [16]. In adults, integrins maintain tissue integrity by mediating cell-ECM interactions that are essential for tissue function and resilience. For example, in muscle tissue, integrins are involved in linking muscle fibers to the ECM, enabling force transmission and contraction.

Dysregulation of integrin function or expression is implicated in various pathological conditions, including cancer metastasis, inflammatory diseases, and cardiovascular disorders. Abnormal integrin signaling can promote tumor cell invasion and metastasis by enhancing cell adhesion to ECM proteins in distant tissues. Inflammatory responses can be exacerbated by integrin-mediated leukocyte adhesion to endothelial cells lining blood vessels, contributing to tissue damage and disease progression.

In biomedical research and clinical practice, integrins are important targets for therapeutic interventions aimed at modulating cell-ECM interactions and treating integrin-related diseases. Strategies include developing integrin-specific inhibitors to block pathological integrin signaling in cancer or inflammation, as well as using integrin-targeting biomaterials in tissue engineering to promote cell adhesion and regeneration.

2.3.2 Discoidin Domain Receptors (DDRs)

Discoidin domain receptors (DDRs) are a unique family of receptor tyrosine kinases that play crucial roles in mediating cellular interactions with the Extracellular Matrix (ECM). Unlike integrins, which are heterodimeric proteins, DDRs are single-pass transmembrane proteins characterized by their discoidin homology domain, a conserved motif involved in ligand binding. DDRs are activated by binding to collagen types I, II, III, and IV, as well as other ECM components, making them key players in collagen-mediated signaling pathways. DDRs consist of an extracellular discoidin domain, a transmembrane domain, and an intracellular tyrosine kinase domain. Upon collagen binding, DDRs undergo autophosphorylation of tyrosine residues within their cytoplasmic domain, leading to the recruitment and activation of downstream signaling molecules such as Src family kinases and focal adhesion kinase (FAK). These signaling events regulate cellular processes including adhesion, migration, proliferation, and survival, thereby influencing tissue development, homeostasis, and response to injury. DDR1 and DDR2 are the two main isoforms of DDRs identified in humans, each with distinct expression patterns and functions. DDR1 is primarily expressed in epithelial cells, endothelial cells, and certain types of cancer cells, where it regulates cell–matrix interactions and promotes epithelial-mesenchymal transition (EMT) in response to ECM remodeling. DDR2, on the other hand, is predominantly found in mesenchymal cells such as fibroblasts and osteoblasts, and is implicated in processes like bone development, wound healing, and tissue repair [17]. In addition to their roles in normal physiological processes, DDRs have been implicated in various pathological conditions. Dysregulation of

DDR signaling is associated with fibrotic diseases, cancer progression, and cardiovascular disorders. For example, aberrant DDR1 signaling in fibroblasts contributes to excessive collagen deposition and tissue fibrosis in conditions like pulmonary fibrosis and liver cirrhosis. In cancer, DDRs promote tumor cell invasion and metastasis by enhancing cell adhesion, migration, and survival in response to collagen-rich ECM environments.

Research into DDR biology has uncovered potential therapeutic strategies targeting DDR signaling pathways. Small molecule inhibitors and monoclonal antibodies that block DDR activation or downstream signaling are being investigated for their potential to suppress fibrosis, inhibit cancer metastasis, and improve outcomes in patients with DDR-related diseases [18]. Furthermore, understanding the intricate mechanisms by which DDRs interact with the ECM and regulate cellular responses provides insights into fundamental aspects of cell-ECM interactions and tissue dynamics.

Discoidin domain receptors (DDRs) represent a specialized family of receptor tyrosine kinases that mediate cellular responses to collagen and other ECM components. Their roles in regulating cell adhesion, migration, proliferation, and survival are essential for tissue development, maintenance of tissue integrity, and adaptive responses to environmental cues. Dysregulation of DDR signaling contributes to various diseases, highlighting their potential as therapeutic targets for treating fibrosis, cancer, and other DDR-associated disorders.

2.3.3 Syndecans

Syndecans are a family of transmembrane proteoglycans that play essential roles in mediating cell-Extracellular Matrix (ECM) interactions and cell signaling processes. They are characterized by a short cytoplasmic domain, a single-pass transmembrane domain, and an extracellular domain that binds to ECM components and growth factors. Syndecans are unique among proteoglycans because they can bind both glycosaminoglycan (GAG) chains, such as heparan sulfate and chondroitin sulfate, and specific ECM proteins, thereby acting as coreceptors for various ligands. There are four mammalian syndecan isoforms (Syndecan-1 to -4), each with distinct tissue distributions and functions. Syndecan-1 is predominantly expressed in epithelial cells and is involved in regulating cell adhesion, migration, and signaling during tissue repair and inflammation. Syndecan-2 is found in mesenchymal cells, where it participates in ECM organization and cell–matrix interactions critical for development and wound healing. Syndecan-3 is mainly expressed in neurons and skeletal muscle cells, playing roles in synaptic function and muscle development. Syndecan-4 is widely distributed and is crucial for cell adhesion and cytoskeletal organization in various cell types, including fibroblasts and endothelial cells.

The extracellular domain of syndecans can bind to growth factors such as fibroblast growth factors (FGFs), vascular endothelial growth factor (VEGF), and transforming growth factor-beta (TGF-β). By binding to these growth factors, syndecans regulate their bioavailability, signaling potency, and spatial distribution within tissues [19]. This interaction facilitates diverse cellular responses, including cell proliferation, differentiation, and migration, thereby influencing tissue development, homeostasis, and repair.

Syndecans also participate in cell signaling pathways through their cytoplasmic domains, which can interact with intracellular signaling molecules and adaptor proteins. For example, syndecan-4 interacts with protein kinase C (PKC), Src family kinases, and focal adhesion kinase (FAK) to regulate actin cytoskeleton dynamics, cell adhesion, and migration in response to ECM cues. This integrative signaling role makes syndecans critical mediators of cellular responses to environmental changes and developmental cues. In addition to their physiological roles, syndecans are implicated in various pathological conditions when dysregulated. Abnormal syndecan expressions or function is associated with diseases such as cancer, inflammation, and fibrosis. For instance, altered syndecan expression in cancer cells can promote tumor growth and metastasis by enhancing cell adhesion, migration, and survival in response to ECM remodeling and growth factor signaling. Similarly, dysregulation of syndecan-mediated inflammation pathways can contribute to chronic inflammatory diseases like rheumatoid arthritis and inflammatory bowel disease. Therapeutically, syndecans and their interactions with ECM components and growth factors represent promising targets for intervention. Strategies to modulate syndecan expression, activity, or downstream signaling pathways are being explored for their potential to treat diseases characterized by aberrant ECM remodeling, inflammation, and tissue dysfunction. Moreover, understanding the complex roles of syndecans in health and disease continues to drive research aimed at uncovering new therapeutic approaches and improving patient outcomes in diverse clinical settings.

Syndecans are versatile transmembrane proteoglycans that integrate ECM interactions with growth factor signaling to regulate cellular behaviors crucial for tissue development, repair, and disease pathogenesis. Their ability to modulate cell adhesion, migration, and signaling pathways underscores their importance in physiological processes and positions them as significant targets for therapeutic intervention in various ECM-related disorders.

Exercise 1

Fill Ups
(MAPK/ERK, Matrix metalloproteinases (MMPs), Integrins, Fibronectin, Elastin)

1. .... are transmembrane receptors that mediate cell attachment to the ECM and play a critical role in cell signaling.

2. The ECM protein .... binds to integrins and facilitates cell adhesion, migration, and wound healing.

3. The interaction between integrins and the ECM can activate the .... signaling pathway, which influences cell proliferation and survival.

4. .... is the ECM component that primarily provides elasticity to tissues and is crucial in tissues like the skin and lungs.

5. Cell-ECM interactions are often modulated by enzymes such as ...., which degrade ECM components to allow for tissue remodeling.

2.4 Mechanisms of Cellular Adhesion

Cellular adhesion refers to the complex mechanisms by which cells attach to and interact with surfaces, including other cells and the Extracellular Matrix (ECM). This fundamental process is essential for maintaining tissue integrity, mediating cell signaling, and facilitating various physiological functions such as embryonic development, immune response, wound healing, and tissue homeostasis. Key molecules involved in cellular adhesion include integrins, cadherins, selectins, and immunoglobulin superfamily proteins, each of which mediates specific interactions depending on the cellular context and the nature of the adhesion substrate [20]. These adhesion molecules not only anchor cells to their surroundings but also transmit mechanical and biochemical signals across the cell membrane, influencing cellular behavior, differentiation, and response to external stimuli. Understanding the intricate mechanisms of cellular adhesion is crucial for elucidating basic biological processes and developing strategies to intervene in diseases characterized by aberrant cell adhesion and migration, such as cancer metastasis and inflammatory disorders.

2.4.1 Focal Adhesions

Focal adhesions are complex protein structures that mediate the physical and functional linkage between cells and the Extracellular Matrix (ECM). These adhesion complexes are crucial for maintaining tissue integrity, transmitting mechanical signals, and regulating cellular behaviors such as adhesion, migration, and differentiation. Focal adhesions are dynamic structures composed of integrins, which span the plasma membrane and bind to ECM proteins like fibronectin and collagen, linking the ECM to the actin cytoskeleton inside the cell. Focal adhesions consist of clusters of proteins including integrins, focal adhesion kinase (FAK), talin, vinculin, paxillin, and other adaptor proteins. Integrins within focal adhesions bind to specific ECM ligands, initiating a cascade of signaling events that lead to the recruitment and activation of FAK and Src family kinases. These kinases phosphorylate downstream targets involved in cytoskeletal organization, actin dynamics, and gene expression, thereby regulating cell adhesion strength and motility. Focal adhesions serve as mechanosensitive signaling hubs that convert mechanical forces from the ECM into biochemical signals within the cell [21]. For example, when cells adhere to a stiff ECM substrate, focal adhesions become larger and more numerous, activating signaling pathways that promote cell spreading and migration. This mechano transduction process is crucial for processes such as tissue development, wound healing, and immune response, where cells must dynamically respond to changes in their mechanical environment. In addition to their role in mechanical sensing and force transmission, focal adhesions play important roles in regulating cell proliferation, survival, and differentiation. Through their interactions with growth factor receptors and other signaling molecules, focal adhesions modulate cellular responses to extracellular cues, influencing tissue morphogenesis and homeostasis. Disruption of focal adhesion dynamics or signaling pathways can lead to various diseases, including cancer metastasis, where aberrant focal adhesion turnover promotes tumor cell invasion and dissemination to distant sites.

Research into focal adhesions has revealed their involvement in diverse cellular processes and their dysregulation in disease states. For instance, focal adhesion proteins like FAK and vinculin are overexpressed or hyperactivated in certain cancers, contributing to enhanced tumor cell adhesion, survival, and metastatic potential. Targeting focal adhesion proteins and signaling pathways has emerged as a promising strategy for cancer therapy, aiming to inhibit tumor cell migration and invasion by disrupting their ability to form stable adhesion complexes with the ECM.

Focal adhesions are integral to tissue engineering and regenerative medicine approaches. Biomaterials designed to mimic the ECM environment often incorporate specific ligands that engage integrins and promote focal adhesion formation, facilitating cell attachment, proliferation, and differentiation. Understanding the molecular mechanisms underlying focal adhesion assembly and regulation is therefore essential for optimizing biomaterial design and enhancing tissue regeneration strategies. Focal adhesions are multifunctional complexes that link cells to the ECM, regulate cellular responses to mechanical and biochemical signals, and play critical roles in tissue development, homeostasis, and disease progression. Their dynamic nature and integration with signaling pathways make focal adhesions a focal point of research in cell biology, biomedical engineering, and therapeutic development, offering opportunities to better understand and manipulate cellular behaviors in health and disease contexts.

2.4.2 Hemidesmosomes

Hemidesmosomes are specialized adhesive structures found in epithelial cells that anchor these cells to the underlying basement membrane, a component of the Extracellular Matrix (ECM). They play a crucial role in maintaining tissue integrity and stability by providing strong adhesion between epithelial cells and the underlying ECM proteins such as laminin and collagen IV. Hemidesmosomes are particularly abundant in tissues subjected to mechanical stress, such as the skin, where they contribute to the resilience and structural integrity of epithelial layers. Hemidesmosomes resemble half-desmosomes (hemi = half), with their assembly initiated by integrin family members, particularly $\alpha6\beta4$ integrins, on the basal surface of epithelial cells. These integrins bind to laminin-332 (formerly known as laminin-5), a major component of the basement membrane. Inside the cell, integrins are linked to intermediate filaments such as keratin filaments via adaptor proteins including plectin and BP230 (bullous pemphigoid antigen 230), forming a stable anchorage complex. The assembly and maintenance of hemidesmosomes are tightly regulated processes involving a network of proteins. Key components include integrin $\alpha6\beta4$, BP180 (bullous pemphigoid antigen 180), BP230, plectin, and various cytoskeletal proteins. Integrin signaling within hemidesmosomes not only anchors epithelial cells to the basement membrane but also regulates cell polarity, differentiation, and survival through activation of signaling pathways such as the PI3K (phosphoinositide 3-kinase)/Akt pathway. Hemidesmosomes are critical for epithelial cell function and tissue integrity, as evidenced by genetic disorders and autoimmune diseases that affect their structure or function. Mutations in genes encoding hemidesmosome components can lead to diseases such as epidermolysis bullosa, a group of inherited skin disorders characterized by blistering and skin fragility due to defective anchoring of epidermal cells to the basement membrane. Autoimmune diseases like pemphigoid and pemphigus involve autoantibodies targeting hemidesmosome proteins, leading to blistering and tissue damage. In addition to their adhesive function, hemidesmosomes play roles in cell signaling and mechano transduction [22]. They sense mechanical forces exerted on epithelial tissues and transmit signals that influence cellular responses to ECM stiffness and tension. This mechanosensitive property is crucial for processes like wound healing, where epithelial cells must adhere firmly to the basement membrane and respond to mechanical cues to migrate and proliferate effectively.

Research into hemidesmosomes is advancing our understanding of their roles in development, tissue homeostasis, and disease. Strategies aimed at modulating hemidesmosome function hold promise for therapeutic interventions in skin diseases and wound healing. Enhancing our knowledge of hemidesmosome biology may also inform efforts to engineer biomaterials that promote epithelial cell adhesion and regeneration, offering potential solutions for treating skin injuries and disorders associated with defective hemidesmosome function. Hemidesmosomes are essential adhesive complexes that anchor epithelial cells to the basement membrane, contributing to tissue stability, mechanical strength, and cellular signaling in epithelial tissues.

2.4.3 Cadherin-Mediated Adhesion

Cadherins are a family of calcium-dependent cell adhesion molecules that play fundamental roles in tissue organization, development, and maintenance of cell–cell contacts in multicellular organisms. These transmembrane glycoproteins are characterized by extracellular cadherin repeats that facilitate homophilic binding with cadherins on adjacent cells. The cytoplasmic tail of cadherins interacts with cytoskeletal proteins such as catenins, linking the cadherin-catenin complex to the actin cytoskeleton, which is crucial for the stabilization and regulation of cell–cell adhesion. The specificity and strength of cadherin-mediated adhesion are critical for establishing and maintaining tissue architecture and function. Different cadherin subtypes are expressed in a tissue-specific manner, contributing to the diversity and specificity of cell–cell interactions in various tissues. For example, E-cadherin (epithelial cadherin) is primarily found in epithelial tissues, where it mediates strong adhesion between adjacent epithelial cells, essential for forming epithelial barriers and maintaining tissue integrity. Cadherin-mediated adhesion is dynamically regulated during development, tissue morphogenesis, and wound healing processes. Changes in cadherin expression, localization, or binding affinity can influence cell sorting, tissue patterning, and collective cell migration [23]. For instance, during embryonic development, differential expressions of cadherins help establish tissue boundaries and organize cell layers, ensuring proper tissue architecture and function.

Beyond their adhesive function, cadherins are involved in signaling pathways that regulate cell proliferation, differentiation, and migration. Adhesive interactions mediated by cadherins can activate intracellular signaling cascades, including those involving Rho GTPases and MAP kinases, which influence cytoskeletal dynamics and gene expression. These signaling events are crucial for coordinating cellular behaviors in response to extracellular cues and for maintaining tissue homeostasis. Dysregulation of cadherin-mediated adhesion is implicated in various diseases, particularly cancer metastasis and tissue fibrosis. Reduced expression or loss of E-cadherin, for example, is associated with increased invasiveness and metastatic potential in many epithelial cancers, where disruption of cell–cell adhesion allows tumor cells to dissociate from primary tumors and invade surrounding tissues. In fibrotic diseases, aberrant cadherin signaling can contribute to excessive tissue remodeling and scar formation, impacting organ function and integrity. Research into cadherin biology continues to uncover new insights into their roles in health and disease, driving efforts to develop therapeutic strategies targeting cadherin-mediated adhesion. Strategies aimed at restoring or modulating cadherin function in cancer therapy, for instance, include approaches to enhance cell–cell adhesion and inhibit tumor cell invasion. Similarly, understanding the mechanisms by which cadherins regulate tissue morphogenesis and wound healing informs strategies for tissue engineering and regenerative medicine. Cadherin-mediated adhesion is a fundamental mechanism for establishing and maintaining cell–cell contacts in tissues, essential for tissue organization, development, and physiological function.

Exercise 2

Fill Ups
(ECM turnover, Homeostasis, Matrix metalloproteinases (MMPs), Fibronectin, Fibrosis)

1. .... are enzymes that degrade ECM components, facilitating tissue remodeling and repair.

2. Excessive ECM deposition during remodeling can lead to the formation of ...., which can impair tissue function.

3. The balance between ECM synthesis and degradation is critical for maintaining .... during tissue remodeling.

4. .... is an ECM glycoprotein that plays a key role in wound healing and is involved in the remodeling of the provisional matrix.

5. ....is a process by which old or damaged ECM is broken down and replaced with new ECM, ensuring tissue integrity and function.

2.5 ECM Remodeling

Extracellular Matrix (ECM) remodeling refers to the dynamic process through which the composition, organization, and mechanical properties of the ECM are modified in response to various physiological and pathological cues. The ECM, comprising a complex network of proteins, glycoproteins, proteoglycans, and other macromolecules, provides structural support to tissues and organs while regulating cell behavior through biochemical and biomechanical signaling. ECM remodeling plays crucial roles in tissue development, wound healing, inflammation, and disease progression by influencing cell adhesion, migration, proliferation, and differentiation [24]. This process involves enzymatic degradation of ECM components such as collagen and elastin by matrix metalloproteinases (MMPs) and other proteases, as well as synthesis and deposition of new ECM components by resident cells like fibroblasts and endothelial cells.

2.5.1 Matrix Metalloproteinases (MMPs)

Matrix metalloproteinases (MMPs) are a family of zinc-dependent endopeptidases that play critical roles in the remodeling of the Extracellular Matrix (ECM) under normal physiological conditions and in various pathological processes. These enzymes are involved in the degradation and turnover of ECM components such as collagen, elastin, fibronectin, and laminin, thereby regulating tissue remodeling, wound healing, angiogenesis, and immune responses. MMPs are synthesized as inactive zymogens (pro-MMPs) and are activated in response to specific stimuli, including cytokines, growth factors, and ECM proteins, highlighting their tight regulation and role in maintaining tissue homeostasis. MMPs are classified into several subgroups based on their substrate specificity and domain organization. Common types include collagenases (e.g., MMP-1, MMP-8, MMP-13) that degrade collagen fibers, gelatinases (e.g., MMP-2, MMP-9) that hydrolyze gelatin and collagen fragments, and stromelysins (e.g., MMP-3, MMP-10) that degrade proteoglycans, laminin, and fibronectin. Each MMP subtype exhibits distinct substrate preferences and cellular localization, allowing for precise regulation of ECM remodeling in different tissues and biological contexts. The activity of MMPs is tightly controlled at multiple levels to prevent excessive ECM degradation and maintain tissue integrity. Regulation occurs through transcriptional control, post-translational modifications, tissue inhibitors of metalloproteinases (TIMPs), and interactions with other ECM components and cell surface receptors. TIMPs, in particular, act as endogenous inhibitors that bind to active MMPs, blocking their enzymatic activity and maintaining ECM homeostasis [25]. Imbalances in MMP/TIMP ratios have been implicated in numerous diseases, including cancer, cardiovascular diseases, arthritis, and neurological disorders, highlighting the importance of MMP regulation in health and disease.

In addition to ECM degradation, MMPs also regulate the bioavailability and activity of growth factors, cytokines, and cell surface receptors through proteolytic cleavage (shedding). For example, MMP-mediated shedding of cell adhesion molecules (CAMs) and growth factor receptors can modulate cell–cell interactions, migration, and signaling pathways involved in inflammation, immunity, and tissue repair. This dual role of MMPs in ECM remodeling and signal transduction underscores their multifaceted functions in cellular physiology and pathology. Dysregulation of MMP activity is implicated in various pathological conditions, particularly in cancer metastasis and tissue fibrosis. In cancer, elevated MMP expression promotes tumor invasion and metastasis by facilitating ECM degradation and remodeling of the tumor microenvironment. Similarly, in fibrotic diseases such as pulmonary fibrosis and liver cirrhosis, excessive MMP activity contributes to ECM deposition and tissue scarring, impairing organ function and integrity. Targeting MMPs and their regulatory pathways has emerged as a promising therapeutic strategy for modulating ECM remodeling and mitigating disease progression in these contexts.

Strategies aimed at selectively targeting MMPs involved in specific disease processes while preserving essential ECM functions hold potential for improving clinical outcomes in cancer therapy, cardiovascular disease, and fibrotic disorders. Moreover, understanding the complex interactions between MMPs, ECM components, and cellular signaling pathways provides opportunities for developing personalized medicine approaches tailored to individual disease profiles and patient responses.

2.5.2 Tissue Inhibitors of Metalloproteinases (TIMPs)

Tissue inhibitors of metalloproteinases (TIMPs) are a family of proteins that regulate the activity of matrix metalloproteinases (MMPs), key enzymes involved in the turnover and remodeling of the Extracellular Matrix (ECM). TIMPs function as endogenous inhibitors by binding directly to active MMPs, thereby blocking their enzymatic activity and preventing excessive degradation of ECM components such as collagen, elastin, and proteoglycans. This regulatory mechanism is crucial for maintaining ECM homeostasis and tissue integrity under normal physiological conditions. There are four known members of the TIMP family: TIMP-1, TIMP-2, TIMP-3, and TIMP-4. Each TIMP isoform exhibits distinct tissue-specific expression patterns and differential affinities for different MMPs, allowing for precise regulation of MMP activity in various tissues and biological contexts. For example, TIMP-1 and TIMP-2 are widely expressed in many tissues and play essential roles in inhibiting broad-spectrum MMPs such as MMP-2 and MMP-9, which are involved in ECM remodeling and tissue repair processes. TIMPs have diverse functions in cellular physiology and pathology. Beyond ECM regulation, TIMPs influence cell proliferation, apoptosis, differentiation, and immune responses through MMP-independent mechanisms. For instance, TIMP-1 has been shown to promote cell survival and growth factor signaling by interacting with cell surface receptors and modulating intracellular signaling pathways. The balance between MMPs and TIMPs is critical for maintaining tissue homeostasis, and dysregulation of this balance is implicated in various pathological conditions [26]. Imbalances favoring MMP activity over TIMP inhibition are associated with diseases such as cancer metastasis, arthritis, cardiovascular diseases, and neurodegenerative disorders. In cancer, elevated MMP expression and reduced TIMP levels contribute to increased ECM degradation, facilitating tumor invasion, angiogenesis, and metastasis to distant organs. Therapeutically, targeting the MMP/TIMP balance has emerged as a strategy for treating diseases characterized by excessive ECM remodeling. Strategies include developing synthetic MMP inhibitors or enhancing TIMP expression to restore ECM homeostasis and inhibit disease progression. However, the complexity of MMP/TIMP interactions and their context-dependent roles in different tissues necessitate careful consideration in therapeutic approaches to avoid unintended consequences on normal tissue function.

Advances in understanding TIMP biology may lead to novel therapeutic interventions aimed at modulating ECM remodeling and mitigating pathological conditions associated with dysregulated MMP activity. Further exploration of TIMP functions beyond MMP inhibition holds promise for uncovering new avenues for therapeutic intervention and personalized medicine approaches tailored to specific disease contexts.

2.5.3 Role in Tissue Repair and Pathology

The role of tissue repair and pathology is intricately tied to the dynamic processes of tissue remodeling, which involve a complex interplay of cellular responses, Extracellular Matrix (ECM) interactions, and biochemical signaling pathways (Fig. 2.3). Tissue repair is a fundamental physiological process essential for maintaining tissue integrity and function in response to injury, infection, or disease. It encompasses a series of coordinated events, including inflammation, ECM deposition and remodeling, cell proliferation, and tissue remodeling, aimed at restoring tissue structure and function to pre-injury levels. At the cellular level, tissue repair involves the activation and recruitment of various cell types, including immune cells, fibroblasts, and endothelial cells. Inflammatory responses initiate the repair process by removing debris and pathogens from the injured site, creating a microenvironment conducive to tissue regeneration. This phase is characterized by the release of cytokines, growth factors, and chemokines that stimulate cell migration, proliferation, and differentiation. ECM remodeling plays a crucial role in tissue repair by providing structural support and scaffolding for migrating cells, regulating their behavior, and influencing the biochemical and mechanical properties of the healing tissue [27]. Components of the ECM, such as collagen, elastin, fibronectin, and proteoglycans, undergo dynamic changes during repair, orchestrated by enzymes like matrix metalloproteinases (MMPs) and their inhibitors (TIMPs). These enzymes degrade and reorganize the ECM, facilitating cell migration and tissue restructuring necessary for wound closure and tissue regeneration. In pathological conditions, however, dysregulated tissue repair processes can lead to chronic inflammation, fibrosis, and impaired healing. Persistent activation of inflammatory pathways or excessive ECM deposition can disrupt normal tissue architecture and function, contributing to diseases such as chronic wounds, fibrotic disorders, and autoimmune diseases. For example, aberrant ECM remodeling characterized by excessive collagen deposition and scar formation is a hallmark of diseases like pulmonary fibrosis and liver cirrhosis, where tissue repair mechanisms fail to resolve injury effectively.

Understanding the molecular mechanisms underlying tissue repair and pathology is crucial for developing therapeutic strategies aimed at enhancing healing processes and mitigating disease progression. Targeting key mediators of tissue repair, such as growth factors, cytokines, ECM-modifying enzymes, and cell signaling pathways, holds promise for promoting effective wound healing and preventing pathological tissue remodeling. Strategies may include the use of biomaterials that mimic the ECM environment, modulation of inflammatory responses, and targeted delivery of therapeutic agents to promote tissue regeneration while limiting fibrosis and scar formation. Advances in regenerative medicine and tissue engineering offer innovative approaches to stimulate tissue repair and regeneration. Techniques such as stem cell therapy, gene editing, and 3D bioprinting hold potential for repairing damaged

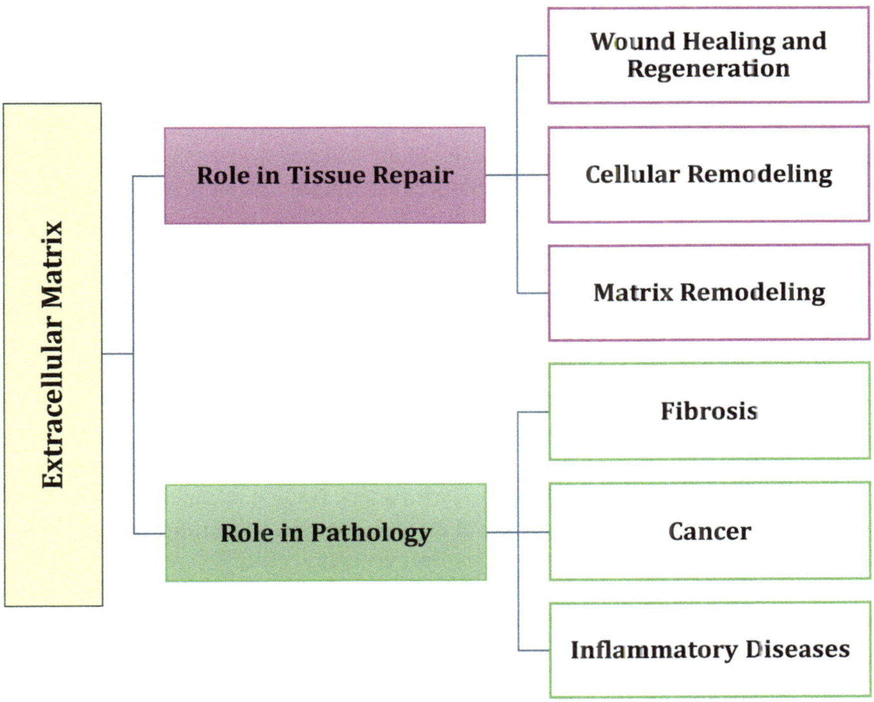

Fig. 2.3 Role of ECM in tissue repair and pathology

tissues and organs by enhancing cellular responses and rebuilding functional ECM structures.

2.6 Cellular Signaling Through ECM

Cellular signaling through the Extracellular Matrix (ECM) represents a pivotal mechanism by which cells communicate with their microenvironment to regulate various physiological processes, including cell adhesion, migration, proliferation, differentiation, and survival. The ECM, composed of a complex network of proteins, proteoglycans, and glycoproteins, not only provides structural support to tissues but also serves as a reservoir for growth factors, cytokines, and other bioactive molecules that modulate cellular behavior. Integrins and other cell surface receptors interact with specific ECM components, initiating signaling cascades that transmit mechanical and biochemical signals across the cell membrane and into the cell interior. This dynamic interplay between cells and the ECM plays crucial roles in tissue development, homeostasis, wound healing, and disease pathogenesis, highlighting the significance of ECM-mediated cellular signaling in orchestrating biological responses in health and disease contexts.

2.6.1 Growth Factors and Cytokines

Growth factors and cytokines are key signaling molecules that play essential roles in regulating cellular processes, including proliferation, differentiation, migration, and survival. These molecules are involved in diverse physiological and pathological processes, ranging from embryonic development and tissue repair to immune responses and cancer progression. Growth factors are typically small proteins or peptides that bind to specific cell surface receptors, activating intracellular signaling pathways that control cellular behaviors. Cytokines, on the other hand, are a broad category of small proteins that mediate communication between cells of the immune system and other cell types, influencing immune responses, inflammation, and tissue repair. Growth factors act through autocrine, paracrine, or endocrine mechanisms to regulate cellular activities in a highly specific manner. For instance, epidermal growth factor (EGF) and fibroblast growth factors (FGFs) stimulate cell proliferation and migration during tissue development and wound healing. Transforming growth factor-beta (TGF-β) family members regulate cell differentiation, immune function, and ECM synthesis [14]. Insulin-like growth factors (IGFs) play critical roles in tissue growth and metabolism, while nerve growth factor (NGF) is essential for neuronal survival and differentiation.

Cytokines function primarily in the immune system to modulate immune responses and inflammation. They include interleukins (ILs), tumor necrosis factors (TNFs), interferons (IFNs), and chemokines, each exerting specific effects on immune cells and other cell types. For example, IL-1β and TNF-α are pro-inflammatory cytokines that promote immune cell activation and inflammation during infection or injury. Interferons stimulate antiviral responses and modulate immune cell functions, while chemokines regulate immune cell migration and positioning within tissues. The signaling pathways activated by growth factors and cytokines are highly diverse and context-dependent, involving receptor tyrosine kinases (RTKs), Janus kinases (JAKs), and signal transducer and activator of transcription proteins (STATs), among others. Upon ligand binding, these pathways trigger phosphorylation events, protein–protein interactions, and gene expression changes that coordinate cellular responses to environmental stimuli. Dysregulation of growth factors and cytokine signaling pathways is implicated in various diseases, including cancer, autoimmune disorders, and metabolic syndromes, highlighting their importance as therapeutic targets.

In cancer, aberrant growth factor signaling often leads to uncontrolled cell proliferation, survival, and metastasis. Growth factors such as vascular endothelial growth factor (VEGF) promote angiogenesis, facilitating tumor growth and metastatic spread. Similarly, dysregulated cytokine signaling can contribute to chronic inflammation and autoimmune diseases, where persistent immune activation damages tissues and disrupts normal physiological functions. Therapeutically, targeting growth factors and cytokines has become a cornerstone of modern medicine. Biologics, including monoclonal antibodies and recombinant proteins, are designed to block or mimic the actions of specific growth factors or cytokines, offering precision therapies for diseases like cancer, rheumatoid arthritis, and inflammatory bowel disease. Moreover, understanding the intricate roles of growth factors and cytokines in cellular signaling networks continues to drive advancements in personalized medicine and drug development, aiming to optimize therapeutic efficacy and minimize side effects in diverse patient populations.

Growth factors and cytokines are essential regulators of cellular behavior, mediating critical processes in development, tissue repair, immunity, and disease. Their ability to orchestrate complex signaling pathways underscores their significance in maintaining tissue homeostasis and responding to environmental cues.

2.6.2 Signal Transduction Pathways

Signal transduction pathways are intricate mechanisms by which cells convert extracellular signals into specific intracellular responses, governing fundamental cellular processes such as growth, differentiation, metabolism, and apoptosis. These pathways are essential for cells to interpret and respond to environmental cues, including hormones, growth factors, cytokines, and neurotransmitters. The process typically begins when a ligand binds to a specific receptor on the cell surface, triggering a cascade of biochemical events that relay the signal from the receptor to the nucleus or other cellular compartments. Several major classes of signal transduction pathways exist, each characterized by specific components and mechanisms. Receptor tyrosine kinase (RTK) pathways, for example, are initiated by growth factors binding to RTKs on the cell surface, leading to receptor dimerization, autophosphorylation, and activation of downstream signaling molecules such as Ras and MAP kinases. These pathways regulate cell growth, proliferation, and survival by modulating gene expression and cytoplasmic activities. Another prominent class of pathways involves G protein-coupled receptors (GPCRs), which transmit signals across the plasma membrane through heterotrimeric G proteins. Upon ligand binding, GPCRs undergo conformational changes that activate G proteins, leading to the dissociation of Gα subunits and activation of downstream effectors such as adenylyl cyclase or phospholipase C. These pathways regulate diverse cellular functions, including neurotransmission, hormone secretion, and immune responses [28]. In addition to RTKs and GPCRs, other signal transduction pathways include cytokine receptors, nuclear hormone receptors, and integrin-mediated pathways, each tailored to respond to specific extracellular signals and modulate distinct cellular responses. For instance, cytokine receptors activate JAK-STAT pathways, regulating immune responses and inflammation, while nuclear hormone receptors such as steroid receptors modulate gene expression in response to hormone binding.

Signal transduction pathways are characterized by their complexity and specificity, often involving intricate networks of protein–protein interactions, post-translational modifications, and feedback loops to ensure precise regulation of cellular responses. For example, phosphorylation cascades mediated by protein kinases and phosphatases play critical roles in signal amplification and termination, allowing cells to respond rapidly to changing environmental conditions while maintaining homeostasis. Dysregulation of signal transduction pathways is associated with various diseases, including cancer, metabolic disorders, and neurodegenerative diseases. Mutations or alterations in pathway components can lead to aberrant signaling, promoting uncontrolled cell growth, inflammation, or impaired cellular function. Consequently, understanding the molecular mechanisms of signal transduction pathways is essential for identifying therapeutic targets and developing interventions to correct signaling defects in disease states.

Advances in molecular biology and systems biology have facilitated the elucidation of complex signal transduction networks, providing insights into how cells integrate and process signals to coordinate physiological responses. Techniques such as pharmacological inhibitors, genetic manipulation, and computational modeling are used to study signal transduction pathways and their roles in health and disease, paving the way for innovative therapies and personalized medicine approaches tailored to target specific pathway dysfunctions in patients.

2.7 Cell Migration in ECM

Cell migration within the Extracellular Matrix (ECM) is a fundamental process essential for various physiological and pathological events, including embryonic development, tissue repair, immune responses, and cancer metastasis (Fig. 2.4). The ECM

Fig. 2.4 Different types of migration within the extracellular matrix

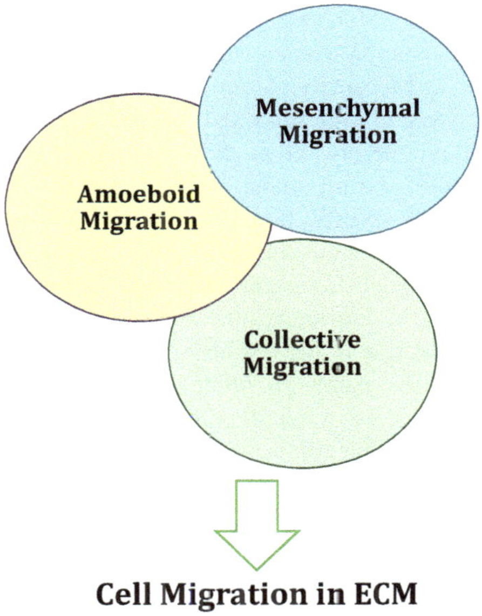

provides a three-dimensional network of structural support and biochemical cues that guide and regulate the movement of cells through complex mechanisms [29]. Cell migration involves coordinated interactions between cells and ECM components such as collagen, fibronectin, and proteoglycans, facilitated by cell surface receptors like integrins and proteoglycan receptors. These receptors bind to specific ECM ligands, triggering signaling pathways that control cytoskeletal dynamics, adhesion formation, and cell polarity necessary for directional migration.

2.7.1 Mesenchymal Migration

Mesenchymal migration, a crucial process in development, wound healing, and disease progression, involves the movement of mesenchymal cells through the Extracellular Matrix (ECM) to specific locations within tissues (Fig. 2.4). Mesenchymal cells are a diverse group of multipotent cells found in connective tissues that possess the ability to migrate and differentiate into various cell types, including fibroblasts, osteoblasts, and adipocytes. This migration is essential during embryogenesis for tissue patterning and organogenesis, where mesenchymal cells migrate along predetermined paths to form structures such as bones, cartilage, and blood vessels. The process of mesenchymal migration is regulated by a complex interplay of molecular signals and ECM interactions. Mesenchymal cells express integrins and other cell surface receptors that bind to ECM proteins such as fibronectin, collagen, and laminin, providing anchorage points for cell adhesion and traction during migration. Through these interactions, cells generate contractile forces via actin cytoskeleton rearrangements, allowing them to pull themselves forward through the ECM. Mesenchymal migration can occur through several distinct modes, including single-cell migration and collective migration. Single-cell migration involves individual mesenchymal cells moving independently through the ECM, guided by chemotactic gradients of growth factors or cytokines that direct cell movement towards specific targets. Collective migration, on the other hand, involves groups of mesenchymal cells migrating together in a coordinated manner, maintaining cell–cell contacts through adherens junctions and gap junctions while navigating through the ECM. During wound healing and tissue repair, mesenchymal migration plays a pivotal role in the recruitment of fibroblasts and other precursor cells to the site of injury. In response to inflammatory signals and growth factors released from damaged tissues, mesenchymal cells migrate towards the wound site, where they participate in ECM deposition, tissue remodeling, and angiogenesis to facilitate healing. This process is tightly regulated to ensure proper tissue regeneration and restoration of structural integrity [30]. In pathological contexts such as cancer, mesenchymal migration contributes to tumor invasion and metastasis, allowing cancer cells to disseminate from the primary tumor site to distant organs. Cancer cells undergoing epithelial-to-mesenchymal transition (EMT) acquire mesenchymal characteristics, including enhanced migratory and invasive abilities enabled by changes in cell adhesion molecules and cytoskeletal dynamics. Mesenchymal migration in cancer involves complex interactions with the ECM and stromal cells, facilitating tumor cell dissemination through blood vessels or lymphatic channels to establish secondary tumors.

Understanding the mechanisms of mesenchymal migration is critical for developing strategies to intervene in diseases where the aberrant cell movement contributes to pathology. Research efforts focus on elucidating the molecular pathways and regulatory mechanisms that govern mesenchymal migration, including signaling pathways mediated by growth factors, cytokines, and ECM receptors.

2.7.2 Amoeboid Migration

Amoeboid migration is a distinctive mode of cell movement characterized by its reliance on dynamic changes in cell shape and high adaptability to diverse microenvironments (Fig. 2.4). This mode of migration is observed in various cell types, including immune cells like leukocytes and cancer cells, allowing them to navigate through complex tissues and reach specific destinations. Amoeboid migration contrasts with mesenchymal migration in that it does not heavily rely on proteolytic ECM degradation or long-term adhesion to the substrate but rather on rapid changes in cell shape and cytoskeletal dynamics. Cells undergoing amoeboid migration exhibit a rounded or irregular shape, facilitated by the contractility of the actomyosin cytoskeleton and the absence of stable focal adhesions with the ECM. Instead of forming long-lasting attachments, amoeboid cells use transient interactions with

ECM components such as proteoglycans and glycoproteins to propel themselves forward through the tissue matrix. This mode of migration is particularly advantageous in environments with dense ECM or confined spaces where cells must squeeze through narrow gaps without degrading the surrounding ECM extensively. The cytoskeletal machinery underlying amoeboid migration is predominantly driven by actin polymerization and myosin II-mediated contractility [31]. Actin filaments at the leading edge of the cell form dynamic protrusions known as blebs, which expand and contract in a pulsatile manner to push the cell forward. Myosin II, a molecular motor protein, generates contractile forces that squeeze the cell nucleus and propel the cell body forward through the ECM. These contractile forces are regulated by Rho GTPases, which control actin dynamics and cell polarity during migration. Amoeboid migration is highly adaptable and responsive to external cues such as chemotactic gradients of cytokines or growth factors, guiding cell movement towards specific targets or sites of injury. For example, immune cells like neutrophils and macrophages use amoeboid migration to infiltrate inflamed tissues in response to chemokines released by injured cells or pathogens. Similarly, cancer cells undergoing amoeboid migration can disseminate through tissues and invade blood vessels or lymphatic channels, facilitating metastasis to distant organs.

In cancer metastasis, amoeboid migration allows cancer cells to evade the immune system and invade surrounding tissues by exploiting the dynamic nature of amoeboid movement. This mode of migration enables cancer cells to traverse dense ECM barriers and migrate collectively or individually through confined spaces within tissues. The plasticity of amoeboid migration also presents challenges in therapeutic targeting, as traditional approaches aimed at inhibiting cell adhesion or proteolytic ECM degradation may be less effective against cells employing amoeboid movement. Understanding the molecular mechanisms and regulatory pathways of amoeboid migration is crucial for developing targeted therapies to inhibit pathological cell migration in diseases such as cancer. Researchers are exploring strategies to interfere with signaling pathways involved in actin dynamics, Rho GTPase activity, and chemotactic responses to disrupt amoeboid migration and inhibit cancer metastasis effectively.

2.7.3 Role in Development and Disease

The role of cell migration in development and disease is pivotal, influencing fundamental processes from embryogenesis to tissue homeostasis and pathology. During embryonic development, orchestrated cell migration patterns are essential for shaping tissues and organs. Cells migrate to specific locations to form structures such as neural networks, blood vessels, and musculoskeletal systems. This precise spatial organization is crucial for establishing functional tissues and organs, highlighting the importance of regulated cell migration in developmental biology. In disease, aberrant cell migration contributes significantly to pathology, including cancer metastasis, chronic inflammation, and autoimmune disorders. Cancer metastasis, for example, involves cancer cells migrating from the primary tumor site to distant organs through blood vessels or lymphatic channels. This process relies on migratory mechanisms such as amoeboid or mesenchymal migration, allowing cancer cells to invade surrounding tissues and disseminate throughout the body. Understanding the molecular mechanisms underlying cancer cell migration is critical for developing targeted therapies to inhibit metastasis and improve patient outcomes. Cell migration also plays a crucial role in wound healing and tissue repair processes. Following injury, cells migrate to the wound site to remove debris, deposit ECM components, and facilitate tissue regeneration. Mesenchymal cells such as fibroblasts and immune cells like macrophages undergo directed migration towards the injury site, guided by chemotactic signals released from damaged tissues [32]. Dysregulated migration in chronic wounds or fibrotic disorders can impair healing processes, leading to prolonged inflammation and tissue damage. In the immune system, cell migration is essential for immune surveillance and responses to infection or injury. Immune cells, including neutrophils, macrophages, and lymphocytes, migrate to sites of infection or inflammation in response to chemotactic signals produced by pathogens or damaged tissues. This migration allows immune cells to engulf pathogens, clear cellular debris, and modulate inflammatory responses to restore tissue homeostasis. Dysregulated immune cell migration can contribute to chronic inflammation and autoimmune diseases characterized by immune cell infiltration and tissue damage. Neurological disorders also involve aberrant cell migration processes that impact brain development and function. During neurodevelopment, neurons migrate from germinal zones to specific brain regions to form neuronal circuits and synaptic connections. Defects in neuronal migration can lead to neurodevelopmental disorders such as lissencephaly and neuronal migration disorders, affecting cognitive function and motor skills.

Cell migration plays diverse and critical roles in both normal development and disease pathology. The ability of cells to migrate in response to external cues is essential for tissue morphogenesis, immune responses, wound healing, and neurological function. Dysregulated cell migration contributes to various diseases, including cancer metastasis, chronic inflammation, and neurodevelopmental disorders.

2.8 ECM and Stem Cell Niche

The Extracellular Matrix (ECM) serves as a critical component of the stem cell niche, playing a pivotal role in regulating stem cell behavior and fate determination. The stem cell niche refers to the specialized microenvironment within tissues that supports and maintains stem cell populations, influencing their self-renewal, differentiation, and interaction with neighboring cells. Within this niche, the ECM provides structural support and biochemical signals that govern stem cell adhesion, migration, and response to growth factors and cytokines. Through dynamic interactions with ECM proteins such as collagen, fibronectin, and laminin, stem cells receive cues that modulate their proliferation and differentiation states, thereby contributing to tissue homeostasis and repair processes.

2.8.1 Role of ECM in Stem Cell Differentiation

The role of the Extracellular Matrix (ECM) in stem cell differentiation (Fig. 2.5):

- **Structural Support and Adhesion**: The ECM provides a three-dimensional scaffold that supports stem cell adhesion and spatial organization within the tissue microenvironment. ECM proteins such as collagen, fibronectin, and laminin interact with cell surface receptors, including integrins, to anchor stem cells and maintain their structural integrity. This physical interaction influences cell shape, polarity, and cytoskeletal organization, which are critical for initiating differentiation processes.
- **Biochemical Signaling**: ECM components serve as reservoirs for growth factors, cytokines, and other bioactive molecules that regulate stem cell behavior. Binding of these molecules to ECM proteins can trigger intracellular signaling pathways that promote or inhibit stem cell differentiation. For instance, specific ECM-bound growth factors like Transforming Growth Factor-beta (TGF-β) or Bone Morphogenetic Proteins (BMPs) can induce lineage-specific differentiation by activating downstream transcription factors and gene expression programs.
- **Mechanical Properties**: The mechanical properties of the ECM, including stiffness and elasticity, play a significant role in stem cell fate determination. Stem cells can sense and respond to mechanical cues from the ECM through mechano transduction pathways, influencing their differentiation into specific cell types. For example, soft ECM substrates may promote neuronal differentiation, while stiff substrates may favor osteogenic differentiation. This mechanosensitive behavior underscores the importance of ECM biomechanics in directing stem cell fate.

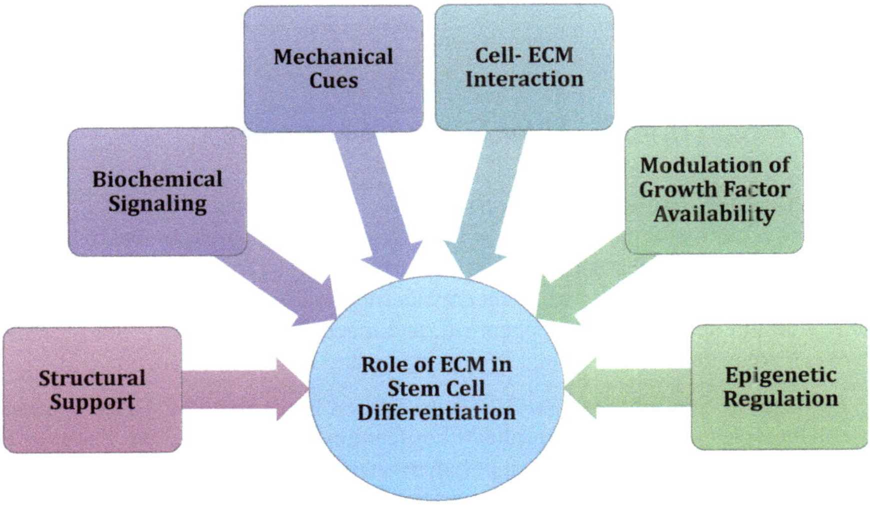

Fig. 2.5 Various roles of ECM in stem cell differentiation

- **Remodeling and Proteolytic Activity**: Dynamic ECM remodeling by enzymes such as matrix metalloproteinases (MMPs) and tissue inhibitors of metalloproteinases (TIMPs) regulate stem cell differentiation. MMPs can cleave ECM proteins to release bioactive fragments or modulate the availability of growth factors bound to the ECM, thereby influencing stem cell responses. TIMPs counterbalance MMP activity, ensuring spatial and temporal regulation of ECM remodeling processes critical for controlled differentiation events.
- **Cell-ECM Interactions in Developmental Niches**: Stem cells reside in specialized niches where ECM composition and organization dictate their differentiation potential. These niches provide spatial cues and signaling gradients that guide stem cell behavior during tissue development and homeostasis. For instance, the basement membrane ECM in neural stem cell niches regulates neurogenesis by maintaining stem cell quiescence or promoting neuronal differentiation in response to niche-derived signals.
- **Applications in Regenerative Medicine**: Understanding ECM interactions with stem cells has significant implications for regenerative medicine and tissue engineering. Biomimetic scaffolds designed to mimic ECM properties can be tailored to promote desired stem cell differentiation outcomes for therapeutic purposes. By modulating ECM composition, stiffness, and biochemical signaling cues, researchers aim to enhance tissue regeneration and repair damaged tissues effectively.

These points underscore the multifaceted role of the ECM in regulating stem cell differentiation through physical, biochemical, and mechanical cues [33]. Elucidating the intricate interactions between stem cells and their ECM microenvironment is crucial for advancing our understanding of developmental processes, disease mechanisms, and therapeutic strategies aimed at harnessing stem cell potential for clinical applications.

2.8.2 Engineering ECM for Stem Cell Therapies

Engineering the Extracellular Matrix (ECM) for stem cell therapies involves designing biomimetic scaffolds that mimic the native ECM environment to support and direct stem cell behavior for therapeutic applications. This approach integrates principles from materials science, bioengineering, and cell biology to create scaffolds with tailored physical, biochemical, and mechanical properties that promote stem cell adhesion, proliferation, differentiation, and tissue regeneration.

- **Biomimetic Scaffold Design**: Biomaterials used for engineering ECM for stem cell therapies are designed to replicate key features of the native ECM. These scaffolds can be composed of natural polymers such as collagen, fibrin, or hyaluronic acid, which resemble ECM components, or synthetic polymers like polyethylene glycol (PEG) or polylactic-co-glycolic acid (PLGA), which offer tunable mechanical properties and degradation kinetics. The choice of biomaterial influences scaffold architecture and functionality, aiming to create environments that optimize stem cell interactions and therapeutic outcomes.

- **Physical and Mechanical Properties**: The physical and mechanical properties of ECM-engineered scaffolds play a critical role in directing stem cell behavior. These properties, including scaffold porosity, stiffness, and elasticity, can be tailored to mimic the specific ECM characteristics of target tissues. For example, scaffolds designed for bone regeneration may possess high stiffness to support osteogenic differentiation, whereas scaffolds for cardiac tissue engineering may require elasticity to mimic myocardial tissue properties and promote cardiomyocyte differentiation.
- **Biochemical Signaling**: Incorporating biochemical cues into ECM-engineered scaffolds can enhance stem cell differentiation and tissue regeneration. Growth factors, cytokines, and ECM-derived peptides can be immobilized within the scaffold or delivered in a controlled manner to modulate stem cell responses. For instance, growth factors like BMPs or VEGF can be tethered to scaffolds to induce osteogenic or angiogenic differentiation of stem cells, promoting tissue-specific regeneration in vivo.
- **ECM Remodeling and Degradation**: ECM-engineered scaffolds are designed to undergo controlled degradation and remodeling to facilitate tissue integration and remodeling by host cells. Enzymatically degradable linkers or materials with tunable degradation rates allow for gradual release of bioactive molecules and ECM remodeling processes that mimic natural tissue turnover. This feature is crucial for ensuring scaffold biocompatibility, minimizing inflammatory responses, and promoting seamless integration with surrounding native tissues.
- **Cell-ECM Interactions**: Engineering ECM scaffolds consider the dynamic interactions between stem cells and their microenvironment. Scaffolds may incorporate cell-adhesive ligands or ECM-mimetic peptides that promote cell adhesion, spreading, and proliferation. Additionally, ECM-engineered scaffolds can modulate cellular mechano transduction pathways by providing cues that influence cytoskeletal organization, gene expression, and differentiation of encapsulated or seeded stem cells.
- **Clinical Applications and Translation**: ECM-engineered scaffolds hold promise for a range of clinical applications in regenerative medicine and tissue engineering. These scaffolds can be used to repair or replace damaged tissues, enhance wound healing, and restore organ function. Clinical translation of ECM-engineered therapies involves rigorous preclinical testing to validate safety, efficacy, and biocompatibility in relevant animal models, followed by clinical trials to evaluate therapeutic outcomes in human patients.

Engineering the ECM for stem cell therapies represents a multidisciplinary approach aimed at creating biomimetic scaffolds that recapitulate native tissue environments to guide and enhance stem cell behaviors for therapeutic purposes. Advances in biomaterials science and cell biology continue to drive innovation in

ECM scaffold design, offering promising strategies to address clinical challenges and improve patient outcomes in regenerative medicine.

2.9 Future Perspectives

Looking ahead, the future of ECM engineering for stem cell therapies holds exciting prospects with potential transformative impacts in regenerative medicine and tissue engineering. Advances in biomaterials science, nanotechnology, and 3D printing are poised to revolutionize the design and fabrication of ECM-mimetic scaffolds with unprecedented precision and complexity. These innovations enable the creation of scaffolds that more closely mimic the intricate architecture and biochemical composition of native ECM, enhancing their ability to guide stem cell differentiation and tissue regeneration. The integration of advanced imaging techniques and computational modeling allows for real-time monitoring and optimization of scaffold-cell interactions and tissue development. This interdisciplinary approach facilitates personalized medicine applications, where ECM-engineered scaffolds can be customized to match patient-specific needs, optimizing therapeutic outcomes and reducing the risk of immune rejection. The development of smart scaffolds capable of dynamic response to physiological cues holds promise for enhancing the efficacy and longevity of ECM-based therapies. These scaffolds can deliver bioactive molecules in response to specific environmental stimuli or modulate their mechanical properties to adapt to changing tissue conditions during healing and regeneration processes. Such advancements are crucial for overcoming current limitations in tissue repair and replacing lost or damaged tissues with functional equivalents. In parallel, ongoing research into the fundamental mechanisms underlying stem cell-ECM interactions continues to uncover novel insights that will inform the design of next-generation scaffolds. By elucidating the intricate signaling pathways and biomechanical cues that regulate stem cell fate decisions within the ECM microenvironment, researchers aim to develop more effective strategies for manipulating stem cell behavior and enhancing tissue-specific differentiation. This knowledge-driven approach promises to accelerate the translation of ECM-engineered therapies from bench to bedside, paving the way for a new era in regenerative medicine where complex tissue regeneration becomes a clinical reality.

2.10 Summary

In summary, the chapter on the Extracellular Matrix (ECM) and its interactions with stem cells provides a comprehensive exploration of how ECM dynamics influence stem cell behavior in both physiological and pathological contexts. The ECM serves as a dynamic scaffold that not only provides structural support but also regulates biochemical signaling pathways critical for stem cell adhesion, migration, proliferation, and differentiation. Through interactions with ECM components such as collagen, fibronectin, and proteoglycans, stem cells receive spatial and biochemical cues that orchestrate their responses and fate decisions within tissue microenvironments. The role of ECM-engineered scaffolds in stem cell therapies emerges as a promising avenue for regenerative medicine, offering tailored environments that mimic native ECM conditions to enhance tissue repair and regeneration. These scaffolds are designed with precise physical, mechanical, and biochemical properties to optimize stem cell interactions and therapeutic outcomes. Advances in biomaterials science and bioengineering continue to drive innovation in scaffold design, enabling researchers to create sophisticated platforms capable of supporting complex tissue structures and guiding stem cell differentiation towards desired cell lineages. Future perspectives highlight the transformative potential of ECM-based therapies, driven by ongoing advancements in nanotechnology, 3D printing, and personalized medicine approaches. Smart scaffolds that respond dynamically to physiological cues and integrate with patient-specific needs hold promises for addressing current challenges in tissue engineering and enhancing the long-term success of regenerative treatments. As research uncovers deeper insights into the molecular mechanisms governing stem cell-ECM interactions, opportunities abound to refine therapeutic strategies and expand the clinical applications of ECM-engineered therapies across diverse medical fields.

The integration of ECM biology with stem cell research represents a pivotal area of study with profound implications for understanding tissue development, disease progression, and therapeutic intervention. By harnessing the intricate interplay between stem cells and their ECM microenvironment, researchers are poised to unlock new therapeutic paradigms that may revolutionize the treatment of degenerative diseases, traumatic injuries, and congenital defects, ultimately improving patient outcomes and quality of life.

Exercise 3

Multiple Choice Questions

1. Which of the following is the primary function of the extracellular matrix (ECM) in cellular interaction?
A) Energy production
B) Genetic information storage
C) Structural support and signaling
D) Waste removal

2. Which cell surface receptor primarily mediates the interaction between cells and the ECM?
A) G-protein-coupled receptors (GPCRs)
B) Integrins
C) Ion channels
D) Receptor tyrosine kinases (RTKs)

3. The rigidity of the ECM influences cell differentiation. Which type of cell is most likely to differentiate in response to a stiff ECM?
A) Adipocytes (fat cells)
B) Chondrocytes (cartilage cells)
C) Neurons (nerve cells)
D) Osteoblasts (bone cells)

4. Which ECM component is primarily responsible for providing tensile strength to tissues?
A) Laminin
B) Collagen
C) Fibronectin
D) Elastin

5. How do matrix metalloproteinases (MMPs) influence cellular interactions with the ECM?
A) They degrade ECM components, facilitating tissue remodeling.
B) They synthesize ECM components.
C) They inhibit cell-ECM adhesion.
D) They bind to integrins, enhancing cell-ECM adhesion.

6. Fibronectin is a key ECM protein. What is one of its primary roles in cellular interaction?
A) Promoting apoptosis (cell death)
B) Facilitating cell adhesion and migration
C) Inhibiting angiogenesis
D) Binding to collagen only

7. Which signaling pathway is commonly activated by integrins upon binding to ECM components?
A) Wnt/β-catenin pathway
B) MAPK/ERK pathway
C) JAK/STAT pathway
D) Notch pathway

8. What role does the ECM play in the context of cancer?
A) It always prevents tumor growth.
B) It can be remodeled to facilitate tumor invasion and metastasis.
C) It has no effect on cancer progression.
D) It exclusively supports immune cell infiltration to eliminate tumors.

9. Which ECM component is particularly important in the basement membrane and influences cell differentiation?
A) Elastin
B) Laminin
C) Hyaluronic acid
D) Fibrinogen

Points to Remember

- **The ECM undergoes continuous remodeling processes orchestrated by enzymes like matrix metalloproteinases (MMPs) and tissue inhibitors of metalloproteinases (TIMPs).**
- **Stem cells are sensitive to mechanical cues from the ECM, such as substrate stiffness and elasticity. This mechano transduction signaling pathway plays a critical role in directing stem cell differentiation into specific lineages.**
- **The ECM serves as a reservoir for bioactive molecules, including growth factors, cytokines, and ECM-derived peptides. These molecules are crucial for regulating stem cell proliferation, migration, and differentiation in response to tissue-specific needs and environmental cues.**
- **Different stem cell niches exhibit distinct ECM compositions and organization tailored to support specific tissue functions. For example, the basement membrane ECM in neural stem cell niches regulates neurogenesis, while the osteogenic niche ECM in bone marrow promotes osteoblast differentiation and bone formation.**
- **ECM-engineered scaffolds are designed to replicate the biochemical and structural properties of native ECM, offering precise control over stem cell behavior for tissue regeneration.**
- **Advances in ECM-based therapies hold promise for clinical applications in treating degenerative diseases, traumatic injuries, and congenital defects. Smart scaffolds capable of dynamic response to physiological cues and integration with patient-specific needs are being developed to improve tissue repair and regeneration outcomes.**

References

1. Sainio A, Järveläinen H (2020) Extracellular matrix-cell interactions: focus on therapeutic applications. Cell Signal 66:109487. https://doi.org/10.1016/j.cellsig.2019.109487
2. Rosso F, Giordano A, Barbarisi M, Barbarisi A (2004) From cell-ECM interactions to tissue engineering. J Cell Physiol 199(2):174–180. https://doi.org/10.1002/jcp.10471
3. Liu AP, Chaudhuri O, Parekh SH (2017) New advances in probing cell-extracellular matrix interactions. Integr Biol Quant Biosci Nano Macro 9(5):383–405. https://doi.org/10.1039/c6ib00251j
4. Li J, Jung W, Nam S, Chaudhuri O, Kim T (2020) Roles of interactions between cells and extracellular matrices for cell migration and matrix remodeling. In: Zhang Y (eds) Multiscale extracellular matrix mechanics and mechanobiology. Studies in mechanobiology, tissue engineering and biomaterials, vol 23. Springer, Cham. https://doi.org/10.1007/978-3-030-20182-1_8
5. Theocharis AD, Skandalis SS, Gialeli C, Karamanos NK (2016) Extracellular matrix structure. Adv Drug Deliv Rev 97:4–27. https://doi.org/10.1016/j.addr.2015.11.001

6. Kular JK, Basu S, Sharma RI (2014) The extracellular matrix: Structure, composition, age-related differences, tools for analysis and applications for tissue engineering. J Tissue Eng. https://doi.org/10.1177/2041731414557112
7. Karamanos NK, Theocharis AD, Piperigkou Z, Manou D, Passi A, Skandalis SS, Vynios DH, Orian-Rousseau V, Ricard-Blum S, Schmelzer CEH, Duca L, Durbeej M, Afratis NA, Troeberg L, Franchi M, Masola V, Onisto M (2021) A guide to the composition and functions of the extracellular matrix. FEBS J 288(24):6850–6912. https://doi.org/10.1111/febs.15776
8. Wang K, Meng X, Guo Z (2021) Elastin structure, synthesis, regulatory mechanism and relationship with cardiovascular diseases. Front Cell Dev Biol 9:596702. https://doi.org/10.3389/fcell.2021.596702
9. Trębacz H, Barzycka A (2023) Mechanical properties and functions of Elastin: an overview. Biomolecules 13(3):574. https://doi.org/10.3390/biom13030574
10. Yue B (2014) Biology of the extracellular matrix: an overview. J Glaucoma 23(8 Suppl 1):S20–S23. https://doi.org/10.1097/IJG.0000000000000108
11. Cerezo-Magaña M, Bång-Rudenstam A, Belting M (2023) Proteoglycans: a common portal for SARS-CoV-2 and extracellular vesicle uptake. Am J Physiol Cell Physiol 324(1):C76–C84. https://doi.org/10.1152/ajpcell.00453.2022
12. Jones FS, Jones PL (2000) The tenascin family of ECM glycoproteins: structure, function, and regulation during embryonic development and tissue remodeling. Dev Dyn Off Publ Am Assoc Anatom 218(2):235–259. https://doi.org/10.1002/(SICI)1097-0177(200006)218:2%3c235::AID-DVDY2%3e3.0.CO;2-G
13. Barallobre-Barreiro J, Baig F, Fava M, Yin X, Mayr M (2017) Glycoproteomics of the extracellular matrix: a method for intact glycopeptide analysis using mass spectrometry. J Vis Exp JoVE 122:55674. https://doi.org/10.3791/55674
14. Kim SH, Turnbull J, Guimond S (2011) Extracellular matrix and cell signalling: the dynamic cooperation of integrin, proteoglycan and growth factor receptor. J Endocrinol 209(2):139–151. https://doi.org/10.1530/JOE-10-0377
15. Chastney MR, Kaivola J, Leppänen VM, Ivaska J (2024) The role and regulation of integrins in cell migration and invasion. Nat Rev Molecul Cell Biol https://doi.org/10.1038/s41580-024-00777-1
16. Danen EHJ. Integrins: an overview of structural and functional aspects. In: Madame curie bioscience database [Internet]. Landes Bioscience, Austin, 2000–2013. Available from: https://www.ncbi.nlm.nih.gov/books/NBK6259/
17. Itoh Y (2018) Discoidin domain receptors: Microenvironment sensors that promote cellular migration and invasion. Cell Adh Migr 12(4):378–385. https://doi.org/10.1080/19336918.2018.1460011
18. Borza CM, Pozzi A (2014) Discoidin domain receptors in disease. Matrix Biol J Int Soc Matrix Biol 34:185–192. https://doi.org/10.1016/j.matbio.2013.12.002
19. Jang B, Kim A, Hwang J, Song HK, Kim Y, Oh ES (2020) Emerging role of syndecans in extracellular matrix remodeling in cancer. J Histochem Cytochem Off J Histochem Soc 68(12):863–870. https://doi.org/10.1369/0022155420930112
20. Mishra YG, Manavathi B (2021) Focal adhesion dynamics in cellular function and disease. Cell Signal 85:110046. https://doi.org/10.1016/j.cellsig.2021.110046
21. Kuo JC (2013) Mechanotransduction at focal adhesions: integrating cytoskeletal mechanics in migrating cells. J Cell Mol Med 17(6):704–712. https://doi.org/10.1111/jcmm.12054
22. Te Molder L, de Pereda JM, Sonnenberg A (2021) Regulation of hemidesmosome dynamics and cell signaling by integrin α6β4. J Cell Sci 134(18):jcs259004. https://doi.org/10.1242/jcs.259004
23. Leckband DE, le Duc Q, Wang N, de Rooij J (2011) Mechanotransduction at cadherin-mediated adhesions. Curr Opin Cell Biol 23(5):523–530. https://doi.org/10.1016/j.ceb.2011.08.003
24. Lu P, Takai K, Weaver VM, Werb Z (2011) Extracellular matrix degradation and remodeling in development and disease. Cold Spring Harb Perspect Biol 3(12):a005058. https://doi.org/10.1101/cshperspect.a005058

25. Wang X, Khalil RA (2018) Matrix Metalloproteinases, vascular remodeling, and vascular disease. Adv Pharmacol (San Diego, Calif.) 81:241–330. https://doi.org/10.1016/bs.apha.2017.08.002
26. Bourboulia D, Stetler-Stevenson WG (2010) Matrix metalloproteinases (MMPs) and tissue inhibitors of metalloproteinases (TIMPs): positive and negative regulators in tumor cell adhesion. Semin Cancer Biol 20(3):161–168. https://doi.org/10.1016/j.semcancer.2010.05.002
27. Mohammed Butt A, Rupareliya V, Hariharan A, Kumar H (2023) Building a pathway to recovery: targeting ECM remodeling in CNS injuries. Brain Res 1819:148533. https://doi.org/10.1016/j.brainres.2023.148533
28. Novoseletskaya ES, Evdokimov PV, Efimenko AY (2023) Extracellular matrix-induced signaling pathways in mesenchymal stem/stromal cells. Cell Commun Signal 21(1):244. https://doi.org/10.1186/s12964-023-01252-8
29. Trepat X, Chen Z, Jacobson K (2012) Cell migration. Compr Physiol 2(4):2369–2392. https://doi.org/10.1002/cphy.c110012
30. Heo JS, Kim S, Yang CE, Choi Y, Song SY, Kim HO (2021) Human adipose mesenchymal stem cell-derived exosomes: a key player in wound healing. Tissue Eng Regener Med 18(4):537–548. https://doi.org/10.1007/s13770-020-00316-x
31. Talkenberger K, Cavalcanti-Adam EA, Voss-Böhme A et al (2017) Amoeboid-mesenchymal migration plasticity promotes invasion only in complex heterogeneous microenvironments. Sci Rep 7:9237. https://doi.org/10.1038/s41598-017-09300-3
32. Bonnans C, Chou J, Werb Z (2014) Remodelling the extracellular matrix in development and disease. Nat Rev Mol Cell Biol 15:786–801. https://doi.org/10.1038/nrm3904
33. Novoseletskaya ES, Grigorieva OA, Efimenko AY, Kalinina NI (2019) Extracellular matrix in the regulation of stem cell differentiation. Biochem Biokhimiia 84(3):232–240. https://doi.org/10.1134/S0006297919030052

Chapter 3
Mechanisms of ECM Remodeling

3.1 Introduction

The extracellular matrix (ECM) is a complex network of molecules that surrounds and supports cells within tissues and organs in multicellular organisms. It provides structural and biochemical support to the cellular components embedded within it, playing crucial roles in tissue development, homeostasis, and disease processes. Composed primarily of proteins such as collagen, elastin, fibronectin, and laminin, the ECM also includes glycosaminoglycans (GAGs) and proteoglycans, which contribute to its diverse structural and functional properties [1]. Functionally, the ECM serves several essential roles. It acts as a scaffold that maintains the structural integrity of tissues, providing mechanical strength and elasticity. This scaffold not only supports cells but also regulates their behavior through biochemical and biomechanical cues. For instance, ECM molecules can bind growth factors and cytokines, controlling their availability and signaling to cells. ECM acts as a reservoir for various bioactive molecules, including growth factors, enzymes, and signaling molecules, influencing processes such as cell proliferation, migration, and differentiation. Beyond its structural and biochemical roles, the ECM is dynamic and undergoes constant remodeling in response to physiological cues and pathological conditions. Cells within tissues actively modify the ECM through the secretion of enzymes like matrix metalloproteinases (MMPs) and tissue inhibitors of metalloproteinases (TIMPs), which regulate ECM turnover and remodeling [2]. This dynamic interaction between cells and their ECM microenvironment is critical for tissue morphogenesis, wound healing, and tissue repair processes. In addition to its local effects within tissues, the ECM also plays systemic roles in processes such as embryonic development, organogenesis, and immune responses. It provides spatial guidance for cell migration and tissue organization during development, influencing cell fate determination and tissue patterning. Alterations in ECM composition and organization are implicated in various diseases, including fibrosis, cancer, and cardiovascular disorders, highlighting its significance in both health and disease.

D. Suhag, *Extracellular Matrix biomimicking for Theranostics*, Biomedical Materials for Multi-functional Applications, https://doi.org/10.1007/978-981-95-7900-6_3

The extracellular matrix is a complex and dynamic network of molecules that surrounds cells within tissues, providing structural support, biochemical signaling, and regulatory cues essential for tissue function and homeostasis. Its composition, organization, and remodeling dynamics are tightly regulated, influencing cellular behavior and tissue physiology across development, health, and disease contexts.

3.2 Importance of ECM Remodeling in Physiology

Extracellular matrix (ECM) remodeling plays a crucial role in maintaining tissue homeostasis and supporting physiological functions across various organ systems. One key importance lies in tissue development and morphogenesis. During embryonic development, ECM remodeling guides cell migration, differentiation, and tissue patterning. It provides spatial cues that direct the formation of complex tissue structures and organs, ensuring proper organogenesis and functional integration of different cell types. For instance, ECM components like collagen and laminin form specialized matrices in specific tissues such as the basement membrane, critical for organ function and integrity. In adult tissues, ECM remodeling continues to be essential for tissue repair and regeneration. After injury or during normal turnover, cells interact with the ECM to initiate repair processes. The ECM not only provides a scaffold for new tissue formation but also regulates the behavior of reparative cells. For example, fibroblasts and immune cells interact with ECM components to orchestrate wound healing responses, including inflammation, matrix deposition, and tissue remodeling [3]. Dysregulation of these processes can lead to impaired healing, chronic wounds, or fibrotic tissue formation. The ECM acts as a reservoir for growth factors, cytokines, and signaling molecules that regulate cell behavior and tissue function. These bioactive molecules are often sequestered within the ECM, released in response to physiological cues or tissue damage. This spatial and temporal control of signaling molecules by the ECM is critical for regulating processes such as cell proliferation, differentiation, and survival. In this way, ECM remodeling influences cellular activities within tissues and organs, maintaining proper function and adaptation to changing physiological demands.

In addition to its structural and signaling roles, the ECM contributes to mechanical support and integrity of tissues. Collagen and elastin fibers within the ECM provide tensile strength, elasticity, and resilience to tissues subjected to mechanical forces. This mechanical property of the ECM is crucial in organs like the skin, blood vessels, and lungs, where it helps maintain structural integrity and function under varying physical stresses [4]. ECM remodeling ensures that tissues can withstand mechanical challenges while maintaining flexibility and responsiveness to physiological cues.

ECM remodeling is implicated in regulating cell behavior in health and disease. Changes in ECM composition, stiffness, or organization can alter cellular responses, contributing to pathological conditions such as fibrosis, cancer progression, and cardiovascular diseases. Understanding the molecular mechanisms underlying ECM remodeling is thus essential for developing therapeutic strategies to modulate tissue

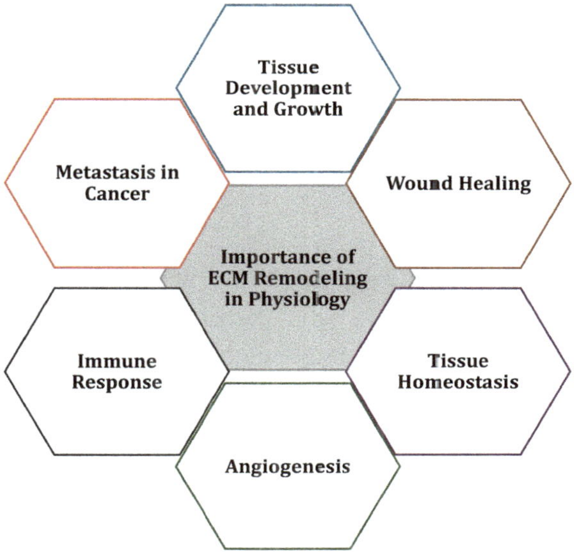

Fig. 3.1 Importance of ECM remodeling in physiology

repair, regeneration, and disease progression effectively. ECM remodeling is indispensable in physiology, influencing tissue development, repair, mechanical support, and cellular signaling. Its dynamic interactions with cells and bioactive molecules regulate tissue function and adaptation, highlighting its fundamental role in maintaining tissue homeostasis and responding to physiological challenges throughout life (Fig. 3.1).

3.3 Cellular Components Involved in ECM Remodeling

Cellular components involved in extracellular matrix (ECM) remodeling play pivotal roles in maintaining tissue structure, function, and repair processes (Fig. 3.2). Key players include fibroblasts, myofibroblasts, and immune cells such as macrophages. Fibroblasts are primary contributors to ECM synthesis and degradation, secreting collagen, elastin, and proteoglycans while producing enzymes like matrix metalloproteinases (MMPs) that facilitate ECM turnover. Myofibroblasts, which exhibit characteristics of both fibroblasts and smooth muscle cells, play crucial roles in wound contraction and scar formation by generating mechanical tension within the ECM [5]. Macrophages, on the other hand, contribute to ECM remodeling through phagocytosis of ECM debris and secretion of cytokines that regulate fibroblast activity and inflammation. Together, these cellular components orchestrate ECM dynamics, influencing tissue architecture and function in both physiological and pathological contexts.

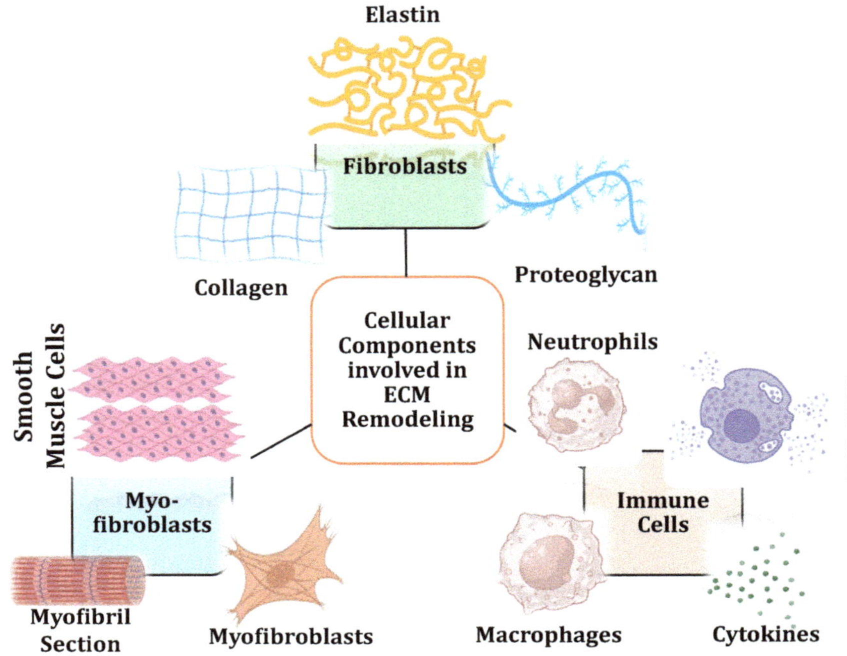

Fig. 3.2 Cellular components involved in ECM remodeling

3.3.1 Fibroblasts and Myofibroblasts

Fibroblasts and myofibroblasts are crucial cellular components involved in the dynamic process of extracellular matrix (ECM) remodeling within tissues. Fibroblasts are connective tissue cells found in various organs throughout the body, primarily responsible for synthesizing and maintaining the ECM. They play essential roles in tissue homeostasis, wound healing, and repair mechanisms. Fibroblasts produce structural proteins such as collagen, elastin, and fibronectin, which form the framework of the ECM and provide mechanical support to surrounding cells. During tissue injury or inflammation, fibroblasts become activated and undergo phenotypic changes to myofibroblasts. Myofibroblasts are contractile cells characterized by the expression of α-smooth muscle actin (α-SMA), which enables them to generate contractile forces within the ECM. This contractile ability is crucial for wound contraction and closure, contributing to tissue remodeling and scar formation [6]. Myofibroblasts also secrete ECM proteins and enzymes like MMPs, facilitating ECM turnover and remodeling processes. The transition of fibroblasts into myofibroblasts is regulated by various cytokines, growth factors, and mechanical cues present in the local microenvironment. Transforming growth factor-beta (TGF-β) is a key regulator known to stimulate fibroblast differentiation into myofibroblasts, promoting ECM synthesis and fibrotic responses in tissues. Mechanical tension within the ECM, such as during tissue stretching or injury, can also induce fibroblast activation into myofibroblasts, highlighting their role in responding to mechanical cues and maintaining tissue integrity. In pathological conditions such as fibrosis, chronic inflammation, or excessive tissue repair, dysregulated activation of myofibroblasts can lead to aberrant ECM deposition and tissue stiffening. This contributes to organ dysfunction and impaired physiological processes. Understanding the molecular mechanisms underlying fibroblast and myofibroblast behavior is crucial for developing therapies to modulate ECM remodeling in disease contexts effectively.

Research into fibroblast and myofibroblast biology continues to uncover new insights into their roles beyond ECM synthesis and contraction. Recent studies have highlighted their involvement in immune responses, angiogenesis, and tissue regeneration processes [7]. For example, fibroblasts can interact with immune cells to regulate inflammation and tissue repair, influencing the overall healing process. Moreover, their ability to modulate ECM composition and stiffness can impact cell migration, differentiation, and signaling within tissues, thereby influencing tissue architecture and function in health and disease.

Fibroblasts and myofibroblasts are integral to ECM remodeling processes essential for tissue maintenance, repair, and adaptation. Their diverse functions in synthesizing ECM components, responding to mechanical cues, and modulating tissue architecture highlight their pivotal roles in physiological and pathological conditions.

Exercise 1

Fill Ups
(Myofibroblasts, Cell migration, Matrix metalloproteinases, Collagen, Tissue Inhibitors of Metalloproteinases)

1. Fibroblasts are key cellular components that produce and secrete enzymes such as ________, which degrade ECM components.

2. The process of ECM remodeling is heavily influenced by ________, which are proteins that help organize and structure the ECM.

3. Integrins are cell surface receptors that mediate the attachment between a cell and the ECM, facilitating the process of ____________

4. The balance between ECM synthesis and degradation is regulated by the activity of enzymes like ________ and their inhibitors.

5. Cells called ________ are involved in the remodeling of the ECM by secreting both ECM components and enzymes that degrade the matrix.

3.3.2 Macrophages and Immune Cells

Macrophages and other immune cells play crucial roles in maintaining tissue homeostasis, orchestrating immune responses, and contributing to extracellular matrix (ECM) remodeling within various physiological and pathological contexts. Macrophages, derived from monocytes, are versatile cells found in almost all tissues where they fulfill diverse functions beyond their traditional role in immune defense. In tissues, macrophages act as scavengers, phagocytosing cellular debris and pathogens, thereby contributing to tissue cleanup and maintaining ECM integrity. Moreover, they secrete cytokines and growth factors that influence the behavior of neighboring cells, including fibroblasts and endothelial cells, thereby regulating ECM synthesis, degradation, and remodeling processes. In the context of inflammation and wound healing, macrophages are crucial mediators of ECM remodeling. During the initial inflammatory phase, they secrete pro-inflammatory cytokines such as tumor necrosis factor-alpha (TNF-α) and interleukin-1 (IL-1), which stimulate fibroblast activation and ECM deposition. This sets the stage for subsequent tissue repair processes. As inflammation resolves, macrophages transition to an anti-inflammatory phenotype, promoting ECM turnover and tissue remodeling by secreting matrix metalloproteinases (MMPs) and tissue inhibitors of metalloproteinases (TIMPs). This balance between pro-inflammatory and anti-inflammatory macrophage functions is essential for proper wound healing and tissue repair [8]. Beyond their roles in acute inflammation and wound healing, macrophages also contribute to chronic inflammatory conditions and tissue remodeling in diseases such as fibrosis and cancer. In fibrotic diseases, prolonged activation of macrophages can lead to excessive ECM deposition and tissue stiffening, impairing organ function. Similarly, in cancer, macrophages within the tumor microenvironment promote ECM remodeling that supports tumor growth, invasion, and metastasis. Targeting macrophage-driven ECM remodeling pathways has thus become a focus of therapeutic strategies aimed at modulating disease progression and improving patient outcomes.

The interaction between macrophages and the ECM is bidirectional and dynamic. ECM components such as fibronectin and collagen can influence macrophage behavior and activation state through integrin-mediated signaling pathways. In turn, activated macrophages remodel the ECM by secreting proteases, influencing its composition and mechanical properties. This reciprocal interaction between macrophages and the ECM underscores their collaborative roles in maintaining tissue architecture, function, and adaptive responses to physiological and pathological challenges. In addition to macrophages, other immune cells, including lymphocytes and dendritic cells, also contribute to ECM remodeling and tissue homeostasis. Lymphocytes, such as T cells and B cells, regulate immune responses and secrete cytokines that can modulate fibroblast activity and ECM turnover [9]. Dendritic cells play roles in antigen presentation and immune surveillance within tissues, influencing local immune responses and tissue remodeling processes. Together, these immune cells form a complex network that integrates immune surveillance with tissue repair and remodeling, ensuring proper tissue function and adaptive responses to environmental changes. Macrophages and other immune cells are integral to ECM remodeling processes that maintain tissue integrity, regulate inflammation, and support wound healing. Their diverse functions in phagocytosis, cytokine secretion, and ECM modulation highlight their importance in both physiological tissue homeostasis and pathological conditions.

3.4 Biochemical Processes in ECM Remodeling

Biochemical processes involved in extracellular matrix (ECM) remodeling are essential mechanisms that regulate tissue architecture, function, and adaptive responses in physiological and pathological conditions (Fig. 3.3). ECM remodeling encompasses a dynamic interplay of biochemical events that regulate the synthesis, degradation, and modification of ECM components such as collagen, elastin, proteoglycans, and glycosaminoglycans (GAGs). Key players in these processes include enzymes like matrix metalloproteinases (MMPs), which degrade ECM proteins, and tissue inhibitors of metalloproteinases (TIMPs), which regulate MMP activity [10]. These biochemical processes are tightly regulated to maintain ECM homeostasis, support tissue repair and regeneration, and influence cellular behaviors such as migration, proliferation, and differentiation.

3.4.1 Matrix Metalloproteinases (MMPs)

Matrix metalloproteinases (MMPs) are a family of zinc-dependent endopeptidases involved in the turnover and remodeling of the extracellular matrix (ECM) in various physiological and pathological processes. This family comprises over 20 members, each with distinct substrate specificities and roles in ECM degradation, remodeling, and cell signaling. MMPs are synthesized as inactive zymogens (pro-MMPs) and

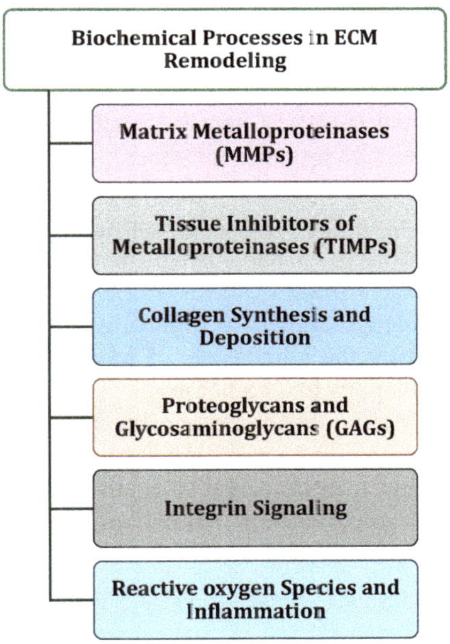

Fig. 3.3 Biochemical processes in ECM remodeling

are activated in response to specific stimuli such as cytokines, growth factors, and mechanical stress, highlighting their regulatory control in tissue homeostasis and repair mechanisms. MMPs play critical roles in tissue remodeling by cleaving ECM components such as collagen, elastin, fibronectin, and laminin. This enzymatic degradation is essential for physiological processes like embryogenesis, tissue morphogenesis, and wound healing. For instance, during tissue repair, MMPs facilitate the breakdown of damaged ECM, allowing for the migration of fibroblasts and other reparative cells to the wound site [11]. This initial degradation phase is followed by ECM synthesis to rebuild tissue structure, highlighting the balanced action of MMPs in tissue remodeling dynamics.

In addition to their role in ECM degradation, MMPs participate in regulating cell behavior and signaling through the cleavage of non-ECM substrates such as growth factors, cytokines, and cell surface receptors. By releasing bioactive fragments from these substrates, MMPs influence cell proliferation, migration, differentiation, and apoptosis, thereby impacting tissue development and pathological conditions. Dysregulation of MMP activity is implicated in various diseases, including arthritis, cancer, cardiovascular diseases, and neurological disorders, underscoring their importance as therapeutic targets. The activity of MMPs is tightly controlled at multiple levels to prevent excessive ECM degradation and maintain tissue integrity. Regulation occurs through transcriptional control, post-translational modifications, and interaction with tissue inhibitors of metalloproteinases (TIMPs), which bind to MMPs and inhibit their enzymatic activity. The balance between MMPs and TIMPs is crucial for ECM turnover and remodeling processes, ensuring proper tissue function and adaptation to physiological changes [12]. Different MMPs exhibit tissue-specific expression patterns and functions, reflecting their roles in specialized physiological processes. For example, MMP-2 and MMP-9 are gelatinases involved in the breakdown of collagen type IV in basement membranes, crucial for angiogenesis and tissue remodeling during development and wound healing. MMP-3 (stromelysin-1) and MMP-7 (matrilysin) are involved in the degradation of proteoglycans and non-ECM proteins, influencing cell migration and tissue repair processes. Advances in research have elucidated the multifaceted roles of MMPs in health and disease, prompting interest in developing MMP inhibitors as potential therapeutics. Targeting MMP activity has shown promise in preclinical studies and clinical trials for diseases characterized by excessive ECM degradation, fibrosis, and tumor invasion. However, challenges remain in achieving specificity and minimizing off-target effects, underscoring the complexity of MMP regulation and their diverse functions in tissue physiology and pathology.

Matrix metalloproteinases (MMPs) are key regulators of ECM remodeling, playing critical roles in tissue development, repair, and disease progression. Their ability to degrade ECM components and modulate cellular behavior underscores their importance in maintaining tissue homeostasis and responding to physiological and pathological cues.

3.4.2 Tissue Inhibitors of Metalloproteinases (TIMPs)

Tissue inhibitors of metalloproteinases (TIMPs) are a family of proteins that regulate the activity of matrix metalloproteinases (MMPs), thereby playing crucial roles in the maintenance of extracellular matrix (ECM) homeostasis, tissue remodeling, and various physiological processes. There are four known members of the TIMP family (TIMP-1, TIMP-2, TIMP-3, and TIMP-4), each with distinct tissue-specific expression patterns and functions. TIMPs primarily function by binding to the active site of MMPs, inhibiting their proteolytic activity and preventing excessive ECM degradation. The balance between MMPs and TIMPs is critical for ECM turnover and remodeling. Under normal physiological conditions, TIMPs help maintain ECM integrity by tightly regulating the activity of MMPs [13]. This balance ensures proper tissue function, structural stability, and adaptation to physiological changes such as growth, development, and wound healing. Dysregulation of TIMP expression or function can lead to pathological conditions characterized by abnormal ECM turnover, tissue fibrosis, and impaired repair mechanisms. In addition to their role as MMP inhibitors, TIMPs have diverse functions beyond ECM regulation. TIMPs can interact with cell surface receptors and other ECM proteins, influencing cell proliferation, survival, and differentiation. For example, TIMP-1 has been shown to promote cell survival through receptor-mediated signaling pathways independent of MMP inhibition. These non-inhibitory functions highlight the multifaceted roles of TIMPs in cellular processes and tissue biology. The expression and activity of TIMPs are tightly regulated at transcriptional, post-translational, and epigenetic levels. Factors such as growth factors, cytokines, and ECM components can modulate TIMP expression in a tissue-specific manner, reflecting their roles in tissue development and adaptive responses. For instance, TIMP-3 is known for its unique ability to bind directly to ECM proteins like heparan sulfate and glycosaminoglycans, influencing ECM organization and function in tissues. TIMPs are involved in various physiological processes, including tissue development, angiogenesis, and immune responses. During embryogenesis, TIMPs contribute to tissue morphogenesis and organogenesis by regulating ECM dynamics and cell behavior. In adult tissues, TIMPs play roles in maintaining tissue homeostasis, responding to injury, and supporting tissue repair mechanisms. Their ability to modulate ECM remodeling and cell signaling pathways underscores their importance in tissue physiology and adaptive responses to environmental changes. The dysregulation of TIMPs has been implicated in numerous pathological conditions, including cancer, cardiovascular diseases, and neurological disorders [14]. In cancer, altered TIMP expression levels can affect tumor progression, invasion, and metastasis by influencing ECM degradation and remodeling processes. Conversely, in cardiovascular diseases, decreased TIMP expression may lead to excessive ECM degradation and vessel wall instability, contributing to vascular pathology.

Therapeutically, targeting TIMPs or manipulating their expression levels has shown potential in modulating ECM remodeling and tissue repair in disease contexts. Strategies aimed at enhancing TIMP activity or delivery may help mitigate ECM degradation in conditions such as arthritis or promote tissue regeneration following injury. However, the complexity of TIMP functions and their interactions with MMPs and other ECM regulators present challenges in developing precise therapeutic interventions.

Tissue inhibitors of metalloproteinases (TIMPs) are essential regulators of ECM homeostasis, tissue remodeling, and cellular processes in health and disease. Their ability to inhibit MMP activity, influence cell behavior, and modulate tissue responses underscores their critical roles in maintaining tissue integrity and function throughout the body.

3.5 Biophysical Mechanisms of ECM Remodeling

Biophysical mechanisms of extracellular matrix (ECM) remodeling encompass the physical forces and mechanical properties that govern the dynamic changes in ECM structure and function. Unlike biochemical mechanisms that involve enzymatic degradation and synthesis of ECM components, biophysical mechanisms focus on how mechanical cues such as tension, compression, shear stress, and stiffness influence ECM organization and remodeling. These mechanical forces are sensed and transduced by cells within tissues, leading to cellular responses that modulate ECM composition, alignment, and stiffness. Understanding biophysical mechanisms is essential for unraveling how mechanical stimuli regulates tissue development, homeostasis, wound healing, and disease processes, highlighting their pivotal role in maintaining tissue integrity and function.

Exercise 2

Fill Ups
(Mechanotransduction, Mechanical, Proteolytic enzymes, Collagen, Traction forces)

1. The mechanical properties of the ECM, such as stiffness, influence cell behavior through a process known as ________.

2. Tensional forces generated by cells contribute to ECM remodeling by contracting and reorganizing ________ fibers.

3. ECM degradation is facilitated by the generation of local ________ that assist in the breakdown of ECM components.

4. The alignment of ECM fibers is influenced by cellular activities such as ________, which involve the pulling and rearrangement of ECM structures.

5. The process of ECM remodeling is often regulated by changes in the local ________ environment, which can affect the activity of remodeling enzymes.

3.5.1 Mechanical Forces and ECM Adaptation

Mechanical forces play a fundamental role in shaping the structure and function of the extracellular matrix (ECM), influencing tissue architecture, cellular behavior, and physiological responses. The ECM is not merely a passive scaffold but actively responds to mechanical cues through processes collectively known as mechano transduction. This process involves the conversion of mechanical stimuli into biochemical signals that regulate cellular activities and ECM remodeling [15]. One key aspect of mechanical forces on ECM is tension. Cells exert forces on the ECM through interactions with integrins and other cell adhesion molecules, leading to ECM stretching and alignment. This mechanical tension is crucial in tissues such as muscles, tendons, and blood vessels, where it contributes to structural integrity and functional properties. For example, in skeletal muscle, mechanical tension regulates the alignment of collagen fibers within the ECM, influencing muscle contraction and force generation. Compression is another mechanical force that impacts ECM adaptation. In tissues subject to compressive loads, such as cartilage and bone, ECM components like proteoglycans and collagen fibers provide resistance to compression while maintaining tissue hydration and elasticity. The balance between compressive forces and ECM resilience is critical for maintaining joint function and preventing tissue degeneration in conditions like osteoarthritis.

Shear stress, generated by fluid flow or movement across endothelial surfaces, also influences ECM remodeling and cellular responses. In blood vessels, shear stress from blood flow stimulates endothelial cells to release nitric oxide and regulate vascular tone. This mechanical signal not only affects vascular function but also influences ECM composition and remodeling in response to hemodynamic changes [16]. The ability of the ECM to adapt to mechanical forces is mediated by cellular processes such as matrix remodeling and synthesis. Fibroblasts and other ECM-producing cells sense mechanical cues through mechanosensitive receptors and cytoskeletal rearrangements. In response, they modulate ECM production, secretion of matrix-degrading enzymes like matrix metalloproteinases (MMPs), and deposition of ECM proteins to strengthen or remodel the tissue in accordance with mechanical demands. Bio-mechanical properties of the ECM, including stiffness and viscoelasticity, also play crucial roles in tissue function and adaptation. ECM stiffness, for instance, affects cell behavior and differentiation in tissues like the lungs and heart. Changes in ECM stiffness can influence cellular responses such as fibroblast activation and differentiation into myofibroblasts, contributing to tissue fibrosis and pathological remodeling.

Mechanical forces are integral to ECM adaptation and remodeling, influencing tissue structure, function, and cellular behavior. The ability of the ECM to sense and respond to mechanical cues through mechano transduction mechanisms is essential for maintaining tissue homeostasis, supporting wound healing, and adapting to physiological challenges.

3.5.2 Role of Cytoskeletal Dynamics

Cytoskeletal dynamics play a crucial role in mediating cellular responses to extracellular signals, including those involved in the remodeling of the extracellular matrix (ECM). The cytoskeleton, composed of microfilaments (actin), intermediate filaments, and microtubules, provides structural support and mechanical integrity to cells while also participating in cell motility, division, and signaling processes. In the context of ECM remodeling, cytoskeletal dynamics are essential for transmitting mechanical forces from the ECM to the cell interior and for regulating cell shape and motility. Actin filaments, a major component of the cytoskeleton, play a central role in cell adhesion and migration processes that are critical for ECM remodeling. Actin polymerization and depolymerization dynamics allow cells to form focal adhesions with the ECM, enabling them to sense and respond to mechanical cues. These focal adhesions serve as sites for force transmission and signaling, linking the ECM to the cytoskeleton and orchestrating cellular responses such as migration, contraction, and ECM remodeling. Intermediate filaments provide mechanical stability and resistance to mechanical stress within cells, contributing to tissue integrity and adaptation to mechanical forces [17]. In tissues subjected to mechanical strain, such as skin and muscle, intermediate filaments help distribute mechanical forces and maintain structural integrity, thereby supporting ECM remodeling processes during development, wound healing, and tissue repair.

Microtubules, another component of the cytoskeleton, participate in intracellular trafficking and organization of organelles, influencing cell polarity and directional migration. Microtubule dynamics regulate the transport of vesicles containing ECM components, such as collagen and fibronectin, to the cell surface for secretion and ECM assembly. Microtubules are involved in signaling pathways that modulate cell adhesion, migration, and ECM remodeling through interactions with focal adhesion proteins and cytoskeletal regulators. Cytoskeletal dynamics also play a role in mechano transduction, the process by which cells convert mechanical stimuli into biochemical signals. Mechanical forces exerted on the ECM are transmitted through integrins and focal adhesions to the cytoskeleton, where they induce cytoskeletal rearrangements and signaling cascades. These mechano transductive signals regulate gene expression, enzyme activity, and ECM remodeling processes, influencing tissue development, homeostasis, and response to injury. Dysregulation of cytoskeletal dynamics can contribute to pathological conditions characterized by abnormal ECM remodeling and tissue dysfunction. For example, mutations in cytoskeletal proteins or alterations in signaling pathways regulating cytoskeletal dynamics can impair cell adhesion, migration, and ECM assembly, leading to tissue fibrosis, impaired wound healing, or developmental abnormalities. Understanding the role of cytoskeletal dynamics in ECM remodeling is critical for developing therapeutic strategies that target mechanical aspects of tissue repair and regeneration. Approaches aimed at modulating cytoskeletal function, such as pharmacological agents targeting actin dynamics or microtubule stabilization, hold promise for enhancing ECM remodeling in diseases characterized by impaired tissue integrity or mechanical dysfunction.

Cytoskeletal dynamics are essential for mediating cellular responses to mechanical cues from the ECM and for regulating processes such as cell adhesion, migration, and ECM remodeling. The interplay between cytoskeletal components—actin filaments, intermediate filaments, and microtubules—provides cells with the structural support and mechanical responsiveness necessary for tissue homeostasis, repair, and adaptation to physiological and pathological conditions involving ECM remodeling.

3.6 Role of Growth Factors and Signaling Pathways

The role of growth factors and signaling pathways is pivotal in orchestrating cellular responses and extracellular matrix (ECM) remodeling processes essential for tissue development, homeostasis, and adaptive responses to environmental changes (Fig. 3.4). Growth factors, such as transforming growth factor-beta (TGF-β), fibroblast growth factors (FGFs), and vascular endothelial growth factor (VEGF), act as potent regulators of cell proliferation, differentiation, and ECM synthesis within tissues. These molecules exert their effects through specific cell surface receptors, initiating intracellular signaling cascades that modulate gene expression, enzyme activity, and ECM remodeling dynamics [18]. By coordinating cellular behaviors and ECM composition, growth factors and their associated signaling pathways play critical roles in physiological processes like embryogenesis, wound healing, and tissue repair, as well as in pathological conditions such as fibrosis, cancer progression, and cardiovascular disease.

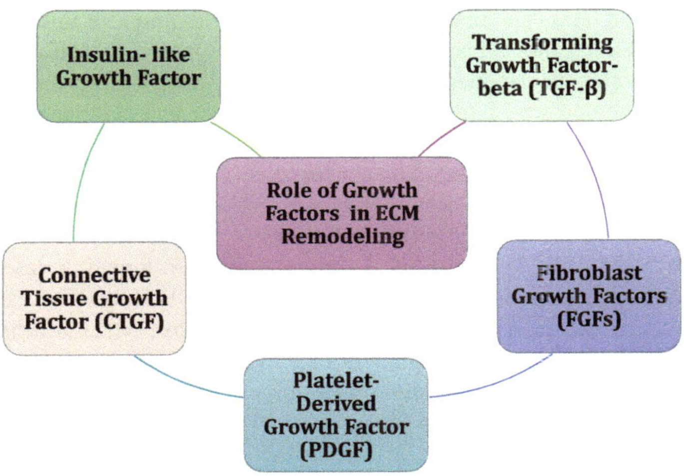

Fig. 3.4 Role of growth factors in ECM remodeling

3.6.1 Transforming Growth Factor-Beta (TGF-B)

Transforming growth factor-beta (TGF-β) is a multifunctional cytokine that plays crucial roles in regulating cellular processes, tissue development, and extracellular matrix (ECM) remodeling. It is part of a large family of growth factors that includes three isoforms: TGF-β1, TGF-β2, and TGF-β3. TGF-β is synthesized and secreted by various cell types, including immune cells, fibroblasts, and endothelial cells, and exerts its effects through specific cell surface receptors known as TGF-β receptors (TβRs). One of the primary functions of TGF-β is its role in ECM synthesis and remodeling. TGF-β stimulates the production of ECM proteins such as collagen, fibronectin, and proteoglycans by fibroblasts and other mesenchymal cells. This contributes to the formation and maintenance of the ECM framework essential for tissue architecture and mechanical support. In pathological conditions, dysregulated TGF-β signaling can lead to excessive ECM deposition, fibrosis, and tissue stiffness, as seen in diseases like pulmonary fibrosis and liver cirrhosis. TGF-β also regulates cellular responses involved in wound healing and tissue repair processes. During tissue injury, TGF-β promotes inflammation and activates fibroblasts and myofibroblasts, leading to ECM deposition and wound closure. It modulates the expression of matrix metalloproteinases (MMPs) and tissue inhibitors of metalloproteinases (TIMPs), thereby influencing ECM turnover and remodeling phases crucial for effective wound healing. In addition to its role in ECM remodeling, TGF-β plays a critical role in regulating immune responses and inflammation. It modulates the function of immune cells such as macrophages, T cells, and dendritic cells, influencing their activation, differentiation, and cytokine production [19]. TGF-β is known for its immunosuppressive effects, contributing to immune tolerance and preventing excessive inflammation and tissue damage in response to injury or infection.

TGF-β signaling pathways are involved in cellular processes that affect cell proliferation, differentiation, and apoptosis. TGF-β can act as both a tumor suppressor and a promoter of tumorigenesis depending on the context and cellular environment. In normal tissues, TGF-β inhibits cell cycle progression and promotes apoptosis to suppress tumor formation. However, in cancer cells, TGF-β signaling pathways may become dysregulated, leading to tumor growth, invasion, and metastasis through effects on ECM remodeling and cellular plasticity.

The effects of TGF-β are mediated through complex signaling cascades that involve Smad-dependent and Smad-independent pathways. Upon binding to TβRs, TGF-β receptors phosphorylate downstream Smad proteins, which translocate into the nucleus to regulate the transcription of target genes involved in ECM synthesis, cell cycle regulation, and immune responses. Smad-independent pathways, such as MAPK (mitogen-activated protein kinase) and PI3K/Akt (phosphatidylinositol-3-kinase/protein kinase B) pathways, also contribute to the diverse biological effects of TGF-β. Dysregulation of TGF-β signaling is implicated in numerous diseases, including fibrosis, cancer, autoimmune disorders, and cardiovascular diseases. Therapeutic strategies targeting TGF-β signaling pathways are actively being explored to modulate ECM remodeling, suppress inflammation, and inhibit tumor progression.

These approaches aim to harness the therapeutic potential of TGF-β while minimizing its adverse effects, highlighting the importance of understanding the complex roles of TGF-β in health and disease.

3.6.2 *Integrin Signaling*

Integrins are a family of cell surface receptors that play essential roles in mediating cell adhesion to the extracellular matrix (ECM) and transmitting signals bidirectionally between the ECM and the intracellular environment. They are heterodimeric transmembrane proteins composed of α and β subunits, which together form a diverse repertoire of integrin receptors. Integrins bind to specific ECM ligands such as collagen, fibronectin, and laminin, establishing physical connections that anchor cells to the ECM and provide mechanical support. Integrin signaling is initiated upon engagement of integrin receptors with ECM ligands, a process known as outside-in signaling. This interaction induces conformational changes in the integrin cytoplasmic domains, leading to the recruitment and activation of intracellular signaling proteins. These include focal adhesion kinase (FAK), Src family kinases, and various adaptor proteins such as talin and vinculin. These signaling complexes serve as hubs for integrating mechanical and biochemical signals from the ECM, regulating processes such as cell migration, proliferation, differentiation, and survival [20]. Beyond outside-in signaling, integrins also transmit signals from the intracellular environment to the ECM, a process known as inside-out signaling. Inside-out signaling involves the regulation of integrin affinity and clustering in response to intracellular cues such as cytoskeletal dynamics and signaling molecules. For example, the binding of small GTPases like Rho and Rac to integrin-associated proteins can modulate integrin activation and promote cell adhesion and migration on the ECM. Integrin signaling is crucial for tissue development, homeostasis, and repair processes. During tissue morphogenesis, integrins regulate cell-ECM interactions that guide cell migration, polarity, and differentiation essential for tissue organization. In adult tissues, integrins maintain tissue integrity and function by mediating cell adhesion to the ECM, thereby supporting mechanical stability and responsiveness to environmental cues.

The dynamic regulation of integrin signaling is essential for physiological processes such as immune responses and wound healing. Integrins on immune cells facilitate adhesion to endothelial cells and migration across tissues during inflammation and immune surveillance. In wound healing, integrin-mediated signaling coordinates the recruitment and activation of fibroblasts and other reparative cells to the wound site, facilitating ECM remodeling and tissue repair processes. Dysregulation of integrin signaling is implicated in various diseases, including cancer, cardiovascular disorders, and autoimmune diseases [21]. Alterations in integrin expression, activation, or downstream signaling pathways can promote tumor cell invasion and metastasis, disrupt tissue integrity, and impair immune cell function. Therapeutic strategies targeting integrin signaling pathways are being investigated to modulate ECM interactions, inhibit disease progression, and promote tissue repair in pathological conditions.

Approaches that aim to manipulate integrin function, block specific integrin-ligand interactions, or modulate downstream signaling pathways hold promise for treating diseases characterized by abnormal ECM remodeling, tissue dysfunction, and impaired cellular responses.

3.7 Pathological ECM Remodeling

Pathological extracellular matrix (ECM) remodeling refers to aberrant changes in the composition, structure, and function of the ECM that contribute to the progression of various diseases and conditions. Unlike physiological ECM remodeling involved in tissue repair and adaptation, pathological ECM remodeling often involves excessive deposition, degradation, or abnormal composition of ECM components. This dysregulation can disrupt tissue architecture, impair organ function, and promote disease progression [22]. Examples of pathological ECM remodeling include fibrosis, where excessive collagen deposition leads to tissue stiffness and organ dysfunction, and cancer metastasis, where ECM remodeling facilitates tumor invasion into surrounding tissues (Fig. 3.5).

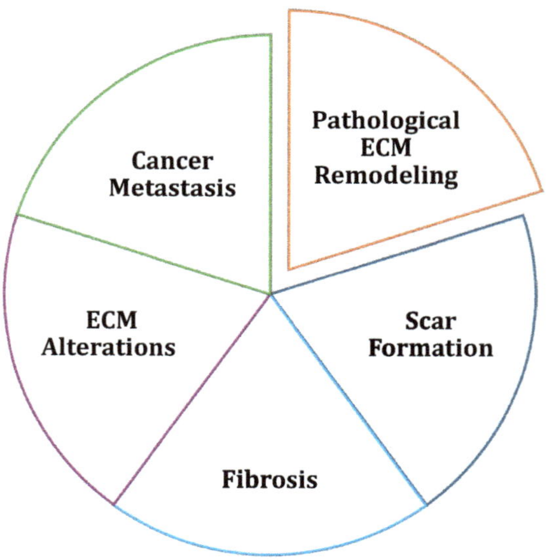

Fig. 3.5 Various examples of pathological ECM remodeling

3.7.1 Fibrosis and Scar Formation

Fibrosis and scar formation are complex processes characterized by excessive deposition of extracellular matrix (ECM) components, primarily collagen, leading to tissue stiffening and functional impairment. These processes occur in response to chronic injury, inflammation, or persistent tissue damage and are common features of various diseases affecting different organs, including the liver, lungs, heart, and skin. The initial phase of fibrosis involves tissue injury or inflammation, which triggers the activation of fibroblasts and other ECM-producing cells. These activated fibroblasts, often referred to as myofibroblasts due to their increased contractile properties, migrate to the site of injury and begin synthesizing and depositing ECM proteins, particularly collagen types I and III. This results in the formation of a provisional ECM matrix that provides structural support but lacks the organization and mechanical properties of normal tissue. As fibrosis progresses, there is a persistent and dysregulated accumulation of collagen and other ECM components, leading to the formation of a dense fibrotic scar. This scar tissue replaces normal parenchymal cells and disrupts tissue architecture and function. In organs such as the liver, fibrosis can progress to cirrhosis, characterized by extensive fibrotic scarring that compromises liver function and leads to complications such as portal hypertension and liver failure [23]. The molecular mechanisms driving fibrosis involve a complex interplay of signaling pathways, including transforming growth factor-beta (TGF-β) and connective tissue growth factor (CTGF), which stimulate ECM synthesis and myofibroblast activation. These cytokines promote fibroblast proliferation, differentiation, and migration, contributing to the sustained production of collagen and ECM proteins. In addition, inflammatory mediators such as cytokines (e.g., interleukin-6, tumor necrosis factor-alpha) and chemokines recruit immune cells to the site of injury, where they further amplify fibrotic responses.

The resolution of fibrosis is hindered by mechanisms that impair ECM degradation and turnover. Matrix metalloproteinases (MMPs), enzymes responsible for ECM degradation, are often inhibited by tissue inhibitors of metalloproteinases (TIMPs) in fibrotic tissues, leading to an imbalance between ECM synthesis and degradation. This imbalance perpetuates the accumulation of ECM proteins and contributes to the progressive nature of fibrotic diseases. Clinically, fibrosis and scar formation have significant implications for disease progression and patient outcomes. In conditions such as idiopathic pulmonary fibrosis (IPF) or systemic sclerosis, fibrotic changes in the lungs and skin, respectively, lead to respiratory insufficiency and impaired mobility. In the heart, post-myocardial infarction fibrosis can impair cardiac function and increase the risk of heart failure. Effective treatments for fibrosis are limited, focusing primarily on managing symptoms and addressing underlying causes such as inflammation or tissue injury. Research efforts are underway to develop therapies that target specific molecular pathways involved in fibrosis, with the aim of inhibiting ECM deposition, promoting ECM degradation, or modulating myofibroblast activation. Potential strategies include anti-fibrotic drugs targeting TGF-β signaling, inhibitors of ECM cross-linking enzymes, and approaches to enhance ECM turnover and repair mechanisms.

3.7.2 Cancer Metastasis and ECM Alterations

Cancer metastasis is a complex and life-threatening process characterized by the spread of cancer cells from the primary tumor to distant organs through the bloodstream or lymphatic system. This metastatic cascade involves multiple steps, including invasion of surrounding tissues, intravasation into blood or lymphatic vessels, survival in circulation, extravasation at distant sites, and colonization to form secondary tumors. Throughout these stages, alterations in the extracellular matrix (ECM) play critical roles in facilitating or hindering cancer cell migration and invasion. The ECM surrounding tumors undergoes significant remodeling during cancer progression. Cancer-associated fibroblasts (CAFs), immune cells, and tumor cells themselves secrete proteolytic enzymes such as matrix metalloproteinases (MMPs), which degrade ECM proteins like collagen, laminin, and fibronectin. This enzymatic degradation creates pathways for cancer cells to invade surrounding tissues and facilitates their entry into blood or lymphatic vessels during metastasis. The physical and biochemical properties of the ECM also influence cancer cell behavior and metastatic potential [24]. Increased ECM stiffness, characteristic of desmoplastic reactions in many solid tumors, promotes cancer cell proliferation, survival, and migration. Changes in ECM composition, such as increased deposition of fibronectin or hyaluronic acid, can enhance cancer cell adhesion, motility, and resistance to anoikis (cell death due to loss of ECM attachment), all of which are critical for metastatic dissemination.

ECM alterations contribute to the establishment of pre-metastatic niches, specialized microenvironments at distant sites that are conducive to cancer cell colonization. Tumor-derived factors, including growth factors (e.g., TGF-β), cytokines, and extracellular vesicles, modulate ECM remodeling in pre-metastatic organs, promoting the recruitment of immune cells, angiogenesis, and the formation of supportive niches for incoming cancer cells.

In addition to facilitating metastasis, ECM alterations can also influence therapeutic responses and disease progression in cancer patients. Abnormal ECM deposition and cross-linking can create physical barriers that limit drug penetration into tumors, contributing to treatment resistance. Interactions between cancer cells and the ECM can activate signaling pathways that promote survival and proliferation, further complicating treatment strategies aimed at targeting tumor cells [25]. Strategies targeting ECM remodeling processes, such as MMP inhibitors or agents that disrupt ECM-stiffening mechanisms, hold promise for mitigating cancer cell invasion and metastatic spread. Biomarkers related to ECM alterations may serve as prognostic indicators or therapeutic targets in personalized cancer treatment approaches.

ECM alterations play integral roles in cancer metastasis by promoting invasive behavior, facilitating intravasation and extravasation, establishing pre-metastatic niches, and influencing therapeutic responses.

3.8 Therapeutic Approaches Targeting ECM Remodeling

Therapeutic approaches targeting extracellular matrix (ECM) remodeling aim to intervene in the dynamic processes that govern ECM composition, structure, and function in various physiological and pathological contexts. The ECM plays critical roles in tissue development, homeostasis, and repair, making it a promising target for treating diseases characterized by abnormal ECM remodeling, such as fibrosis, cancer metastasis, and cardiovascular disorders. Therapeutic strategies include inhibiting enzymes involved in ECM degradation, such as matrix metalloproteinases (MMPs), to prevent excessive ECM breakdown and tissue damage [23]. Alternatively, promoting ECM synthesis and deposition through growth factors or ECM-mimicking scaffolds can enhance tissue repair and regeneration in wound healing or tissue engineering applications.

3.8.1 Anti-fibrotic Therapies

Anti-fibrotic therapies represent a critical area of research and treatment development aimed at mitigating the pathological process of fibrosis, characterized by excessive accumulation of extracellular matrix (ECM) proteins such as collagen, leading to tissue scarring and organ dysfunction. These therapies target various stages of fibrotic disease progression, from inhibiting initial injury and inflammation to preventing ECM deposition and promoting tissue regeneration. One approach to anti-fibrotic therapy involves targeting signaling pathways that drive fibroblast activation and ECM production. Transforming growth factor-beta (TGF-β), a key cytokine involved in fibrosis, stimulates fibroblast proliferation and collagen synthesis. Inhibitors of TGF-β signaling, such as small molecule inhibitors or monoclonal antibodies, are being investigated to suppress fibrotic responses in diseases like pulmonary fibrosis, liver cirrhosis, and systemic sclerosis. By blocking TGF-β-mediated fibroblast activation and ECM deposition, these therapies aim to slow disease progression and preserve organ function.

Another strategy focuses on disrupting ECM cross-linking, a process that contributes to tissue stiffness and fibrotic scar formation. Enzymes like lysyl oxidases (LOX) catalyze the cross-linking of collagen fibers, enhancing ECM stability but also impairing tissue flexibility and function. Inhibitors targeting LOX enzymes or agents that degrade abnormal cross-links are under investigation as potential anti-fibrotic therapies. These approaches aim to restore tissue elasticity and reduce the impact of fibrotic scarring on organ function. In addition to targeting ECM synthesis and cross-linking, anti-fibrotic therapies may involve promoting ECM degradation and turnover. Matrix metalloproteinases (MMPs) are enzymes responsible for ECM degradation, playing essential roles in tissue remodeling and repair. However, in fibrotic diseases, MMP activity may be impaired or insufficient to counterbalance ECM accumulation. Therapeutic strategies aimed at enhancing MMP activity or delivery, such as gene therapy or MMP activators, are being explored to facilitate ECM degradation and promote resolution of fibrotic lesions. Furthermore, anti-fibrotic therapies may involve modulating immune responses and inflammation associated with fibrotic diseases [26]. Immune cells play critical roles in regulating tissue repair and fibrosis, with dysregulated immune responses contributing to chronic inflammation and ECM remodeling. Targeted immunomodulatory therapies, including cytokine inhibitors or immune cell-targeting agents, aim to attenuate inflammatory responses and prevent progressive fibrosis in affected tissues.

Recent advances in tissue engineering and regenerative medicine have also contributed to anti-fibrotic approaches by developing biomaterials and scaffolds that mimic ECM properties and support tissue regeneration. These ECM-mimicking materials can promote cell adhesion, migration, and differentiation while providing mechanical support and structural integrity. By integrating with host tissues and facilitating ECM remodeling, these biomaterial-based therapies aim to repair fibrotic lesions and restore normal tissue function.

Clinical trials evaluating anti-fibrotic therapies are ongoing across various disease indications, with a focus on assessing safety, efficacy, and long-term outcomes in patients. Combination therapies targeting multiple aspects of fibrosis, such as ECM synthesis, cross-linking, degradation, and immune regulation, may offer synergistic benefits in treating complex fibrotic diseases. Personalized medicine approaches, guided by biomarkers of fibrosis severity and patient response to therapy, hold promise for optimizing treatment strategies and improving outcomes for individuals affected by debilitating fibrotic conditions.

Anti-fibrotic therapies encompass a diverse array of strategies aimed at halting or reversing the pathological process of fibrosis. By targeting key mechanisms involved in ECM remodeling, inflammation, and tissue repair, these therapies seek to preserve organ function, improve quality of life, and mitigate disease progression in patients with fibrotic disorders.

3.8.2 ECM Modulation in Regenerative Medicine

Extracellular matrix (ECM) modulation plays a pivotal role in regenerative medicine, where the manipulation of ECM composition, structure, and mechanical properties is leveraged to promote tissue repair and regeneration. The ECM serves as a dynamic scaffold that supports cell adhesion, migration, proliferation, and differentiation—essential processes in tissue development and healing. In regenerative medicine, ECM modulation strategies aim to mimic or enhance natural ECM cues to guide and optimize tissue regeneration. One approach involves using ECM-derived biomaterials or scaffolds to provide a bioactive environment that mimics the native ECM composition and architecture of the target tissue. These biomaterials can be derived from decellularized tissues or synthesized using ECM-mimicking components such as collagen, elastin, and glycosaminoglycans. By recapitulating the biochemical and biomechanical properties of native ECM, these scaffolds support cell attachment and growth while facilitating tissue-specific differentiation and remodeling. ECM modulation in regenerative medicine encompasses strategies to enhance ECM remodeling and turnover within damaged or diseased tissues. This may involve delivering growth factors, such as transforming growth factor-beta (TGF-β) or fibroblast growth factors (FGFs), that stimulate ECM synthesis by resident cells and promote tissue regeneration. These growth factors can be incorporated into biomaterials or delivered locally to enhance ECM deposition and remodeling at the site of injury or defect. In addition to promoting ECM synthesis, regenerative strategies aim to facilitate ECM degradation and turnover, processes critical for tissue remodeling and repair [25]. Matrix metalloproteinases (MMPs) play key roles in ECM degradation and are involved in tissue remodeling during wound healing and regeneration. Strategies to enhance MMP activity or delivery, such as gene therapy or biomaterial-based delivery systems, can promote ECM remodeling and facilitate the integration of newly formed tissue with the host environment. ECM modulation in regenerative medicine includes approaches to modify ECM stiffness and mechanical properties, which influence cellular behaviors and tissue function. In tissues like muscle and heart, ECM stiffness affects cell differentiation and contractility, while in bone and cartilage, ECM stiffness influences matrix deposition and mineralization. Biomaterials with tunable mechanical properties allow researchers to mimic specific tissue microenvironments and optimize conditions for cell growth, differentiation, and tissue integration.

ECM modulation also encompasses the use of cell-based therapies, where stem cells or progenitor cells are seeded onto ECM-mimicking scaffolds to promote tissue regeneration. These cells interact with the ECM scaffold, responding to biochemical and mechanical cues that guide their differentiation into specific cell types and support tissue repair processes. By integrating with host tissues and promoting ECM remodeling, cell-based therapies aim to restore functional tissue architecture and enhance organ function in regenerative medicine applications.

Clinical translation of ECM modulation approaches in regenerative medicine is advancing through preclinical studies and clinical trials across various tissue types and disease conditions. Strategies that combine biomaterial-based scaffolds, growth factors, and cell therapies show promise for treating complex tissue defects, enhancing wound healing, and repairing injured or diseased organs. Personalized approaches guided by patient-specific factors and tissue-specific requirements are paving the way for tailored regenerative therapies that optimize outcomes and improve patient quality of life. ECM modulation in regenerative medicine represents a diverse and innovative approach to promote tissue repair and regeneration. By harnessing the bioactive properties of the ECM and leveraging advanced biomaterials and cell-based therapies, researchers aim to develop effective treatments for restoring functional tissue architecture and improving patient outcomes in a wide range of clinical settings.

Exercise 3

Multiple Choice Questions

1. Which of the following enzymes is primarily responsible for degrading the extracellular matrix (ECM)?
a) Kinases
b) Matrix metalloproteinases (MMPs)
c) Phosphatases
d) Ligases

2. The process of ECM remodeling is crucial in which of the following physiological processes?
a) DNA replication
b) Muscle contraction
c) Wound healing
d) Neuronal signaling

3. Which cell type is most commonly associated with producing ECM components?
a) Neurons
b) Fibroblasts
c) Red blood cells
d) Adipocytes

4. Integrins are important in ECM remodeling because they:
a) Degrade the ECM
b) Regulate cell-ECM adhesion
c) Synthesize ECM proteins
d) Transport ECM components

5. The balance between ECM synthesis and degradation is maintained by which of the following?
a) Growth factors alone
b) MMPs and TIMPs (Tissue Inhibitors of Metalloproteinases)
c) ATP and GTP
d) Hormones like insulin

6. ECM stiffness can influence cell behavior through a process known as:
a) Apoptosis
b) Mechanotransduction
c) Transcription
D) Endocytosis

7. Which of the following is NOT a component of the ECM?
a) Collagen
b) Elastin
c) Actin
d) Fibronectin

8. TGF-β (Transforming Growth Factor-beta) influences ECM remodeling by:
a) Promoting ECM degradation only
b) Inhibiting ECM synthesis
c) Regulating both ECM synthesis and degradation
D) Preventing cell migration

9. The ECM plays a critical role in which of the following cancer-related processes?
a) Metastasis
b) DNA repair
c) Protein synthesis
D) Hormone secretion

10. Which biophysical property of the ECM is most closely linked to cellular differentiation?
a) Thickness
b) Color
c) Stiffness
d) Temperature

3.9 Future Directions and Emerging Research

The future directions and emerging research in ECM modulation and regenerative medicine:

- **Advanced Biomaterials**: Continued development of biomaterials with precise control over ECM mimicking properties, including composition, mechanical stiffness, and bioactivity, to better mimic native tissue environments and enhance regenerative outcomes.
- **Bio fabrication Techniques**: Advancements in 3D printing and bio fabrication methods to create complex tissue scaffolds with spatially controlled ECM properties, enabling personalized tissue engineering and regenerative therapies.
- **Cell-Free Therapies**: Exploration of cell-free approaches, such as ECM-derived vesicles or nanoparticles, for delivering bioactive ECM components and signaling molecules to promote tissue repair and modulate immune responses.
- **Gene Editing and Therapy**: Application of gene editing technologies, such as CRISPR/Cas9, to manipulate ECM-related genes and enhance ECM remodeling capabilities in regenerating tissues.
- **Mechanobiology**: Increasing focus on understanding how mechanical forces and ECM stiffness influence cellular behaviors and tissue regeneration, leading to the development of mechanobiology-based strategies for enhancing regenerative outcomes.
- **Biomedical Imaging**: Integration of advanced imaging techniques, such as multiphoton microscopy and MRI, for non-invasive monitoring of ECM remodeling dynamics and therapeutic efficacy in real-time.
- **Personalized Medicine Approaches**: Implementation of precision medicine strategies to tailor regenerative therapies based on patient-specific factors, including genetic variability, disease stage, and tissue-specific microenvironment.

These points highlight the ongoing and future research directions aimed at harnessing the potential of ECM modulation and regenerative medicine to address complex tissue defects, enhance healing processes, and improve patient outcomes across diverse clinical applications.

3.10 Summary

In this chapter, we have explored the multifaceted roles of extracellular matrix (ECM) in physiological processes, pathological conditions, and therapeutic interventions within the realms of biomedical research and clinical practice. The ECM serves as a dynamic scaffold that not only provides structural support to tissues but also regulates cellular behaviors critical for tissue development, homeostasis, and repair. From its involvement in cell adhesion, migration, and differentiation to its modulation of biochemical signaling pathways, the ECM plays a central role in maintaining tissue integrity and function. Our discussion delved into the intricate mechanisms of ECM remodeling in both health and disease contexts. In physiological conditions, controlled ECM turnover facilitates tissue remodeling and adaptive responses to environmental cues. However, dysregulated ECM remodeling underlies numerous pathological conditions such as fibrosis, cancer metastasis, and cardiovascular diseases. Understanding the molecular and cellular mechanisms driving these processes is essential for developing targeted therapies aimed at restoring ECM homeostasis and improving patient outcomes. We examined various therapeutic approaches targeting ECM modulation, including anti-fibrotic therapies that aim to inhibit ECM deposition, enhance degradation, or disrupt abnormal cross-linking processes. These strategies hold promise for mitigating tissue scarring, preserving organ function, and improving therapeutic outcomes in fibrotic diseases and beyond. Additionally, we explored the application of ECM modulation in regenerative medicine, where advanced biomaterials, cell-based therapies, and bio fabrication techniques are being leveraged to promote tissue repair, regeneration, and functional restoration. Looking forward, emerging research directions such as the development of advanced biomaterials with tailored ECM properties, bio fabrication technologies for complex tissue engineering, and the exploration of mechanobiology and gene editing techniques offer exciting opportunities to further enhance our understanding and capabilities in ECM modulation and regenerative medicine. These innovations pave the way for personalized treatment strategies that harness the therapeutic potential of ECM remodeling to address diverse clinical challenges and improve quality of life for patients worldwide.

Points to Remember

- **ECM remodeling is a dynamic process involving continuous synthesis, degradation, and modification of ECM components.**
- **It involves complex cellular crosstalk between various cell types including fibroblasts, immune cells, and endothelial cells.**
- **MMPs play a crucial role in ECM remodeling by degrading specific ECM components like collagen and elastin.**
- **Dysregulation of ECM remodeling is implicated in various diseases such as cancer metastasis, fibrosis, and cardiovascular disorders.**
- **Proper ECM remodeling is essential for tissue regeneration and wound healing processes.**
- **ECM remodeling also influences mechano transduction, where mechanical cues from the ECM impact cellular behavior and function.**
- **ECM remodeling involves biochemical signaling pathways that regulate matrix turnover and organization.**

References

1. Yue B (2014) Biology of the extracellular matrix: an overview. J Glaucoma 23(8 Suppl 1):S20–S23. https://doi.org/10.1097/IJG.0000000000000108
2. Theocharis AD, Skandalis SS, Gialeli C, Karamanos NK (2016) Extracellular matrix structure. Adv Drug Deliv Rev 97:4–27. https://doi.org/10.1016/j.addr.2015.11.001
3. Bonnans C, Chou J, Werb Z (2014) Remodelling the extracellular matrix in development and disease. Nat Rev Mol Cell Biol 15(12):786–801. https://doi.org/10.1038/nrm3904
4. Walma DAC, Yamada KM (2020) The extracellular matrix in development. Development (Cambridge, England) 147(10):dev175596. https://doi.org/10.1242/dev.175596
5. Naba A (2024) Mechanisms of assembly and remodelling of the extracellular matrix. Nature reviews. Molecul Cell Biol. Advance online publication. https://doi.org/10.1038/s41580-024-00767-3
6. Bowers SLK, Meng Q, Molkentin JD (2022) Fibroblasts orchestrate cellular crosstalk in the heart through the ECM. Nat Cardiovasc Res 1:312–321. https://doi.org/10.1038/s44161-022-00043-7
7. D'Urso M, Kurniawan NA (2020) Mechanical and physical regulation of fibroblast-myofibroblast transition: from cellular mechanoresponse to tissue pathology. Front Bioeng Biotechnol 8:609653. https://doi.org/10.3389/fbioe.2020.609653
8. Zhang YL, Tang TT, Wang B, Wen Y, Feng Y, Yin Q, Jiang W, Zhang Y, Li ZL, Wu M, Wu QL, Song J, Crowley SD, Lan HY, Lv LL, Liu BC (2024) Identification of a novel ECM remodeling macrophage subset in AKI to CKD transition by integrative spatial and single-cell analysis. Adv Sci (Weinheim, Baden-Wurttemberg, Germany) 11(38):e2309752. https://doi.org/10.1002/advs.202309752
9. Vilardi A, Przyborski S, Mobbs C et al (2024) Current understanding of the interplay between extracellular matrix remodelling and gut permeability in health and disease. Cell Death Discov 10:258. https://doi.org/10.1038/s41420-024-02015-1
10. Winkler J, Abisoye-Ogunniyan A, Metcalf KJ et al (2020) Concepts of extracellular matrix remodelling in tumour progression and metastasis. Nat Commun 11:5120. https://doi.org/10.1038/s41467-020-18794-x
11. Wang X, Khalil RA (2018) Matrix metalloproteinases, vascular remodeling, and vascular disease. Adv Pharmacol (San Diego, Calif.) 81:241–330. https://doi.org/10.1016/bs.apha.2017.08.002
12. Cui N, Hu M, Khalil RA (2017) Biochemical and biological attributes of matrix metalloproteinases. Prog Mol Biol Transl Sci 147:1–73. https://doi.org/10.1016/bs.pmbts.2017.02.005
13. Verstappen J, Von den Hoff JW (2006) Tissue inhibitors of metalloproteinases (TIMPs): their biological functions and involvement in oral disease. J Dent Res 85(12):1074–1084. https://doi.org/10.1177/154405910608501202
14. Vanhoutte D, Heymans S (2010) TIMPs and cardiac remodeling: 'embracing the MMP-independent-side of the family.' J Mol Cell Cardiol 48(3):445–453. https://doi.org/10.1016/j.yjmcc.2009.09.013
15. Benmelech S, Le T, McKay M, Nam J, Subramaniam K, Tellez D, Vlasak G, Mak M (2024) Biophysical and biochemical aspects of immune cell–tumor microenvironment interactions. APL Bioeng 8(2). https://doi.org/10.1063/5.0195244
16. Naba A (2024) Mechanisms of assembly and remodelling of the extracellular matrix. Nat Rev Mol Cell Biol 25:865–885. https://doi.org/10.1038/s41580-024-00767-3
17. Azadi S, Tafazzoli Shadpour M (2020) The microenvironment and cytoskeletal remodeling in tumor cell invasion. Int Rev Cell Mol Biol 356:257–289. https://doi.org/10.1016/bs.ircmb.2020.06.003
18. Farooq M, Khan AW, Kim MS, Choi S (2021) The role of fibroblast growth factor (FGF) signaling in tissue repair and regeneration. Cells 10(11):3242. https://doi.org/10.3390/cells10113242

19. Kamato D, Burch ML, Piva TJ, Rezaei HB, Rostam MA, Xu S, Zheng W, Little PJ, Osman N (2013) Transforming growth factor-β signalling: role and consequences of Smad linker region phosphorylation. Cell Signal 25(10):2017–2024. https://doi.org/10.1016/j.cellsig.2013.06.001
20. Danen EHJ. Integrins: an overview of structural and functional aspects. In: Madame Curie bioscience database [Internet]. Landes Bioscience; 2000–2013, Austin. Available from: https://www.ncbi.nlm.nih.gov/books/NBK6259/
21. Pang X, He X, Qiu Z et al (2023) Targeting integrin pathways: mechanisms and advances in therapy. Sig Transduct Target Ther 8:1. https://doi.org/10.1038/s41392-022-01259-6
22. Cox TR, Erler JT (2011) Remodeling and homeostasis of the extracellular matrix: implications for fibrotic diseases and cancer. Dis Model Mech 4(2):165–178. https://doi.org/10.1242/dmm.004077
23. Mohammed Butt A, Rupareliya V, Hariharan A, Kumar H (2023) Building a pathway to recovery: targeting ECM remodeling in CNS injuries. Brain Res 1819:148533. https://doi.org/10.1016/j.brainres.2023.148533
24. Eble JA, Niland S (2019) The extracellular matrix in tumor progression and metastasis. Clin Exp Metas 36(3):171–198. https://doi.org/10.1007/s10585-019-09966-1
25. Prakash J, Shaked Y (2024) The interplay between extracellular matrix remodeling and cancer therapeutics. Cancer Discov 14(8):1375–1388. https://doi.org/10.1158/2159-8290.CD-24-0002
26. Zhao P, Sun T, Lyu C et al (2023) Cell mediated ECM-degradation as an emerging tool for anti-fibrotic strategy. Cell Regen 12:29. https://doi.org/10.1186/s13619-023-00172-9

Chapter 4
ECM in Health and Disease

4.1 Introduction

The Extracellular Matrix (ECM) is a complex and dynamic network of macromolecules that surrounds cells within tissues and organs of multicellular organisms. It provides structural support, regulates biochemical and biomechanical signaling, and influences cellular behavior in both physiological and pathological conditions (Fig. 4.1). The ECM is primarily composed of proteins such as collagen, elastin, fibronectin, and laminin, which are secreted by cells known as fibroblasts, epithelial cells, and others depending on the tissue type [1]. These proteins form a scaffold that anchors cells in place, allowing for tissue integrity and proper functioning. Collagen, the most abundant protein in the ECM, provides tensile strength and resistance to stretching, crucial for maintaining tissue architecture. Elastin, on the other hand, imparts elasticity, allowing tissues like blood vessels and skin to recoil after stretching. Fibronectin and laminin are adhesive glycoproteins that bind cells to the ECM, facilitating cell migration, differentiation, and tissue organization during development and wound healing [2]. Together, these components create a three-dimensional microenvironment that supports cellular activities essential for tissue homeostasis.

The ECM's role extends beyond structural support; it actively participates in regulating cellular processes through biochemical and mechanical signaling pathways. Embedded within the ECM are growth factors, cytokines, and proteases that modulate cell proliferation, differentiation, and migration. For instance, growth factors like transforming growth factor-beta (TGF-β) regulate ECM synthesis and degradation, influencing tissue remodeling in response to physiological cues or injury [3]. Moreover, mechanical cues from the ECM, such as stiffness and topography, can direct cell behavior, impacting processes like stem cell differentiation and cancer progression. In tissues like bone and cartilage, the ECM forms a specialized niche that regulates the maintenance and repair of these structures. In bone, for instance, the mineralized ECM provides a framework for osteoblasts and osteoclasts to remodel bone

D. Suhag, *Extracellular Matrix biomimicking for Theranostics*, Biomedical Materials for Multi-functional Applications, https://doi.org/10.1007/978-981-95-7900-6_4

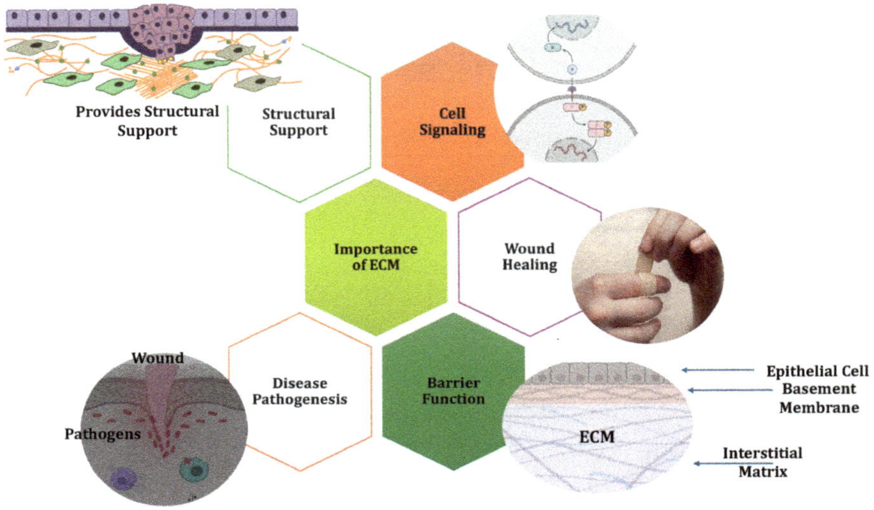

Fig. 4.1 Importance of extracellular cellular membrane [5]

tissue in response to mechanical stress and hormonal signals. Similarly, in cartilage, the ECM's composition of collagen and proteoglycans confers resilience and load-bearing capacity necessary for joint function [4]. Disruptions in ECM composition or function can lead to pathological conditions such as osteoarthritis or osteoporosis, highlighting its critical role in maintaining tissue integrity and function.

The ECM's importance in tissue morphogenesis and organogenesis is evident during embryonic development, where it guides cell migration, tissue patterning, and organ formation. For example, ECM gradients help establish body axes and direct the migration of neural crest cells during neural tube formation. Throughout adulthood, the ECM continues to modulate tissue remodeling in response to physiological demands or injury, orchestrating processes such as wound healing, angiogenesis, and immune responses. Its ability to integrate biochemical and mechanical signals makes the ECM a key regulator of tissue homeostasis and adaptive responses to environmental changes.

The Extracellular Matrix (ECM) represents a fundamental component of tissues and organs, providing structural support, guiding cellular behavior, and influencing tissue function through biochemical and mechanical signaling pathways [5]. Composed of diverse proteins and embedded with regulatory molecules, the ECM maintains tissue integrity, regulates development, and supports adaptive responses to physiological challenge. ECM dynamics is crucial for advancing therapies in regenerative medicine, tissue engineering, and treating diseases associated with ECM dysregulation.

Exercise 1
Fill Ups:

Laminins, Adhesion, Cell Signaling, Glycosaminoglycans (GAGs). Matrix Metalloproteinases (MMPs).

1. Integrins are cell surface receptors that mediate .... between cells and the ECM.
2. .... are enzymes that degrade ECM components, facilitating tumor invasion and metastasis.
3. .... are glycoproteins that connect cells to the ECM and regulate cell behavior.
4. The composition of the ECM can influence .... within the tumor microenvironment.
5. .... are proteoglycans in the ECM that can bind growth factors and regulate their availability to cells.

4.2 ECM Alterations in Cancer

The Extracellular Matrix (ECM) plays a pivotal role in cancer progression and metastasis through its dynamic alterations within the tumor microenvironment. ECM composition and organization undergo significant changes during cancer development, influencing tumor cell behavior, therapeutic responses, and overall prognosis. These alterations involve aberrant deposition of ECM proteins such as collagen, fibronectin, and proteoglycans, as well as remodeling of their spatial distribution and mechanical properties [6]. Such changes not only provide structural support to tumor cells but also create a conducive environment for invasive behavior and metastatic dissemination. In cancer, ECM remodeling is orchestrated by both tumor and stromal cells, leading to the generation of a stiff and dense matrix termed "desmoplasia." This fibrotic ECM promotes tumor progression by enhancing cell survival, facilitating angiogenesis, and impeding drug delivery to malignant cells. Additionally, altered ECM dynamics contribute to the activation of signaling pathways that promote epithelial-mesenchymal transition (EMT), a critical process enabling cancer cells to acquire invasive traits and disseminate to distant sites. Understanding these ECM alterations in cancer not only elucidates mechanisms of disease progression but also informs the development of novel therapeutic strategies targeting the tumor microenvironment.

4.2.1 Role of ECM in Tumor Microenvironment

The tumor microenvironment (TME) is a dynamic ecosystem where cancer cells interact with surrounding non-cancerous cells, soluble factors, and the ECM. The ECM within the TME plays a crucial role in regulating tumor progression, metastasis, and response to therapies (Fig. 4.2). One key aspect of the ECM's importance lies in its structural support for tumor architecture. Aberrant deposition and crosslinking of ECM proteins, such as collagen and fibronectin, lead to the formation of a dense and stiff matrix known as desmoplasia [7]. This altered ECM structure not only physically confines tumor cells but also promotes their survival and proliferation by activating integrin-mediated signaling pathways that enhance cell adhesion and survival signals. Beyond its structural role, the ECM serves as a reservoir for various growth factors, cytokines, and proteases that influence tumor cell behavior and interactions within the TME. For instance, ECM-bound growth factors like transforming growth factor-beta (TGF-β) and vascular endothelial growth factor (VEGF) regulate angiogenesis, promoting the formation of new blood vessels to sustain tumor growth and facilitate metastasis. Moreover, proteases such as matrix metalloproteinases (MMPs) and cathepsins, produced by both tumor and stromal cells, remodel the ECM by degrading matrix components, thereby facilitating tumor invasion into surrounding tissues and metastatic spread to distant organs. The mechanical properties of the ECM also contribute significantly to tumor progression. Increased ECM stiffness alters mechano transduction signaling in tumor cells, promoting aggressive phenotypes such as enhanced proliferation, resistance to apoptosis, and increased motility. Integrins and other cell surface receptors on tumor cells sense and respond to these mechanical cues, activating intracellular pathways that promote cell survival and migration [5]. Additionally, the physical barriers created by the dense ECM hinder drug delivery to tumor cells, contributing to therapy resistance commonly observed in advanced cancers.

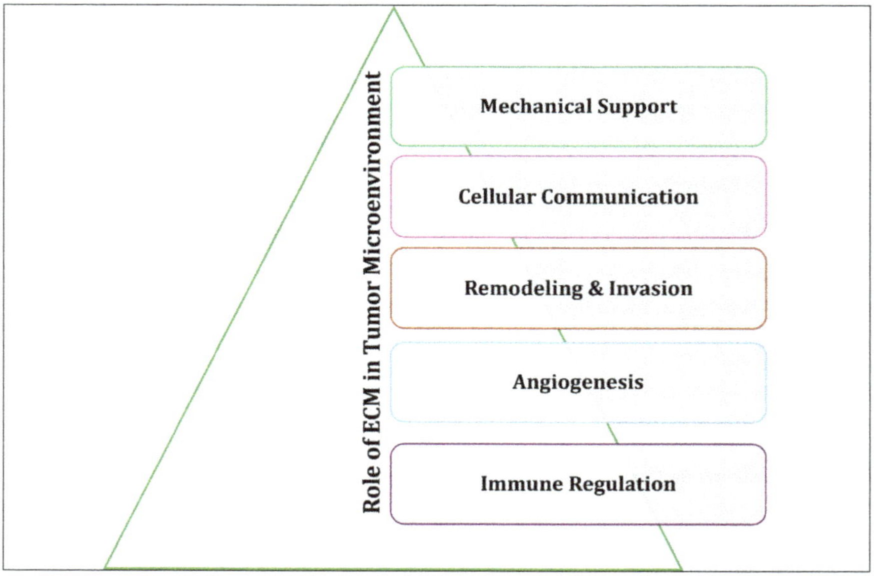

Fig. 4.2 Major role of ECM in tumor microenvironment

In addition to its direct effects on tumor cells, the ECM influences immune cell infiltration and function within the TME. The composition and organization of ECM components regulate immune cell trafficking, activation, and polarization, thereby shaping anti-tumor immune responses. For example, hyaluronic acid-rich ECM promotes an immunosuppressive TME by impairing T cell infiltration and function, while collagen-rich matrices can induce polarization of tumor-associated macrophages towards a pro-tumorigenic phenotype. These interactions between ECM and immune cells create a complex immunosuppressive environment that shields tumor cells from immune surveillance and promotes immune evasion.

ECM-mediated signaling pathways contribute to the induction of epithelial-mesenchymal transition (EMT) in cancer cells, a process characterized by the loss of epithelial characteristics and acquisition of mesenchymal traits associated with increased invasive potential. ECM proteins and their receptors activate EMT-promoting transcription factors such as Snail, Slug, and Twist, facilitating tumor cell dissemination and metastasis. EMT-driven changes in cell morphology and behavior enable cancer cells to breach the basement membrane, invade surrounding tissues, and enter the bloodstream or lymphatic system for dissemination to distant organs, thereby driving metastatic progression [8]. The dynamic remodeling of the ECM in response to biochemical and biomechanical cues within the TME represents a hallmark of cancer progression and metastasis. Tumor-associated fibroblasts (CAFs), endothelial cells, and immune cells contribute to ECM remodeling through the secretion of ECM-modifying enzymes, growth factors, and cytokines. This collaborative process creates a feedforward loop wherein ECM remodeling further enhances tumor cell survival, invasion, and metastasis. Consequently, targeting the ECM and its interactions within the TME has emerged as a promising therapeutic strategy to inhibit tumor progression, enhance drug delivery, and restore anti-tumor immune responses in cancer patients.

The Extracellular Matrix (ECM) within the tumor microenvironment (TME) exerts profound effects on cancer progression by providing structural support, regulating biochemical signaling, influencing immune responses, and promoting metastasis. Alterations in ECM composition, organization, and mechanical properties contribute to tumor cell survival, invasion, and therapy resistance, highlighting its pivotal role in shaping the malignant phenotype. ECM-mediated mechanisms in the TME are crucial for developing effective therapeutic interventions that target the ECM to disrupt tumor-stroma interactions, enhance treatment efficacy, and improve clinical outcomes for cancer patients.

4.2.2 ECM Remodeling and Cancer Progression

Extracellular Matrix (ECM) remodeling is a critical process in cancer progression, characterized by dynamic changes in the composition, organization, and mechanical properties of the ECM within the tumor microenvironment (TME). These alterations play a pivotal role in facilitating tumor cell survival, invasion, metastasis, and resistance to therapies. ECM remodeling involves the deposition of ECM proteins such as collagen, fibronectin, and hyaluronic acid by cancer-associated fibroblasts (CAFs) and other stromal cells, as well as the enzymatic degradation of ECM components by matrix metalloproteinases (MMPs) and other proteases produced by both tumor and stromal cells.

The altered ECM composition in tumors creates a supportive niche that promotes cancer cell survival and proliferation [9]. For example, increased deposition of collagen and fibronectin fibers forms a dense and stiff ECM, termed desmoplasia, which enhances mechanical support for tumor cells and facilitates their anchorage and survival. This stiff ECM also activates integrin-mediated signaling pathways in cancer cells, promoting cell adhesion, migration, and resistance to apoptosis. Moreover, ECM proteins such as laminins and nidogens provide adhesive cues that regulate cell polarity and tissue architecture, contributing to tumor cell organization and invasive behavior within the TME. ECM remodeling is also crucial for facilitating tumor cell invasion and metastasis. Degradation of basement membrane components such as collagen IV and laminin by MMPs and other proteases enables cancer cells to breach physical barriers and invade surrounding tissues. The proteolytic activity of MMPs not only facilitates local invasion but also supports intravasation into blood vessels or lymphatics, enabling cancer cells to disseminate to distant sites and initiate metastatic colonies [10]. Furthermore, remodeling of the ECM in premetastatic niches primes these secondary sites for colonization by circulating tumor cells, enhancing metastatic outgrowth and establishing secondary tumors in distant organs.

The mechanical properties of the ECM also play a crucial role in cancer progression. Changes in ECM stiffness and elasticity influence cellular mechano transduction pathways, altering gene expression, cell behavior, and therapeutic responses in cancer cells. High ECM stiffness promotes integrin clustering and focal adhesion kinase (FAK) signaling, which in turn activates downstream pathways involved in cell survival, proliferation, and migration. These mechano transduction signals contribute to the acquisition of aggressive phenotypes in cancer cells, such as enhanced invasiveness and resistance to chemotherapy or targeted therapies. Moreover, ECM remodeling contributes to the establishment of an immunosuppressive TME that shields tumor cells from immune surveillance and promotes immune evasion [11]. Altered ECM composition and organization can impair immune cell infiltration and function within the TME, limiting anti-tumor immune responses and promoting tumor immune escape. For instance, dense collagen networks and hyaluronic acid-rich matrices can physically obstruct T cell infiltration into tumors, while ECM-derived signals may induce immunosuppressive phenotypes in tumor-associated macrophages and other immune cells, further fostering an immunosuppressive environment. The dynamic nature of ECM remodeling provides cancer cells with a survival advantage and facilitates adaptation to evolving microenvironmental cues. Tumor cells and stromal cells constantly interact to modify ECM composition and structure in response to biochemical and biomechanical signals within the

TME. This adaptive remodeling process allows tumors to overcome therapeutic challenges, including resistance to chemotherapy, targeted therapies, and immune checkpoint blockade. Targeting ECM remodeling pathways represents a promising therapeutic strategy to disrupt tumor-stroma interactions, enhance drug delivery to tumors, and restore anti-tumor immune responses, thereby improving clinical outcomes for cancer patients.

ECM remodeling is a fundamental process in cancer progression that regulates tumor cell survival, invasion, metastasis, and therapy resistance. Alterations in ECM composition, organization, and mechanical properties create a supportive microenvironment for tumor growth and dissemination, while also contributing to immune evasion and therapeutic resistance.

4.2.3 Therapeutic Implications Targeting ECM in Cancer Treatment

Therapeutic strategies targeting the Extracellular Matrix (ECM) in cancer treatment have emerged as promising approaches to disrupt tumor-stroma interactions, enhance drug delivery, and overcome therapeutic resistance. One approach involves inhibiting ECM remodeling enzymes such as matrix metalloproteinases (MMPs) and their regulators. MMP inhibitors aim to prevent ECM degradation, thereby reducing tumor cell invasion and metastasis. Additionally, targeting ECM crosslinking enzymes like lysyl oxidase (LOX) can decrease ECM stiffness, impairing mechano transduction signaling in cancer cells and sensitizing them to chemotherapy and immunotherapy. Another therapeutic strategy focuses on disrupting ECM-integrin interactions that promote tumor cell survival and migration [12]. Integrin inhibitors block integrin-mediated signaling pathways involved in cell adhesion, proliferation, and invasion within the ECM. This approach not only impedes tumor progression but also enhances the efficacy of conventional therapies by reducing ECM-mediated resistance mechanisms. Moreover, antibodies targeting specific ECM proteins or glycoproteins, such as fibronectin or hyaluronic acid, can interfere with ECM-mediated signaling and inhibit tumor growth and metastasis.

Modulating ECM stiffness and composition represents another avenue for therapeutic intervention in cancer. Agents that degrade collagen or disrupt collagen crosslinking can reduce ECM stiffness, thereby enhancing drug penetration into tumors and improving therapeutic outcomes. Furthermore, therapies that remodel the ECM to restore normal tissue architecture and function, such as enzymatic degradation of excessive ECM deposits in fibrotic tumors, have shown promise in preclinical studies and clinical trials. These approaches aim to normalize the tumor microenvironment, reduce hypoxia, and improve drug delivery to enhance treatment efficacy. Targeting ECM-mediated immunosuppression is another critical therapeutic strategy in cancer treatment [13]. Agents that disrupt ECM barriers and promote immune cell infiltration into tumors, such as hyaluronidase or anti-fibrotic agents, can enhance anti-tumor immune responses and improve the efficacy of immunotherapy. By altering the immunosuppressive TME, these therapies help overcome immune evasion mechanisms and restore immune surveillance against cancer cells. Additionally, combining ECM-targeted therapies with immune checkpoint inhibitors or adoptive T cell therapies holds promise for synergistic anti-tumor effects.

The advantages of targeting the ECM in cancer treatment extend beyond direct tumor cell effects to encompass broader therapeutic benefits. By disrupting ECM-mediated signaling pathways and remodeling processes, these therapies can sensitize tumor cells to conventional therapies such as chemotherapy and radiation. Enhanced drug delivery resulting from ECM modulation reduces treatment resistance and improves patient outcomes by increasing tumor response rates and prolonging progression-free survival. Moreover, targeting the ECM may mitigate the risk of metastasis by inhibiting ECM-driven processes that facilitate cancer cell dissemination and colonization of secondary sites. Furthermore, ECM-targeted therapies offer the potential for personalized medicine approaches in cancer treatment [14]. By characterizing individual tumor ECM profiles and targeting specific ECM components or pathways, clinicians can tailor treatment strategies to each patient's unique disease characteristics and therapeutic responses. This precision medicine approach maximizes treatment efficacy while minimizing systemic toxicity and adverse effects associated with traditional therapies. Additionally, ECM-targeted therapies have the advantage of being combinable with existing treatment modalities, offering opportunities for synergistic effects and improved overall therapeutic outcomes in multidisciplinary cancer care.

Therapeutic strategies targeting the Extracellular Matrix (ECM) in cancer treatment represent innovative approaches to overcome therapeutic resistance, enhance drug delivery, and improve patient outcomes. By disrupting ECM remodeling processes, inhibiting ECM-mediated signaling pathways, and modulating ECM stiffness and composition, these therapies offer promising avenues for reducing tumor growth, inhibiting metastasis, and enhancing the efficacy of conventional and immunotherapies. Continued research and clinical development in ECM-targeted therapies hold the potential to transform cancer treatment paradigms and improve survival rates for patients with various cancer types.

4.3 ECM and Cardiovascular Diseases

The Extracellular Matrix (ECM) plays a critical role in maintaining the structural integrity and function of cardiovascular tissues, including the heart, blood vessels, and valves. Composed primarily of collagen, elastin, proteoglycans, and glycoproteins, the ECM provides mechanical support, regulates tissue elasticity, and modulates cellular signaling pathways essential for cardiovascular homeostasis. In cardiovascular diseases such as atherosclerosis, hypertension, and heart failure, alterations in ECM composition, organization, and mechanical properties contribute to disease progression and complications [15]. For instance, excessive deposition of collagen in the myocardium leads to fibrosis, impairing cardiac function and promoting arrhythmias. Understanding the dynamic interactions between ECM components and cardiovascular cells is crucial for developing targeted therapies to prevent or mitigate ECM-related pathologies and improve cardiovascular outcomes. ECM remodeling in cardiovascular diseases involves complex interactions between cardiac fibroblasts, endothelial cells, smooth muscle cells, and inflammatory mediators within the vascular and cardiac microenvironments [16]. In response to mechanical stress, inflammatory signals, and neurohormonal factors, these cells modulate ECM synthesis, degradation, and crosslinking processes. Dysregulated ECM remodeling contributes to vascular stiffening, plaque formation, and myocardial fibrosis, exacerbating cardiovascular dysfunction and increasing the risk of adverse cardiovascular events. Therapeutic strategies aimed at restoring ECM homeostasis, reducing fibrosis, and enhancing ECM turnover represent promising approaches to attenuate disease progression and improve cardiac function in patients with cardiovascular diseases.

Exercise 2
Fill Ups:

Valvular heart diseases, Contractility, Cell signaling, Proliferation, Survival, Reducing fibrosis, Cell adhesion, Migration, Differentiation, Improving tissue regeneration.

1. The ECM provides structural support and biochemical cues for …………, …………, and ………… in cardiovascular tissues.
2. Dysregulation of ECM remodeling can lead to …………… in heart valves.
3. Integrins are cell surface receptors that mediate cell-ECM interactions, influencing …………., ……………, and …………………. .
4. ECM stiffness affects cardiac …………….. and function.
5. Targeting ECM components like collagen may offer therapeutic strategies for ……………, and ………………………… in cardiovascular diseases.

4.3.1 ECM Composition in Cardiovascular Tissues

Extracellular matrix (ECM) composition in cardiovascular tissues undergoes significant changes in both structure and composition compared to non-cancerous tissues. The ECM in cardiovascular tissues plays crucial roles in providing structural support, maintaining tissue integrity, and regulating cellular behaviors such as proliferation, migration, and differentiation [17]. In healthy cardiovascular tissues, the ECM primarily consists of collagen, elastin, glycoproteins like fibronectin and laminin, proteoglycans, and various growth factors (Fig. 4.3). Collagen fibers, particularly types I and III, form a scaffold that imparts tensile strength to the tissue, while elastin fibers confer elasticity and resilience. Glycoproteins like fibronectin and laminin facilitate cell adhesion and migration, crucial for tissue repair and maintenance of vascular integrity. In contrast, ECM composition in cancerous cardiovascular tissues undergoes profound alterations due to the pathological changes induced by cancer. These changes often involve excessive deposition of ECM components,

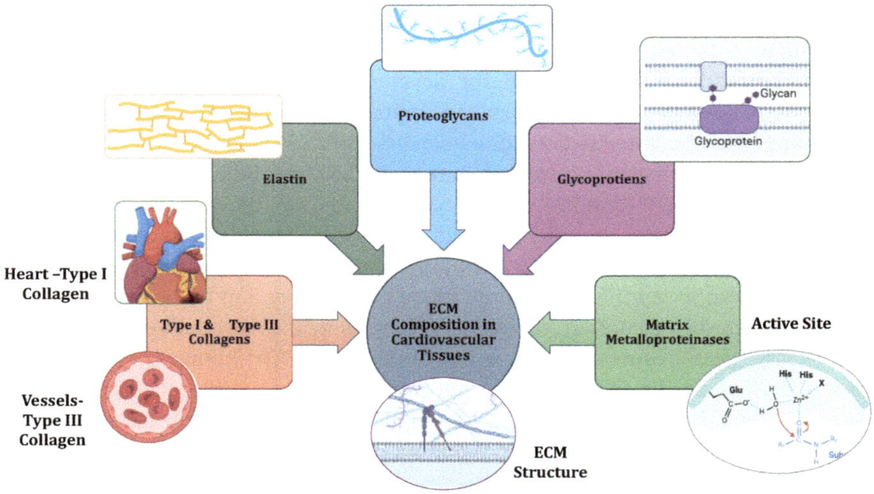

Fig. 4.3 The composition ECM in cardiovascular tissues (created using bio render)

disorganized collagen and elastin fibers, and altered expression of proteoglycans and growth factors. Cancer-associated ECM remodeling creates a microenvironment that promotes tumor growth, invasion, and metastasis.

One notable change in cancerous ECM is the increased deposition of fibrillar collagens, especially collagen type I, which contributes to the stiffness and rigidity of the tissue (Table 4.1). This altered mechanical environment not only supports cancer cell survival but also facilitates their migration through tissue barriers [17]. Additionally, there is an increase in proteoglycans such as versican and decorin, which can modulate signaling pathways involved in cancer progression and metastasis.

Table 4.1 Major differences between cancerous and non-cancerous ECM

Aspect of comparison	Non-cancerous ECM	Cancerous ECM
Composition	Predominantly collagen (types I and III), elastin, glycoproteins (fibronectin, laminin), proteoglycans	Increased collagen type I, disorganized collagen and elastin fibers, altered proteoglycan expression
Structural organization	Organized collagen and elastin fibers	Disorganized collagen and elastin fibers
Mechanical properties	Provides tensile strength and elasticity	Increased stiffness and rigidity
Angiogenic potential	Maintains vascular integrity, minimal angiogenic factors	Enhanced angiogenesis (increased VEGF, FGF expression)
Functional implications	Supports tissue structure and cellular functions	Promotes tumor growth, invasion and metastasis

Furthermore, cancer-associated ECM often exhibits enhanced angiogenesis, driven by upregulated expression of angiogenic factors like vascular endothelial growth factor (VEGF) and fibroblast growth factor (FGF). This neo angiogenesis supports tumor growth by ensuring an adequate blood supply, promoting tumor cell proliferation, and facilitating metastatic spread through the bloodstream. The dysregulated ECM in cancerous cardiovascular tissues also promotes immune evasion and resistance to therapy. ECM components like hyaluronic acid and tenascin-C can create a physical barrier that restricts immune cell infiltration into the tumor microenvironment. Moreover, ECM-mediated signaling pathways can induce tumor cell survival mechanisms and resistance to chemotherapy or radiation therapy.

The ECM composition in cardiovascular tissues undergoes profound changes in cancerous conditions compared to non-cancerous states. These alterations include increased deposition of fibrillar collagens, enhanced expression of proteoglycans and growth factors, altered mechanical properties, and enhanced angiogenesis.

4.3.2 ECM Remodeling in Heart Disease

Extracellular matrix (ECM) remodeling plays a critical role in the pathogenesis of heart disease, encompassing various processes that alter the composition, structure, and mechanical properties of the cardiac ECM. This dynamic remodeling is driven by a combination of mechanical stress, biochemical signaling, and cellular responses, contributing significantly to both adaptive and maladaptive changes in heart tissue. In healthy conditions, ECM remodeling in the heart is tightly regulated to maintain cardiac structure and function. Cardiomyocytes, fibroblasts, endothelial cells, and immune cells interact with the ECM, continuously synthesizing and degrading matrix proteins to support myocardial contraction, electrical conduction, and mechanical integrity. Collagens, particularly types I and III, along with elastin and glycoproteins like fibronectin and laminin, form the scaffold that provides structural support and elasticity essential for cardiac function. In heart disease, such as myocardial infarction (heart attack), hypertension, or heart failure, ECM remodeling becomes dysregulated [18]. Ischemic injury during a heart attack leads to the release of inflammatory mediators and oxidative stress, triggering an acute inflammatory response. This inflammatory cascade activates fibroblasts to proliferate and differentiate into myofibroblasts, which secrete excessive amounts of collagen and other ECM proteins as part of the wound healing process.

Chronic conditions like hypertension or chronic ischemia further exacerbate ECM remodeling by causing sustained mechanical stress on the heart. Increased blood pressure or reduced blood flow stimulates cardiac fibroblasts to deposit more collagen, leading to myocardial fibrosis. Fibrotic changes in the ECM stiffen the myocardium, impairing its ability to contract and relax properly, which contributes to diastolic dysfunction and eventual heart failure. Moreover, ECM remodeling in heart disease disrupts normal signaling pathways that regulate cellular behavior. Growth factors such as TGF-β and CTGF are upregulated in response to injury, promoting fibroblast activation and ECM deposition. This leads to the formation of scar tissue in the myocardium, which can disrupt electrical conduction and increase the risk of arrhythmias. The altered ECM composition and stiffness also affect cardiomyocyte function and survival. Changes in matrix elasticity can activate mechano transduction pathways within cardiomyocytes, altering gene expression and cellular responses. In pathological conditions, these changes can contribute to cardiomyocyte hypertrophy, apoptosis, and impaired contractility, further compromising cardiac function.

Understanding the mechanisms of ECM remodeling in heart disease is crucial for developing therapeutic strategies. Interventions aimed at modulating ECM turnover, such as targeting MMPs or inhibiting TGF-β signaling, hold promise for preventing or reversing fibrotic changes in the heart. Additionally, therapies that promote ECM degradation or enhance ECM repair mechanisms may help mitigate the adverse effects of myocardial remodeling and improve outcomes for patients with heart disease.

ECM remodeling in heart disease is a complex process driven by cellular responses to mechanical and biochemical cues [19]. Dysregulation of ECM turnover and deposition contributes to myocardial fibrosis, impaired cardiac function, and increased susceptibility to arrhythmias and heart failure. Efforts to unravel the molecular mechanisms underlying ECM remodeling are essential for developing effective treatments to preserve cardiac structure and function in patients with heart disease.

4.3.3 Impact of ECM on Vascular Health

The extracellular matrix (ECM) plays a crucial role in vascular health by providing structural support and biochemical cues essential for the maintenance and functionality of blood vessels. Composed primarily of proteins such as collagen, elastin, and fibronectin, the ECM forms a scaffold that not only maintains tissue integrity but also regulates cellular behavior through interactions with integrins and other cell surface receptors [20]. This intricate network of molecules not only supports vascular cells but also influences processes like cell adhesion, migration, proliferation, and differentiation, all of which are vital for vascular development, repair, and function. One of the key roles of ECM in vascular health is in maintaining the mechanical strength and elasticity of blood vessels. For instance, collagen fibers in the ECM provide tensile strength, ensuring that blood vessels can withstand the pressure exerted by blood flow without rupturing. Meanwhile, elastin fibers confer elasticity, allowing vessels to expand and contract in response to changes in blood pressure, thereby regulating blood flow and optimizing circulation. The ECM serves as a reservoir for growth factors, cytokines, and other signaling molecules that regulate vascular cell behavior. These bioactive molecules are stored within the ECM and released in response to physiological cues or injury, thereby orchestrating processes such as angiogenesis (the formation of new blood vessels), vasculogenesis (the formation of blood vessels from precursor cells), and tissue remodeling. This dynamic interaction between ECM components and signaling molecules ensures the proper development, repair, and adaptation of blood vessels in response to various physiological demands and pathological conditions.

Importantly, disruptions in ECM composition or function can lead to vascular diseases such as atherosclerosis, hypertension, and aneurysms [21]. For example, excessive deposition of ECM proteins can contribute to the thickening of arterial walls in atherosclerosis, narrowing blood vessels and impairing blood flow. Conversely, degradation of ECM components, often mediated by enzymes such as matrix metalloproteinases (MMPs), can weaken vessel walls, increasing the risk of aneurysm formation or vessel rupture.

Advances in understanding the role of ECM in vascular health have led to innovative therapeutic strategies. One approach involves developing biomaterials that mimic the ECM's composition and architecture to support vascular regeneration and repair. These biomaterials can be used as scaffolds for tissue engineering applications or as delivery systems for therapeutic agents that promote vascular healing and regeneration. The ECM is indispensable for maintaining vascular health by providing structural support, regulating cellular behavior, and modulating signaling pathways critical for vascular function. Understanding the complexities of ECM interactions in vascular biology not only sheds light on normal physiological processes but also offers insights into the pathogenesis of vascular diseases and informs the development of novel therapeutic interventions aimed at restoring vascular homeostasis [22]. Thus, the ECM stands as a fundamental determinant of vascular health, offering opportunities for innovative research and clinical applications in cardiovascular medicine.

4.4 Role of ECM in Fibrosis

Extracellular matrix (ECM) plays an important role in the pathogenesis of fibrosis, a complex condition characterized by excessive deposition of ECM components in tissues (Fig. 4.4). The ECM, consisting of proteins such as collagen, fibronectin, and proteoglycans, serves not only as a structural scaffold but also regulates cellular behavior and tissue homeostasis. In fibrotic diseases, dysregulation of ECM turnover and remodeling processes leads to the accumulation of ECM proteins, resulting in tissue stiffness and impaired organ function [23]. Moreover, ECM components interact with various growth factors and cytokines, influencing cell proliferation, differentiation, and inflammatory responses, thereby perpetuating the fibrotic cascade.

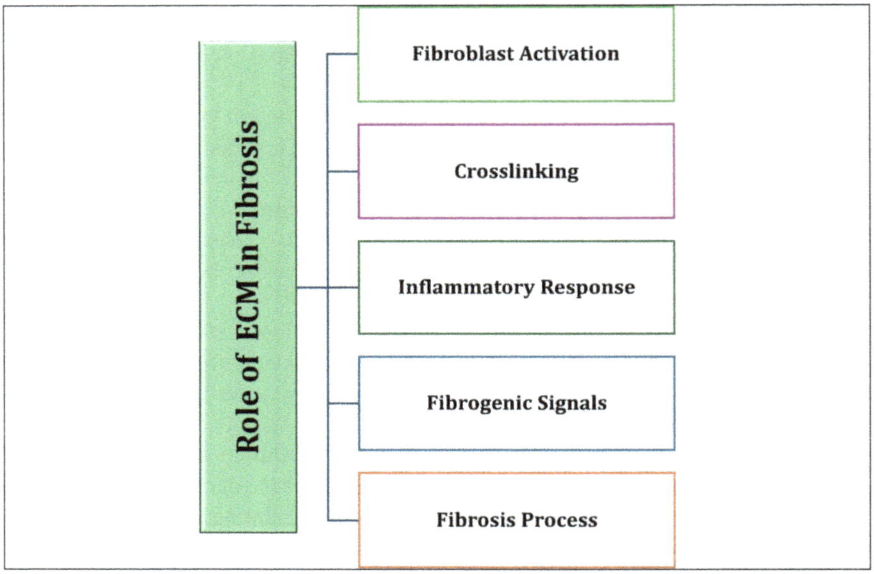

Fig. 4.4 Various major roles of ECM in fibrosis

4.4.1 Mechanisms of ECM Deposition in Fibrotic Diseases

Fibrotic diseases encompass a diverse group of conditions characterized by abnormal deposition of extracellular matrix (ECM) components in tissues. The process of ECM deposition in fibrosis involves intricate cellular and molecular mechanisms that disrupt tissue architecture and function. Understanding these mechanisms is crucial for developing effective therapies to combat fibrotic diseases.

- **Initiation and Activation of Fibroblasts**: Fibroblasts, typically quiescent cells in healthy tissues, become activated in response to various stimuli such as tissue injury, chronic inflammation, or persistent signaling from cytokines and growth factors [24]. Activated fibroblasts undergo phenotypic changes, acquiring contractile and synthetic properties that drive ECM production. Transforming growth factor-beta (TGF-β) is a key cytokine implicated in fibroblast activation, promoting the synthesis and secretion of ECM proteins like collagen and fibronectin.
- **ECM Protein Synthesis and Secretion**: Activated fibroblasts and myofibroblasts are the primary sources of ECM proteins in fibrotic tissues. These cells increase their synthesis of collagen types I and III, fibronectin, and proteoglycans. Collagen, the most abundant ECM protein, forms cross-linked fibrils that contribute to tissue stiffness and impaired function [25]. Fibronectin serves as a linker between ECM components and cell surface receptors, facilitating cell adhesion and ECM organization.

- **Disrupted ECM Turnover**: Normally, ECM turnover is tightly regulated by enzymes such as matrix metalloproteinases (MMPs) and tissue inhibitors of metalloproteinases (TIMPs). In fibrotic diseases, this balance is disrupted, leading to excessive ECM accumulation. Increased expression of TIMPs inhibits MMP activity, impairing ECM degradation and turnover [26]. Consequently, ECM proteins accumulate, contributing to tissue fibrosis and dysfunction.
- **Role of Myofibroblasts**: Myofibroblasts are specialized fibroblast-like cells that play a central role in ECM deposition during fibrosis. These cells exhibit enhanced contractility and ECM synthesis capabilities, driven by persistent TGF-β signaling and mechanical stress within the tissue microenvironment. Myofibroblasts also promote ECM cross-linking through enzymes like lysyl oxidase, further exacerbating tissue stiffness and fibrosis progression.
- **Epithelial-Mesenchymal Transition (EMT)**: In certain fibrotic diseases, epithelial cells can undergo EMT, a process where epithelial cells lose their polarity and cell–cell adhesion properties, acquiring a mesenchymal-like phenotype. EMT-derived cells contribute to ECM production and fibrotic tissue remodeling. This transition is regulated by various growth factors, cytokines, and transcription factors, highlighting its role in ECM deposition and fibrosis pathogenesis.
- **Role of Inflammatory Mediators**: Chronic inflammation is a hallmark of many fibrotic diseases and contributes significantly to ECM deposition. Inflammatory cells, such as macrophages and lymphocytes, release cytokines (e.g., TGF-β, interleukins) and chemokines that stimulate fibroblast activation and ECM synthesis. Additionally, inflammatory mediators can directly induce ECM protein expression in resident cells within affected tissues, perpetuating the fibrotic response.
- **Mechanical Stress and ECM Remodeling**: Mechanical stress within fibrotic tissues influences ECM remodeling and deposition. Increased tissue stiffness alters cellular mechano transduction pathways, activating fibroblasts and promoting ECM synthesis [23]. This feedback loop further reinforces ECM deposition and fibrosis progression, highlighting the interplay between mechanical cues and cellular responses in fibrotic diseases.
- **Microenvironmental Factors**: The tissue microenvironment, including oxygen tension, pH levels, and nutrient availability, profoundly influences ECM deposition in fibrotic diseases. Hypoxia, common in fibrotic tissues due to vascular abnormalities, can activate fibroblasts and promote ECM production through hypoxia-inducible factors (HIFs) and other signaling pathways. Similarly, altered nutrient metabolism and oxidative stress contribute to ECM remodeling and fibrosis development.

ECM deposition in fibrotic diseases results from complex interactions involving activated fibroblasts, myofibroblasts, inflammatory mediators, mechanical stress, and microenvironmental factors. These processes disrupt normal tissue homeostasis, leading to excessive ECM accumulation, tissue fibrosis, and organ dysfunction. Targeting these mechanisms holds promise for developing therapies that can halt or reverse fibrotic progression and restore tissue function in affected individuals.

4.4.2 Consequences of ECM Dysregulation in Organ Fibrosis

There are Various consequences of ECM dysregulation in organ fibrosis.

- **Tissue Stiffness and Dysfunction**: Excessive deposition of ECM proteins, particularly collagen, leads to increased tissue stiffness, impairing the normal elasticity and flexibility required for organ function. In organs such as the liver, lungs, and kidneys, this stiffness can restrict organ movement or expansion during normal physiological processes, compromising overall function [27]. For example, in pulmonary fibrosis, stiffening of lung tissue reduces the ability to expand and contract during breathing, leading to impaired gas exchange and respiratory function.
- **Impaired Organ Architecture**: The abnormal accumulation of ECM disrupts the intricate architecture of organs. Normally, ECM components provide structural support and organize cells into functional units. In fibrotic diseases, the disorganized deposition of collagen and other matrix proteins alters tissue architecture, leading to loss of specialized cell–cell interactions and tissue compartments. This architectural disruption impairs organ function and can contribute to progressive organ failure over time.
- **Altered Cellular Signaling**: ECM proteins act as signaling molecules, interacting with cell surface receptors such as integrins and proteoglycans to regulate cell behavior. In fibrotic conditions, dysregulated ECM composition and organization alter these signaling pathways. For instance, changes in ECM stiffness can activate intracellular signaling cascades that promote fibroblast activation, ECM production, and inflammation. This altered signaling environment perpetuates fibrotic processes and contributes to sustained tissue injury and dysfunction.
- **Vascular Abnormalities**: ECM dysregulation in fibrosis affects vascular integrity and function. Abnormal ECM deposition around blood vessels can lead to vascular remodeling, narrowing of vessel lumens, and increased vascular resistance. In organs like the liver, disrupted ECM composition contributes to portal hypertension, a common complication of cirrhosis, where increased resistance in the hepatic circulation leads to complications such as variceal bleeding and ascites.
- **Fibrosis-Associated Organ Dysfunction**: Progressive ECM deposition in organs often results in functional impairments specific to the affected organ. For example, in cardiac fibrosis, excessive ECM accumulation around cardiomyocytes compromises myocardial contractility and electrical conduction, predisposing individuals to arrhythmias and heart failure. Similarly, renal fibrosis disrupts the filtration and reabsorption functions of the kidneys, leading to impaired fluid and electrolyte balance and eventual kidney failure if left untreated [28].
- **Compromised Wound Healing**: In fibrotic conditions, the abnormal ECM environment interferes with normal wound healing processes. Excessive ECM deposition creates a barrier to cellular migration and impedes the resolution of inflammation. This delayed wound healing can lead to chronic wounds or ulcers, particularly in conditions such as diabetic nephropathy where renal fibrosis contributes to impaired tissue repair and progressive kidney damage.

- **Increased Risk of Cancer Development**: Chronic inflammation and ECM dysregulation in fibrotic tissues create a microenvironment that promotes tumor initiation and progression. Abnormal ECM composition and mechanical stress can induce genetic mutations, enhance angiogenesis, and impair immune surveillance, facilitating the growth and spread of cancer cells. Liver cirrhosis, characterized by extensive ECM deposition and inflammation, significantly increases the risk of hepatocellular carcinoma, highlighting the link between fibrosis and cancer development.
- **Impact on Organ Transplantation**: ECM dysregulation in fibrotic organs poses challenges for transplantation outcomes. Fibrotic changes alter organ structure and function, increasing the risk of graft rejection, vascular thrombosis, and surgical complications post-transplantation. Strategies to mitigate ECM deposition and promote tissue regeneration are crucial for improving transplant success rates and long-term graft survival in patients with end-stage organ fibrosis.

ECM dysregulation in organ fibrosis contributes to a range of detrimental consequences including tissue stiffness, architectural disruption, altered cellular signaling, vascular abnormalities, organ-specific dysfunction, impaired wound healing, increased cancer risk, and challenges in organ transplantation.

4.4.3 Therapeutic Strategies Targeting ECM in Fibrotic Conditions

Therapeutic strategies targeting ECM in fibrotic diseases encompass a wide range of approaches aimed at inhibiting ECM synthesis, promoting ECM degradation, modulating cross-linking, targeting fibroblast activation, and employing biological and cell-based therapies. These strategies aim to mitigate tissue stiffness, preserve organ function, and improve clinical outcomes for individuals affected by fibrotic conditions. Ongoing research and clinical trials continue to explore novel therapies and combination approaches to effectively manage and treat fibrosis across different organ systems.

- **Inhibition of ECM Synthesis**: Therapeutic approaches aimed at reducing ECM deposition focus on inhibiting the synthesis of collagen and other matrix proteins. Small molecule inhibitors or monoclonal antibodies targeting key enzymes involved in ECM production, such as lysyl oxidase or prolyl hydroxylases, have shown promise in preclinical studies [29]. These inhibitors aim to disrupt fibroblast activation and reduce collagen deposition, thereby slowing down fibrotic progression in affected organs.
- **Promotion of ECM Degradation**: Enhancing ECM turnover and degradation is another strategy to alleviate fibrosis. Agents that stimulate matrix metalloproteinases (MMPs), enzymes responsible for ECM degradation, or inhibit tissue inhibitors of metalloproteinases (TIMPs) can help restore the balance between

ECM synthesis and degradation. For example, MMP activators like GM6001 have been investigated to promote ECM remodeling and improve tissue function in fibrotic diseases such as liver fibrosis and pulmonary fibrosis.

- **Modulation of ECM Cross-Linking**: Cross-linking of ECM proteins contributes to tissue stiffness and impaired function in fibrosis. Therapeutic interventions targeting cross-linking enzymes, such as lysyl oxidase inhibitors (e.g., beta-aminopropionitrile), aim to reduce collagen cross-linking and improve tissue elasticity. By disrupting ECM cross-linking, these agents mitigate fibrotic changes and restore organ function in affected individuals.
- **Anti-fibrotic Drugs**: Several anti-fibrotic drugs targeting ECM-related pathways are under investigation for their potential to treat fibrotic diseases. Pirfenidone and nintedanib, approved for idiopathic pulmonary fibrosis, inhibit fibroblast activation, cytokine production, and ECM synthesis [30]. These drugs demonstrate anti-inflammatory and anti-fibrotic properties, slowing disease progression and improving lung function in patients with pulmonary fibrosis.
- **Targeting Fibroblast Activation**: Activated fibroblasts are major contributors to ECM deposition in fibrotic diseases. Therapies targeting fibroblast activation pathways, such as TGF-β signaling inhibitors (e.g., pirfenidone, galunisertib), aim to suppress fibroblast proliferation and collagen synthesis. By blocking key signaling molecules involved in fibroblast activation, these agents reduce ECM accumulation and mitigate fibrotic tissue changes in various organs.
- **Biological Therapies**: Biological therapies targeting specific cytokines or growth factors involved in ECM regulation offer promising approaches for treating fibrotic conditions. Monoclonal antibodies against TGF-β, PDGF (platelet-derived growth factor), and CTGF (connective tissue growth factor) have been developed to interfere with ECM synthesis and fibroblast activation. These targeted therapies aim to disrupt pro-fibrotic signaling pathways and attenuate ECM deposition in affected tissues.
- **Cell-Based Therapies**: Cell-based therapies, such as mesenchymal stem cell (MSC) transplantation, hold potential for modulating ECM remodeling and reducing fibrosis. MSCs exert anti-inflammatory and anti-fibrotic effects by secreting paracrine factors that inhibit fibroblast activation, promote ECM degradation, and stimulate tissue repair processes. Clinical trials investigating MSC therapy in liver fibrosis, kidney fibrosis, and cardiac fibrosis are ongoing to evaluate their efficacy in reversing ECM changes and improving organ function.
- **Combination Therapies**: Given the multifaceted nature of fibrotic diseases, combination therapies targeting multiple aspects of ECM dysregulation and fibrogenic pathways are increasingly explored. Combinations of anti-fibrotic drugs, biological agents, and cell-based therapies aim to synergistically inhibit ECM synthesis, promote ECM degradation, and modulate fibroblast activity. These integrated approaches offer potential synergistic benefits in halting fibrotic progression and restoring tissue homeostasis in patients with advanced fibrotic conditions.

4.5 ECM in Chronic Inflammatory Conditions

Extracellular matrix (ECM) plays a pivotal role in chronic inflammatory conditions by serving as a dynamic scaffold that regulates cellular behavior and tissue architecture. Inflammation triggers a cascade of events that remodel the ECM, altering its composition and physical properties. This remodeling process is orchestrated by various enzymes, such as matrix metalloproteinases (MMPs), which degrade ECM components like collagen and fibronectin. Concurrently, inflammatory cytokines like tumor necrosis factor-alpha (TNF-α) and interleukins stimulate fibroblasts and other ECM-producing cells to increase matrix synthesis, leading to fibrosis and tissue stiffening [31]. The importance of ECM in chronic inflammation lies in its ability to influence immune cell recruitment, activation, and resolution of inflammation. ECM components serve as reservoirs for chemokines and growth factors that guide immune cells to the site of inflammation. Dysregulation of ECM turnover disrupts these signaling mechanisms, perpetuating chronic inflammation and tissue damage. Moreover, altered ECM composition can create a microenvironment that favors persistent inflammatory responses, contributing to disease progression.

Exercise 3
Fill Ups:

Deposition, Tissue destruction, Collagen & fibrosis, Degradation & remodeling, Inflammation & disease progression.

1. ECM remodeling in chronic inflammation can lead to excessive deposition of .... & ....
2. Inflammatory cytokines such as TNF-alpha and IL-1 can stimulate ECM .... & ....
3. Fibroblasts play a key role in ECM .... during chronic inflammation.
4. ECM degradation products like MMPs contribute to .... in chronic inflammatory diseases.
5. Alterations in ECM composition and stiffness can promote .... & .... in chronic inflammatory conditions.

4.5.1 ECM Involvement in Chronic Inflammation

Chronic inflammation is a complex process characterized by prolonged activation of the immune system, often resulting from unresolved acute inflammation or persistent exposure to harmful stimuli such as pathogens, toxins, or damaged cells. The Extracellular Matrix (ECM) plays a pivotal role in chronic inflammation through various mechanisms that contribute to tissue remodeling, immune cell recruitment, and signaling processes (Fig. 4.5).

Firstly, the ECM serves as a structural scaffold within tissues, providing physical support and organizing cells into functional units. During chronic inflammation, ECM components such as collagen, fibronectin, and elastin undergo dynamic changes

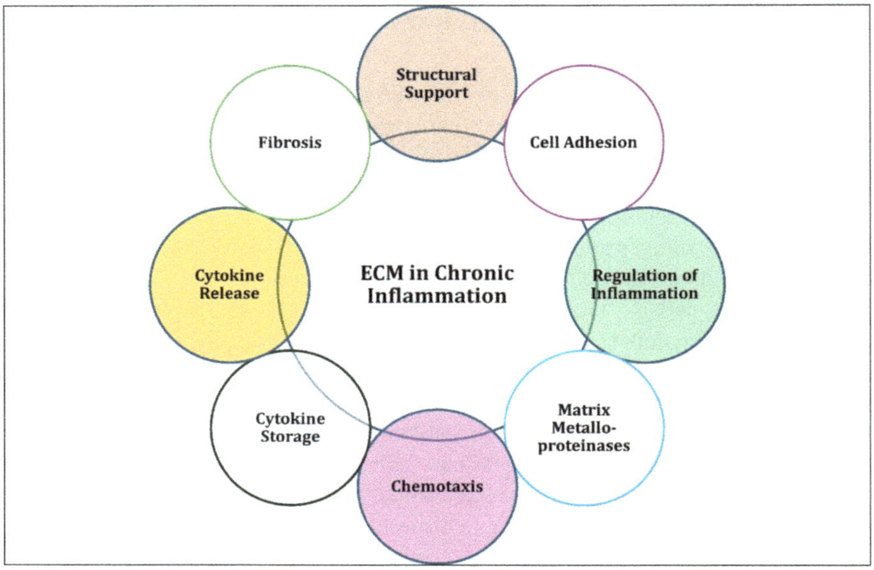

Fig. 4.5 Role of ECM in chronic inflammation

orchestrated by cells like fibroblasts and immune cells [32]. These changes include ECM degradation and deposition, which influence tissue architecture and stiffness, thereby affecting immune cell behavior and inflammatory responses. Moreover, ECM molecules act as signaling molecules, capable of directly influencing immune cell activation and function. For instance, ECM fragments like hyaluronan or proteoglycans can bind to specific receptors on immune cells, triggering signaling pathways that promote inflammation. This direct interaction between ECM components and immune cells modulates the recruitment, adhesion, and activation of leukocytes at the site of inflammation, perpetuating chronic inflammatory processes. The ECM serves as a reservoir for cytokines, chemokines, and growth factors that regulate immune responses. These molecules are stored within the ECM in an inactive form and are released upon degradation or by specific enzymatic cleavage during inflammation. This controlled release of bioactive molecules from the ECM influences the local immune cell microenvironment, promoting cell survival, proliferation, and differentiation, which are crucial aspects of chronic inflammation.

ECM remodeling during chronic inflammation alters tissue architecture and function. Excessive ECM deposition, often associated with fibrosis, can impair tissue function and exacerbate inflammation by creating physical barriers that hinder immune cell migration and tissue repair processes. Conversely, ECM degradation by matrix metalloproteinases (MMPs) and other enzymes contributes to tissue destruction and perpetuates inflammatory responses by releasing bioactive ECM fragments and exposing cryptic binding sites for immune cell receptors. Furthermore, the

ECM modulates angiogenesis and vascular permeability during chronic inflammation. It regulates the formation of new blood vessels (angiogenesis) in response to tissue injury or hypoxia, facilitating immune cell trafficking and nutrient delivery to the inflamed tissue. Dysregulated ECM turnover and deposition disrupt vascular integrity, leading to increased permeability and leakage of inflammatory mediators into surrounding tissues, thereby sustaining chronic inflammation. The ECM interacts with immune cells through integrin-mediated adhesion and signaling pathways [33]. Integrins, which are transmembrane receptors that bind ECM proteins, play a crucial role in immune cell adhesion, migration, and activation. Dysregulated integrin-ECM interactions during chronic inflammation can alter immune cell behavior, promoting sustained inflammatory responses and tissue damage.

The ECM contributes to the formation of specialized microenvironments within tissues, known as niches, which regulate immune cell homeostasis and function. These niches provide a spatial and biochemical framework that supports immune cell survival, proliferation, and differentiation, influencing the balance between pro-inflammatory and anti-inflammatory responses during chronic inflammation. The ECM exerts multifaceted effects on chronic inflammation through its roles in structural support, signaling, cytokine regulation, tissue remodeling, angiogenesis, integrin-mediated interactions, and niche formation.

4.5.2 Crosstalk Between Immune Cells and ECM

The crosstalk between immune cells and the Extracellular Matrix (ECM) is essential for regulating immune responses in health and disease. The ECM, a complex network of proteins and carbohydrates surrounding cells, not only provides structural support but also actively participates in immune cell behavior through various mechanisms. Immune cells interact with ECM components such as collagen, fibronectin, and laminin via cell surface receptors like integrins. These interactions mediate adhesion, migration, and activation of immune cells. Integrins serve as molecular bridges that transmit signals bidirectionally between the ECM and immune cells, influencing their recruitment to inflammatory sites and modulating their activation state. ECM molecules act as reservoirs for cytokines, chemokines, and growth factors that regulate immune cell function [34]. These bioactive molecules are stored within the ECM and released upon enzymatic cleavage or degradation during inflammation, impacting immune cell activation, differentiation, and survival. This process helps to orchestrate complex immune responses and maintain tissue homeostasis. Furthermore, ECM remodeling by immune cells themselves, such as fibroblasts and macrophages, contributes to the dynamic changes observed during immune responses. For example, fibroblasts deposit ECM components during tissue repair, creating a supportive environment for immune cell infiltration and function. Conversely, macrophages and other immune cells produce enzymes like matrix metalloproteinases (MMPs) that degrade ECM proteins, facilitating tissue remodeling and influencing immune cell behavior.

The ECM influences immune cell polarization and function through physical cues and biochemical signals. ECM stiffness and composition can dictate immune cell differentiation towards pro-inflammatory (M1) or anti-inflammatory (M2) phenotypes. This polarization is mediated by ECM-mediated activation of signaling pathways such as NF-κB and MAPK, which regulate cytokine production and immune cell fate. ECM-derived molecules such as hyaluronan and versican can directly engage immune cell receptors, triggering intracellular signaling cascades that modulate immune cell adhesion, migration, and activation. These interactions not only regulate immune cell behavior locally but also impact systemic immune responses and disease progression. The ECM serves as a barrier that restricts immune cell migration or facilitates their movement within tissues. This spatial regulation is critical for controlling immune surveillance, inflammation resolution, and tissue repair processes. Dysregulation of ECM-immune cell interactions can impair immune cell trafficking and function, contributing to chronic inflammation and autoimmune diseases.

Additionally, immune cells can influence ECM turnover and composition through secretion of cytokines, enzymes, and other factors. For instance, inflammatory cytokines like TNF-α and IL-1β stimulate ECM-producing cells to increase deposition of ECM components, promoting tissue fibrosis and exacerbating inflammatory responses [35]. Conversely, anti-inflammatory cytokines and regulatory immune cells can promote ECM remodeling and tissue repair, resolving inflammation and restoring tissue homeostasis.

The crosstalk between immune cells and the ECM is dynamic and context-dependent, shaping immune responses in a spatial and temporal manner. Understanding these interactions is crucial for developing therapeutic strategies that target ECM components or immune cell-ECM interactions to modulate immune responses and treat inflammatory diseases effectively. The crosstalk between immune cells and the ECM is a complex interplay that regulates immune cell recruitment, activation, polarization, and function during health and disease. This reciprocal relationship underscores the importance of ECM dynamics in shaping immune responses and maintaining tissue homeostasis.

4.5.3 ECM-Based Therapies for Chronic Inflammatory Diseases

ECM (Extracellular Matrix) based therapies are emerging as promising treatments for chronic inflammatory diseases, offering novel approaches to managing conditions like rheumatoid arthritis, inflammatory bowel disease, and chronic obstructive pulmonary disease, among others. The ECM, a complex network of proteins and carbohydrates surrounding cells, plays crucial roles in tissue architecture, signaling, and immune modulation. Harnessing ECM components or modifying their interactions presents opportunities to target inflammation at its root causes. One of the key strategies in ECM-based therapies involves leveraging ECM-derived scaffolds to promote tissue repair and modulate immune responses. These scaffolds can be derived from natural sources like decellularized tissues or synthetic materials designed to mimic ECM characteristics. By providing a supportive environment for cell migration, proliferation, and differentiation, these scaffolds aid in tissue regeneration while potentially dampening chronic inflammation. This approach not only addresses structural damage but also targets the underlying inflammatory processes. Another avenue of ECM-based therapy explores the use of ECM-derived bioactive molecules, such as growth factors and cytokines, which can regulate immune cell function and tissue repair mechanisms. These molecules, when administered locally or systemically, have shown promise in preclinical and clinical studies for reducing inflammation and promoting tissue healing [33]. By modulating immune cell behavior and cytokine production, ECM-derived bioactive molecules contribute to restoring tissue homeostasis and mitigating chronic inflammation in affected organs. Furthermore, researchers are exploring the therapeutic potential of ECM-based hydrogels, which can be engineered to release bioactive molecules in a controlled manner. These hydrogels not only provide mechanical support but also serve as reservoirs for therapeutic agents that can exert anti-inflammatory effects locally. This targeted delivery minimizes systemic side effects while maximizing therapeutic efficacy, making ECM-based hydrogels an attractive option for chronic inflammatory diseases.

In addition to their direct effects on tissue repair and immune modulation, ECM-based therapies also influence the microenvironment surrounding inflammatory sites. By altering the ECM composition or stiffness, researchers can influence cell behavior and signaling pathways critical to inflammation resolution. This approach holds promise for diseases where chronic inflammation leads to tissue fibrosis or dysfunction, offering a means to restore normal tissue architecture and function. ECM-based therapies are being investigated for their potential synergies with existing treatments, such as biologics and immunomodulatory drugs. Combining ECM scaffolds or bioactive molecules with conventional therapies may enhance treatment outcomes by addressing multiple aspects of inflammation and tissue damage simultaneously. This integrated approach could lead to more effective management of chronic inflammatory diseases and potentially reduce the need for long-term medication use.

Clinical translation of ECM-based therapies faces challenges such as standardization of ECM materials, optimal delivery methods, and long-term safety profiles. Despite these challenges, ongoing research and clinical trials continue to demonstrate encouraging results, underscoring the potential of ECM-based therapies to revolutionize the treatment landscape for chronic inflammatory diseases.

4.6 ECM in Neurological Disorders

The extracellular matrix (ECM) plays a crucial role in the nervous system, providing structural support and regulating various cellular functions (Fig. 4.6). In the central nervous system (CNS), the ECM is primarily composed of glycoproteins such as laminin, fibronectin, and various types of collagens. These molecules form a complex network that surrounds neurons, glial cells, and blood vessels, contributing to the integrity and function of neural tissues.

In normal physiological conditions, the ECM in the nervous system acts as a scaffold that supports cell adhesion, migration, and differentiation during development and adulthood. It also regulates synaptic plasticity, which is essential for learning and memory processes. Moreover, the ECM molecules interact with cell surface receptors, such as integrins and dystroglycan, thereby influencing signal transduction pathways that govern cell behavior and survival. In neurodegenerative diseases such as Alzheimer's disease, Parkinson's disease, and amyotrophic lateral sclerosis (ALS), significant alterations occur in the ECM composition and organization. These changes contribute to disease progression and pathology [36]. For instance, in Alzheimer's disease, the accumulation of beta-amyloid plaques and tau protein aggregates disrupts the ECM structure, leading to impaired synaptic function and neuronal loss. The altered ECM environment also affects the clearance of toxic metabolites and the inflammatory response in the brain, exacerbating neurodegeneration. In Parkinson's disease, misfolded alpha-synuclein proteins can aggregate

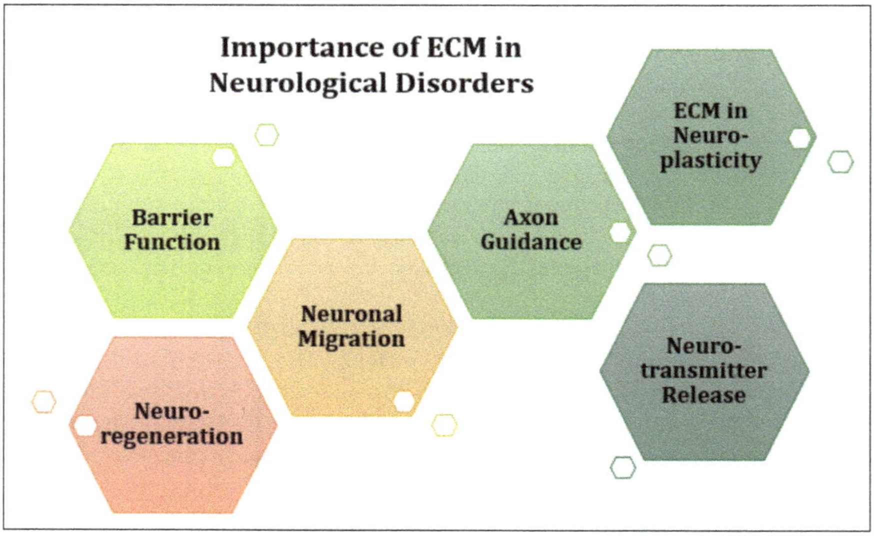

Fig. 4.6 Importance of ECM in neurological disorders

within the ECM, triggering neuroinflammation and impairing cellular communication. These ECM alterations disrupt the balance between neuronal survival and death, contributing to the characteristic motor and cognitive symptoms of the disease.

Similarly, in ALS, mutations in genes encoding ECM proteins like collagen IV and laminin disrupt the integrity of the blood–brain barrier and impair motor neuron function. The resulting ECM abnormalities contribute to the degeneration of motor neurons and the progressive muscle weakness observed in ALS patients.

Exercise 4
Tick The Correct Answer:

1. Which of the following is a function of the ECM in neurological tissues?
a. Intracellular signaling
b. Structural support
c. DNA replication
d. Ribosomal assembly

2. ECM components such as collagen and laminin are primarily involved in:
a. Neurotransmitter release
b. Myelination of neurons
c. Cell adhesion and migration
d. Electrical signal propagation

3. Disruption of ECM integrity in neurological disorders can lead to:
a. Enhanced synaptic transmission
b. Neuronal degeneration
c. Increased blood-brain barrier permeability
d. Accelerated axonal growth

4. Which ECM molecule is crucial for forming the perineuronal net (PNN), involved in synaptic stability and plasticity?
a. Collagen
b. Fibronectin
c. Laminin
d. Hyaluronic acid

5. ECM remodeling in response to neural injury typically involves the activation of:
a. Astrocytes
b. Microglia
c. Oligodendrocytes
d. Neurons

Recent research has highlighted the dynamic nature of the ECM in neurological disorders, revealing its involvement in modulating neuroinflammation and promoting neuroprotection or neurodegeneration. In conditions like multiple sclerosis (MS), ECM remodeling contributes to the formation of scar tissue or gliosis, which impedes neuronal regeneration and repair. The ECM components interact with immune cells and glial cells, influencing the inflammatory response and disease progression. Therapeutic approaches targeting ECM molecules, such as proteoglycans or metalloproteinases involved in ECM degradation, are being explored to mitigate neuroinflammation and promote tissue repair in MS and other inflammatory CNS disorders. Moreover, the ECM in the peripheral nervous system (PNS) plays a critical role in axon regeneration and nerve repair following injury. Injury-induced changes in ECM composition and stiffness at the site of nerve damage influence axonal growth and guidance cues, affecting functional recovery. Understanding these ECM dynamics is crucial for developing interventions that facilitate nerve regeneration and functional restoration after traumatic injuries or peripheral neuropathies. In the context of aging, alterations in ECM turnover and composition contribute to age-related cognitive decline and susceptibility to neurodegenerative diseases [37]. Age-related changes, such as increased deposition of ECM proteins like collagen and fibronectin in the brain, can impair synaptic plasticity and neuronal communication. Addressing ECM-related mechanisms of aging may offer new strategies to maintain cognitive function and delay the onset of age-related neurological disorders, providing opportunities for enhancing healthy brain aging and quality of life in the elderly population.

Understanding the role of ECM in neurological disorders is crucial for developing novel therapeutic strategies aimed at restoring ECM homeostasis and preserving neuronal function. Targeting specific ECM components or their interacting receptors may offer potential avenues for neuroprotective therapies to halt or slow the progression of neurodegenerative diseases. By elucidating the intricate mechanisms underlying ECM dysfunction in these disorders, researchers can pave the way for future advancements in treatment options and improve outcomes for patients affected by neurological conditions.

4.7 Novel Techniques for Studying ECM Dynamics

Studying the dynamics of the Extracellular Matrix (ECM) involves innovative approaches that integrate advanced technologies and methodologies across various scientific disciplines. ECM, a complex network of proteins and carbohydrates surrounding cells, plays crucial roles in tissue architecture, cellular behavior, and signaling. Novel techniques aim to dissect ECM dynamics with unprecedented detail and precision, offering insights into both normal physiological processes and pathological conditions. One pioneering approach utilizes advanced imaging techniques such as super-resolution microscopy. This method surpasses the diffraction limit of conventional light microscopy, enabling researchers to visualize ECM components and their interactions at nanometer scales [38]. By resolving structural details that were previously inaccessible, super-resolution microscopy provides new perspectives on ECM organization and dynamics within tissues. Another cutting-edge technique involves the use of high-throughput screening technologies. These platforms allow researchers to systematically analyze how various ECM components influence cellular behavior and tissue function. By screening large libraries of ECM-modifying molecules or cellular responses to ECM cues, scientists can identify key regulators of ECM dynamics and uncover novel therapeutic targets for diseases involving ECM dysregulation.

Furthermore, biophysical techniques such as atomic force microscopy (AFM) have revolutionized the study of ECM mechanics. AFM enables direct measurement of ECM stiffness, elasticity, and topography at the nanoscale level. This approach not only provides quantitative data on ECM mechanical properties but also offers insights into how these properties influence cellular behavior, migration, and differentiation.

In addition to imaging and biophysical approaches, omics technologies play a pivotal role in ECM research. Genomics, transcriptomics, proteomics, and metabolomics provide comprehensive molecular profiles of ECM composition, turnover, and remodeling. Integrative analysis of omics data allows researchers to uncover ECM-related biomarkers and molecular pathways implicated in ECM dynamics under various physiological and pathological conditions. Bioengineering approaches have also been instrumental in developing biomimetic ECM models. These models replicate key aspects of native ECM architecture and composition, facilitating controlled studies of ECM dynamics in vitro. By systematically varying ECM parameters in engineered models, researchers can elucidate how specific ECM characteristics regulate cell behavior, tissue development, and disease progression.

Lastly, computational modeling and simulation techniques are increasingly employed to predict ECM behaviors and interactions in silico. These models integrate experimental data with mathematical algorithms to simulate ECM dynamics under different conditions [39]. By refining and validating computational predictions against experimental findings, scientists can gain deeper insights into the complex and dynamic nature of ECM regulation. The field of ECM dynamics benefits from a diverse array of novel techniques that span imaging, biophysics, omics, bioengineering, and computational modeling.

4.7.1 Potential Biomarkers Related to ECM Alterations

Identifying biomarkers associated with alterations in the Extracellular Matrix (ECM) holds promise for diagnosing and monitoring various diseases characterized by ECM dysregulation. These biomarkers can serve as indicators of ECM remodeling, fibrosis, or aberrant deposition of ECM components, offering valuable insights into disease progression and treatment response (Fig. 4.7). One potential biomarker is hyaluronic acid (HA), a glycosaminoglycan abundant in the ECM. Elevated levels of HA in bodily fluids, such as serum or urine, are linked to ECM degradation and turnover, reflecting conditions like liver fibrosis, arthritis, or cancer metastasis. Collagens, the main structural proteins of the ECM, also serve as significant biomarkers when altered in disease states. For instance, specific collagen fragments, such as C-terminal telopeptide of type I collagen (CTX-I), can be detected in blood samples and are indicative of bone resorption and ECM degradation in conditions like osteoporosis.

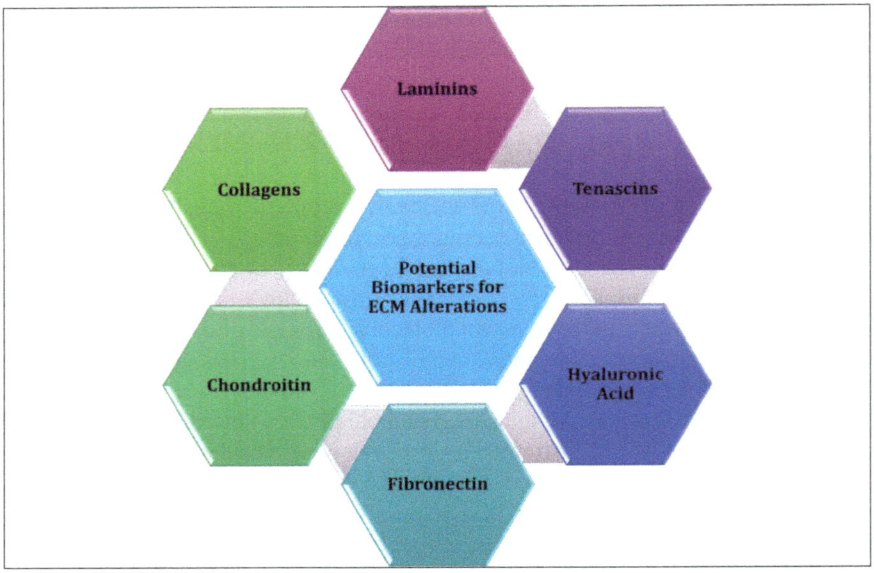

Fig. 4.7 Potential biomarkers for ECM alterations

Similarly, matrix metalloproteinases (MMPs), enzymes responsible for ECM remodeling, serve as biomarkers when their expression and activity are dysregulated [40]. Elevated MMP levels in tissues or bodily fluids correlate with ECM degradation in diseases such as cardiovascular disorders and inflammatory conditions.

Moreover, fibronectin, a glycoprotein involved in cell adhesion and ECM organization, exhibits altered expression patterns in diseases associated with ECM remodeling. Changes in circulating levels or tissue deposition of fibronectin have been linked to conditions like fibrosis and cancer progression. These biomarkers offer clinicians valuable tools for early detection, prognosis, and therapeutic monitoring in diseases where ECM alterations play a pivotal role in pathogenesis.

4.7.2 Translational Opportunities for ECM-Based Therapies

Translational opportunities for Extracellular Matrix (ECM)-based therapies involve a promising frontier in medical research. ECM, the non-cellular component present in all tissues and organs, serves not only as a structural support but also regulates various cellular functions through biochemical and biomechanical cues. This unique biological scaffold has sparked interest in therapeutic applications across several medical fields:

- **Regenerative Medicine**: ECM-based therapies show potential in tissue engineering and regenerative medicine by providing a natural microenvironment that supports cell adhesion, proliferation, and differentiation. Applications range from skin regeneration to complex organ reconstruction.
- **Wound Healing**: ECM-derived products are utilized in wound care to promote healing by enhancing tissue repair processes. They can modulate inflammation, promote angiogenesis, and provide a scaffold for new tissue formation.
- **Orthopedics**: ECM scaffolds are explored for repairing musculoskeletal tissues such as cartilage, ligaments, and tendons. They can improve outcomes in joint repair surgeries and help regenerate damaged tissues.
- **Cardiovascular Medicine**: ECM-based therapies are investigated for repairing heart tissues post-infarction and in treating vascular diseases. They offer potential in enhancing cardiac function and reducing scar formation.
- **Neurology**: ECM-derived materials are being researched for neural tissue repair and regeneration in conditions like spinal cord injuries and neurodegenerative diseases. They can facilitate axonal regrowth and neural network repair.
- **Drug Delivery**: ECM can serve as a carrier for drug delivery systems due to its biocompatibility and ability to protect and release therapeutic agents at specific sites within the body.
- **Cancer Therapeutics**: ECM-based approaches are explored to modulate the tumor microenvironment and inhibit cancer progression [41]. They can be used to deliver anticancer agents or modulate immune responses against tumors.

These translational opportunities underscore the versatility of ECM-based therapies in addressing a wide array of medical challenges, from chronic wounds to complex tissue regeneration. Their biocompatibility, bioactivity, and ability to mimic natural tissue environments make them a promising avenue for future therapeutic innovations.

4.8 Summary

The Extracellular Matrix (ECM) is a complex network of proteins and glycoproteins that surrounds cells within tissues, providing structural support, regulating biochemical signaling, and influencing cellular behavior. In health, the ECM plays a crucial role in maintaining tissue integrity, supporting organ function, and regulating physiological processes such as wound healing and tissue morphogenesis. Composed of collagen, elastin, fibronectin, and other components, the ECM forms a dynamic microenvironment that orchestrates cellular activities through interactions with growth factors, cytokines, and cell surface receptors. In disease contexts, ECM alterations contribute significantly to pathogenesis across various conditions. In cancer, for example, dysregulated ECM remodeling promotes tumor progression, metastasis, and resistance to therapies. Increased ECM stiffness, aberrant deposition of ECM proteins, and altered mechanical signaling pathways within the tumor microenvironment foster aggressive cancer phenotypes and create barriers to effective treatment. Targeting ECM-mediated processes in cancer holds promise for disrupting tumor-stroma interactions, enhancing drug delivery, and improving patient outcomes.

In cardiovascular diseases, ECM remodeling plays a critical role in the pathophysiology of conditions such as atherosclerosis, hypertension, and heart failure. Changes in ECM composition and organization within blood vessels and cardiac tissues contribute to vascular stiffening, plaque formation, and myocardial fibrosis, which impair cardiovascular function and increase the risk of adverse events. Therapeutic strategies aimed at modulating ECM turnover, reducing fibrosis, and restoring ECM homeostasis offer potential avenues to mitigate disease progression and improve cardiovascular outcomes.

Moreover, the ECM's involvement extends to chronic inflammatory conditions and fibrotic diseases, where dysregulated ECM deposition and crosslinking contribute to tissue stiffness, organ dysfunction, and impaired healing responses. Understanding the mechanisms underlying ECM-mediated pathologies in these conditions is essential for developing targeted therapies that address ECM abnormalities and restore tissue function. Approaches such as ECM-modifying enzymes, integrin inhibitors, and ECM-targeted biologics are being explored to intervene in these diseases by restoring ECM balance and promoting tissue regeneration. Future research directions include further elucidating the molecular mechanisms of ECM regulation, exploring novel therapeutic targets within the ECM, and advancing imaging techniques to monitor ECM dynamics in real-time. By harnessing the therapeutic potential of ECM-targeted interventions, researchers aim to translate these discoveries into clinical applications that improve treatment efficacy, reduce disease burden, and enhance patient quality of life across diverse health conditions.

Points to Remember

- ECM provides a scaffold for cells, offering mechanical support and maintaining tissue architecture.
- It influences cell adhesion, migration, proliferation, and differentiation through biochemical and mechanical cues.
- ECM components, like proteoglycans and glycoproteins, can bind growth factors and cytokines, regulating cellular responses.
- ECM helps maintain tissue hydration, pH balance, and ion concentrations critical for cellular function.
- Dysregulation of ECM remodeling is associated with various diseases, including fibrosis, cancer metastasis, and cardiovascular disorders.
- ECM components can interact with immune cells, influencing inflammation and immune responses in tissues.
- Due to its central role in disease progression, ECM components are targets for therapeutic interventions aimed at modifying tissue remodeling and disease progression.

References

1. Yue B (2014) Biology of the extracellular matrix: an overview. J Glaucoma 23(8 Suppl 1):S20–S23. https://doi.org/10.1097/IJG.0000000000000108
2. Frantz C, Stewart KM, Weaver VM (2010) The extracellular matrix at a glance. J Cell Sci 123(Pt 24):4195–4200. https://doi.org/10.1242/jcs.023820
3. Labat-Robert J, Bihari-Varga M, Robert L (1990) Extracellular matrix. FEBS Lett 268(2):386–393. https://doi.org/10.1016/0014-5793(90)81291-u
4. Theocharis AD, Skandalis SS, Gialeli C, Karamanos NK (2016) Extracellular matrix structure. Adv Drug Deliv Rev 97:4–27. https://doi.org/10.1016/j.addr.2015.11.001
5. Walker C, Mojares E, Del Río Hernández A (2018) Role of extracellular matrix in development and cancer progression. Int J Mol Sci 19(10):3028. https://doi.org/10.3390/ijms19103028
6. Yuan Z, Li Y, Zhang S, Wang X, Dou H, Yu X, Zhang Z, Yang S, Xiao M (2023) Extracellular matrix remodeling in tumor progression and immune escape: from mechanisms to treatments. Mol Cancer 22(1):48. https://doi.org/10.1186/s12943-023-01744-8
7. Popova NV, Jücker M (2022) The functional role of extracellular matrix proteins in cancer. Cancers 14(1):238. https://doi.org/10.3390/cancers14010238
8. Neophytou CM, Panagi M, Stylianopoulos T, Papageorgis P (2021) The role of tumor microenvironment in cancer metastasis: molecular mechanisms and therapeutic opportunities. Cancers 13(9):2053. https://doi.org/10.3390/cancers13092053
9. Winkler J, Abisoye-Ogunniyan A, Metcalf KJ, Werb Z (2020) Concepts of extracellular matrix remodelling in tumour progression and metastasis. Nat Commun 11(1):5120. https://doi.org/10.1038/s41467-020-18794-x
10. Kim M, Lee C, Park J (2022) Extracellular matrix remodeling facilitates obesity-associated cancer progression. Trends Cell Biol 32(10):825–834. https://doi.org/10.1016/j.tcb.2022.02.008
11. Kaushik N, Kim S, Suh Y, Lee SJ (2019) Proinvasive extracellular matrix remodeling for tumor progression. Arch Pharmacal Res 42(1):40–47. https://doi.org/10.1007/s12272-018-1097-0
12. Huang J, Zhang L, Wan D, Zhou L, Zheng S, Lin S, Qiao Y (2021) Extracellular matrix and its therapeutic potential for cancer treatment. Signal Transduct Target Ther 6(1):153. https://doi.org/10.1038/s41392-021-00544-0
13. Joyce JA (2005) Therapeutic targeting of the tumor microenvironment. Cancer Cell 7(6):513–520. https://doi.org/10.1016/j.ccr.2005.05.024
14. Chen X, Zhang X, Wu Y, Chen Y, Guo Y, Jana D, Wang D, Yuan W, Zhao Y (2022) Tumor extracellular matrix-targeted nanoscavengers reverse suppressive microenvironment for cocktail therapy. Mater Today 61:78–90. https://doi.org/10.1016/j.mattod.2022.10.018
15. Del Monte-Nieto G, Fischer JW, Gorski DJ, Harvey RP, Kovacic JC (2020) Basic biology of extracellular matrix in the cardiovascular system, Part 1/4. J Am Coll Cardiol 75(17):2169–2188. https://doi.org/10.1016/j.jacc.2020.03.024
16. Bloksgaard M, Lindsey M, Martinez-Lemus LA (2018) Extracellular matrix in cardiovascular pathophysiology. Am J Physiol Hear Circ Physiol 315(6):H1687–H1690. https://doi.org/10.1152/ajpheart.00631.2018
17. Hematti P (2018) Role of extracellular matrix in cardiac cellular therapies. Adv Exp Med Biol 1098:173–188. https://doi.org/10.1007/978-3-319-97421-7_9
18. Berk BC, Fujiwara K, Lehoux S (2007) ECM remodeling in hypertensive heart disease. J Clin Investig 117(3):568–575. https://doi.org/10.1172/JCI31044
19. Hamsho K, Broadwin M, Stone CR, Sellke FW, Abid MR (2024) The current state of extracellular matrix therapy for ischemic heart disease. Med Sci 12(1):8. https://doi.org/10.3390/medsci12010008
20. Chaqour B, Karrasch C (2020) Eyeing the extracellular matrix in vascular development and microvascular diseases and bridging the divide between vascular mechanics and function. Int J Mol Sci 21(10):3487. https://doi.org/10.3390/ijms21103487
21. Xu J, Shi GP (2014) Vascular wall extracellular matrix proteins and vascular diseases. Biochem Biophys Acta 1842(11):2106–2119. https://doi.org/10.1016/j.bbadis.2014.07.008

22. Lin PK, Davis GE (2023) Extracellular matrix remodeling in vascular Disease: defining its regulators and pathological influence. Arterioscler Thromb Vasc Biol 43(9):1599–1616. https://doi.org/10.1161/atvbaha.123.318237
23. Herrera J, Henke CA, Bitterman PB (2018) Extracellular matrix as a driver of progressive fibrosis. J Clin Investig 128(1):45–53. https://doi.org/10.1172/JCI93557
24. Dees C, Chakraborty D, Distler JHW (2021) Cellular and molecular mechanisms in fibrosis. Exp Dermatol 30(1):121–131. https://doi.org/10.1111/exd.14193
25. Wynn TA, Ramalingam TR (2012) Mechanisms of fibrosis: therapeutic translation for fibrotic disease. Nat Med 18(7):1028–1040. https://doi.org/10.1038/nm.2807
26. Cox TR, Erler JT (2011) Remodeling and homeostasis of the extracellular matrix: implications for fibrotic diseases and cancer. Dis Model Mech 4(2):165–178. https://doi.org/10.1242/dmm.004077
27. Zhao P, Sun T, Lyu C, Liang K, Du Y (2023) Cell mediated ECM-degradation as an emerging tool for anti-fibrotic strategy. Cell Regen (London, England) 12(1):29. https://doi.org/10.1186/s13619-023-00172-9
28. Lloyd SM, He Y (2024) Exploring extracellular matrix crosslinking as a therapeutic approach to fibrosis. Cells 13(5):438. https://doi.org/10.3390/cells13050438
29. Isshiki T, Naiel S, Vierhout M, Otsubo K, Ali P, Tsubouchi K, Yazdanshenas P, Kumaran V, Dvorkin-Gheva A, Kolb MRJ, Ask K (2024) Therapeutic strategies to target connective tissue growth factor in fibrotic lung diseases. Pharmacol Ther 253:108578. https://doi.org/10.1016/j.pharmthera.2023.108578
30. Gourdie RG, Dimmeler S, Kohl P (2016) Novel therapeutic strategies targeting fibroblasts and fibrosis in heart disease. Nat Rev Drug Discov 15(9):620–638. https://doi.org/10.1038/nrd.2016.89
31. Marangio A, Biccari A, D'Angelo E, Sensi F, Spolverato G, Pucciarelli S, Agostini M (2022) The study of the extracellular matrix in chronic inflammation: a way to prevent cancer initiation? Cancers 14(23):5903. https://doi.org/10.3390/cancers14235903
32. Sorokin L (2010) The impact of the extracellular matrix on inflammation. Nat Rev Immunol 10(10):712–723. https://doi.org/10.1038/nri2852
33. McQuitty CE, Williams R, Chokshi S, Urbani L (2020) Immunomodulatory role of the extracellular matrix within the liver disease microenvironment. Front Immunol 11:574276. https://doi.org/10.3389/fimmu.2020.574276
34. Kuehn C, Vermette P, Fülöp T (2014) Cross talk between the extracellular matrix and the immune system in the context of endocrine pancreatic islet transplantation. A review article. Pathologie-Biologie 62(2):67–78. https://doi.org/10.1016/j.patbio.2014.01.001
35. Bhattacharjee O, Ayyangar U, Kurbet AS, Ashok D, Raghavan S (2019) Unraveling the ECM-immune cell crosstalk in skin diseases. Front Cell Dev Biol 7:68. https://doi.org/10.3389/fcell.2019.00068
36. Ge X, Xu X, Cai Q, Xiong H, Xie C, Hong Y, Gao X, Yao Y, Bachoo R, Qin Z (2024) Live Mapping of the brain extracellular matrix and remodeling in neurological disorders. Small Methods 8(1):e2301117. https://doi.org/10.1002/smtd.202301117
37. Soleman S, Filippov MA, Dityatev A, Fawcett JW (2013) Targeting the neural extracellular matrix in neurological disorders. Neuroscience 253:194–213. https://doi.org/10.1016/j.neuroscience.2013.08.050
38. Dibus M, Joshi O, Ivaska J (2024) Novel tools to study cell-ECM interactions, cell adhesion dynamics and migration. Curr Opin Cell Biol 88:102355. https://doi.org/10.1016/j.ceb.2024.102355
39. Afratis NA, Sagi I (2019) Novel approaches for extracellular matrix targeting in disease treatment. Methods Mol Biol (Clifton, N.J.) 1952:261–275. https://doi.org/10.1007/978-1-4939-9133-4_21
40. Brassart-Pasco S, Brézillon S, Brassart B, Ramont L, Oudart JB, Monboisse JC (2020) Tumor microenvironment: extracellular matrix alterations influence tumor progression. Front Oncol 10:397. https://doi.org/10.3389/fonc.2020.00397

41. Capella-Monsonís H, Crum RJ, Hussey GS, Badylak SF (2024) Advances, challenges, and future directions in the clinical translation of ECM biomaterials for regenerative medicine applications. Adv Drug Deliv Rev 211:115347 (Advance online publication). https://doi.org/10.1016/j.addr.2024.115347

Chapter 5
Biomimicry of ECM in Medical Applications

5.1 Introduction

The study of the Extracellular Matrix (ECM) represents a pivotal junction in the realms of biology and medical science, tracing its origins to early investigations into tissue structure and function [1]. Historically, the ECM was initially perceived as a passive scaffold providing structural support to cells within tissues. It wasn't until the mid-twentieth century that researchers began to recognize its dynamic role in cellular behavior and tissue homeostasis. Pioneering work by Albert Claude and Keith Porter in the 1940s provided the first electron microscopy images revealing the intricate meshwork of ECM fibers surrounding cells, challenging the prevailing notion of cells as isolated entities and highlighting the ECM's integral role in tissue architecture [2].

The 1970s witnessed a significant shift in ECM research with the identification and characterization of specific ECM components such as collagens, proteoglycans, and glycoproteins. This era marked a turning point in understanding the ECM's biochemical complexity and its diverse functions beyond structural support. The discovery of integrins in the 1980s further revolutionized ECM research by elucidating how cells interact with and respond to ECM cues, influencing processes like cell adhesion, migration, and differentiation [3]. These findings underscored the ECM's active participation in cellular signaling and tissue development, laying the groundwork for subsequent advancements in tissue engineering and regenerative medicine. The late twentieth and early twenty-first centuries witnessed exponential growth in ECM research driven by interdisciplinary collaborations spanning biology, biochemistry, materials science, and engineering. Innovations in imaging technologies, such as confocal microscopy and atomic force microscopy, enabled detailed visualization and analysis of ECM structure and function at unprecedented resolutions. Concurrently, bioinformatics and genomic studies provided insights into the genetic regulation [4] of ECM synthesis and remodeling processes, further expanding our understanding of ECM dynamics in health and disease. Today, ECM research

D. Suhag, *Extracellular Matrix biomimicking for Theranostics*, Biomedical Materials for Multi-functional Applications, https://doi.org/10.1007/978-981-95-7900-6_5

stands at the forefront of biomedical innovation, with profound implications for therapeutic interventions and medical device development. Insights gained from ECM studies have paved the way for biomimetic approaches in tissue engin eering, where synthetic ECM scaffolds and matrices mimic native tissue environments to promote regeneration and repair. Moreover, the integration of ECM-derived biomaterials in drug delivery systems and personalized medicine holds promise for targeted therapies and enhanced patient outcomes. As we continue to unravel the intricate interactions between cells and their ECM microenvironments, the future of ECM research promises to unlock new frontiers in regenerative medicine, disease modeling, and biotechnology.

The historical evolution of ECM research underscores its transformative impact on biomedical sciences, from foundational discoveries elucidating ECM structure and function to contemporary applications driving innovation in regenerative medicine and beyond. By harnessing the principles of biomimicry and advancing technologies, researchers are poised to unlock the full therapeutic potential of ECM-based therapies, heralding a new era in personalized healthcare and biomedical engineering.

5.2 Importance of ECM in Tissue Engineering and Medical Applications

The Extracellular Matrix (ECM) plays a pivotal role in tissue engineering, serving as a foundational scaffold that mimics the natural microenvironment essential for cell growth, differentiation, and tissue regeneration (Fig. 5.1). In tissue engineering applications, ECM-derived biomaterials provide structural support and biochemical cues that guide cellular behavior, fostering tissue repair and regeneration. These biomimetic scaffolds mimic the native ECM's composition of proteins (e.g., collagen, elastin) and glycosaminoglycans, creating an environment conducive to cell adhesion, migration, and proliferation. By closely replicating the ECM's biochemical and biophysical properties, tissue engineers aim to develop implantable constructs capable of integrating seamlessly with host tissues, offering potential solutions for organ replacement and repair [5]. The ECM's importance in tissue engineering extends beyond structural support to include its role in modulating cellular functions and tissue development. ECM components act as reservoirs for growth factors, cytokines, and other bioactive molecules that regulate cell signaling pathways crucial for tissue morphogenesis and homeostasis. By presenting these biochemical cues in a spatially and temporally controlled manner, ECM-based constructs can promote specific cellular responses, such as stem cell differentiation into specialized cell types or the recruitment of progenitor cells to sites of injury. This bioactivity makes ECM-derived biomaterials invaluable tools for engineering complex tissues and organs with functional integrity and physiological relevance. In medical applications, ECM-based therapies have shown promise in enhancing wound healing, particularly in chronic and non-healing wounds where traditional treatments often fall short. ECM dressings and matrices provide a biocompatible platform that supports cellular infiltration and tissue remodeling, accelerating the closure of wounds and reducing scar formation. Moreover, ECM-derived products have been utilized in surgical procedures to reinforce weakened tissues, such as hernia repairs, where they promote tissue integration and reduce the risk of recurrence. These applications highlight the versatility of ECM-based biomaterials in clinical settings, offering solutions that harness the body's natural regenerative capacity to improve patient outcomes.

ECM's inherent biocompatibility and ability to modulate the immune response are critical factors in its success as a biomaterial for medical applications. Unlike synthetic materials, ECM-derived scaffolds exhibit reduced immunogenicity and inflammatory responses, minimizing adverse reactions and supporting long-term tissue integration. This biocompatibility makes ECM-based therapies suitable for a wide range of clinical scenarios, from cardiovascular repair to neurological regeneration, where compatibility with host tissues and functional restoration are paramount [6]. Additionally, the natural degradation and remodeling of ECM scaffolds over time align with the body's healing processes, ensuring seamless integration and functional recovery in treated patients. As research continues to advance, ECM-based biomaterials are poised to revolutionize personalized medicine by offering

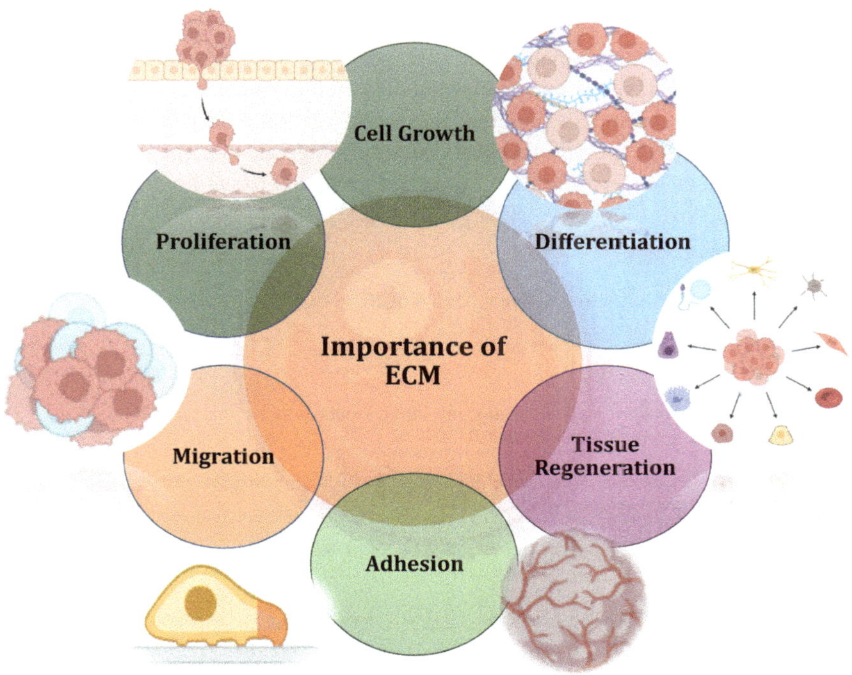

Fig. 5.1 Importance of ECM in tissue engineering and medical applications

tailored therapies that account for individual patient variability and disease complexities. By leveraging advances in biomaterial science and bioengineering, researchers can customize ECM scaffolds to mimic the unique biochemical and biomechanical properties of specific tissues, enhancing treatment efficacy and patient outcomes. These personalized approaches hold promise for addressing unmet clinical needs in areas such as tissue regeneration, organ transplantation, and disease modeling, where traditional therapies have limitations. Ultimately, the integration of ECM-based biomaterials into clinical practice represents a paradigm shift towards regenerative medicine strategies that harness the body's innate healing capacities to restore form and function in diseased or injured tissues.

The ECM's multifaceted role in tissue engineering and medical applications underscores its significance as a versatile biomaterial capable of promoting tissue regeneration, modulating cellular responses, and enhancing therapeutic outcomes.

5.3 Structural and Functional Properties of ECM

The Extracellular Matrix (ECM) serves as a complex, dynamic network that provides structural support and regulates cellular behavior within tissues. Structurally, ECM comprises a diverse array of macromolecules, including collagen, elastin, proteoglycans, and glycoproteins, organized into a three-dimensional architecture that varies across different tissues and developmental stages. This intricate composition not only imparts mechanical resilience to tissues but also facilitates cellular adhesion, migration, and signaling through its biochemical cues [7]. Functionally, ECM components interact with cells via specific receptors such as integrins, transmitting signals that regulate processes like proliferation, differentiation, and tissue homeostasis.

Exercise 1

Fill Ups
(Biochemical, Growth, Mechanical, Stiffness, Flexibility, 3D, Porosity)

1. Biomimetic approaches in ECM design aim to replicate the __________ and __________ properties of natural ECM to promote tissue regeneration.

2. Hydrogels are commonly used in biomimetic ECM design because they provide a __________ environment that supports cell growth and differentiation.

3. One of the main challenges in designing biomimetic ECM is achieving the correct __________ for optimal cellular interaction and tissue integration.

4. Incorporating __________ factors into biomimetic ECM designs can enhance cell signaling and tissue repair processes.

5. Biomimetic ECM scaffolds must balance __________ and __________ to support both structural integrity and cellular function.

5.3.1 *Composition and Functionality*

The composition of the natural Extracellular Matrix (ECM) is a complex interplay of diverse macromolecules, each playing crucial roles in tissue structure, function, and cellular interactions. This intricate matrix primarily consists of proteinaceous components and glycosaminoglycans (GAGs), which together create a supportive and dynamic environment essential for tissue integrity and function.

- **Proteinaceous Components**

Central to the ECM's structural framework are various proteins, with collagen being the most abundant and widely recognized. Collagens, particularly types I, II, and III, form fibrillar structures that provide tensile strength and elasticity to tissues like skin, bone, cartilage, and tendons. These proteins are characterized by their triple-helical structure and serve as the main scaffolding elements of the ECM, ensuring mechanical resilience and tissue organization. In addition to collagens, elastin is another crucial protein that imparts elasticity to tissues, enabling recoil and flexibility in structures such as blood vessels and lungs [8]. Fibronectin and laminins are adhesive glycoproteins that mediate cell attachment to the ECM, facilitating cell migration, differentiation, and tissue development.

- **Glycosaminoglycans (GAGs)**

Complementing the proteinaceous components are glycosaminoglycans (GAGs), long unbranched polysaccharide chains composed of repeating disaccharide units. These molecules are highly hydrophilic due to their negative charges and contribute to the ECM's hydration and viscosity. Among the key GAGs are hyaluronic acid (HA), chondroitin sulfate, dermatan sulfate, heparan sulfate, and keratan sulfate, each with specific roles in different tissues [8]. For instance, HA serves as a lubricant and shock absorber in joints and contributes to the ECM's viscoelastic properties. Chondroitin sulfate and heparan sulfate are involved in cell signaling and growth factor binding, regulating processes such as tissue repair and inflammation.

- **Organization and Localization**

The organization of ECM components varies significantly across tissues and developmental stages, reflecting specific functional requirements. In cartilage, for example, the ECM is rich in proteoglycans like aggrecan, which bind to hyaluronic acid and form aggregates that resist compressive forces. In contrast, bone ECM consists predominantly of collagen fibers mineralized with calcium phosphate, providing hardness and rigidity. The localization of ECM components within tissues is tightly regulated, with specific cells responsible for their synthesis, deposition, and remodeling in response to physiological cues and mechanical stresses.

- **Dynamic Remodeling and Turnover**

A hallmark of ECM biology is its dynamic remodeling and turnover, essential for tissue homeostasis, repair, and adaptation [9]. Cells such as fibroblasts, osteoblasts, chondrocytes, and endothelial cells actively synthesize and degrade ECM components in response to environmental signals and mechanical stimuli. This continuous remodeling process ensures that the ECM maintains structural integrity while accommodating tissue growth, repair, and regeneration. Dysregulation of ECM remodeling is implicated in various pathological conditions, including fibrosis, osteoarthritis, and cardiovascular diseases, highlighting the critical role of balanced ECM turnover in maintaining tissue health.

- **Interactions with Cells and Biochemical Signaling**

Beyond its structural role, the ECM serves as a dynamic interface for cellular communication and biochemical signaling. Cell surface receptors, such as integrins and discoidin domain receptors, interact with specific ECM ligands, transmitting mechanical and biochemical signals that regulate cellular behaviors such as adhesion, migration, proliferation, and differentiation. These interactions are crucial for tissue development, wound healing, immune response modulation, and maintaining tissue-specific functions. Moreover, ECM-bound growth factors and cytokines are sequestered and released in a spatiotemporal manner, influencing cellular responses and tissue repair processes.

- **Clinical Relevance and Biomedical Applications**

The composition and properties of natural ECM have profound implications for biomedical applications, particularly in tissue engineering, regenerative medicine, and therapeutic interventions. Biomaterials derived from ECM components or mimicking their biochemical cues are utilized to develop scaffolds, matrices, and implants that promote tissue repair and regeneration. By recapitulating the native ECM environment, these biomimetic materials enhance cell adhesion, proliferation, and differentiation, facilitating the reconstruction of damaged tissues and organs. Furthermore, ECM-based therapies hold promise for treating chronic wounds, osteoarthritis, myocardial infarction, and other conditions where ECM integrity and function are compromised.

Advancing our understanding of the composition, organization, and dynamics of natural ECM is essential for developing innovative therapies and improving clinical outcomes. Future research efforts aim to decipher ECM heterogeneity across tissues and species, elucidate its role in disease progression, and engineer ECM-based biomaterials with enhanced bioactivity and biocompatibility [10]. Addressing challenges such as scalability, immunogenicity, and regulatory approval will be crucial for translating ECM-based therapies from bench to bedside, ushering in a new era of personalized medicine and regenerative therapies harnessing the innate capabilities of the ECM.

Exercise 2

Fill Ups
(Growth factors, Stiffness, Biochemical, Proteins, Elasticity)

1. The __________ of biomimetic ECM affects how cells sense and respond to their environment.
2. Biochemical cues in biomimetic ECM often involve the incorporation of __________ to regulate cell behavior.
3. The __________ of biomimetic ECM plays a critical role in determining how well it can support tissue regeneration under mechanical stress.
4. Biomimetic ECMs are designed to mimic the natural __________ found in tissues, such as collagen and elastin.
5. __________ properties of biomimetic ECM help guide cellular migration, adhesion, and differentiation.

5.3.2 Functional Roles of ECM

The Extracellular Matrix (ECM) serves multifaceted functional roles in tissues, encompassing structural support, mechanical properties, and signaling through biochemical cues (Fig. 5.2). These functions are crucial for maintaining tissue integrity, regulating cellular behavior, and orchestrating physiological processes essential for organismal health and function.

- **Structural Support**

One of the primary functions of the ECM is to provide structural support to tissues and organs. The ECM's network of collagen and elastin fibers forms a scaffold that determines tissue architecture and mechanical stability. For instance, in connective tissues like tendons and ligaments, collagen fibers align to resist tensile forces, ensuring the integrity of joints and facilitating movement. In bones, the ECM's mineralized matrix of collagen and calcium phosphate confers rigidity and strength, supporting the body's skeletal framework and protecting vital organs.

- **Mechanical Properties**

Beyond structural support, the ECM contributes to tissue biomechanics by imparting mechanical properties such as elasticity, stiffness, and resilience [11]. Elastin fibers, for example, provide tissues like blood vessels and lungs with elasticity, allowing them to stretch and recoil with each heartbeat or breath. The composition and organization of collagen fibers in skin and cartilage determine tissue stiffness and resistance to deformation, crucial for maintaining skin integrity and joint function. These mechanical properties are essential for tissue adaptation to physiological forces and mechanical loading, ensuring functional performance and resilience under varying conditions.

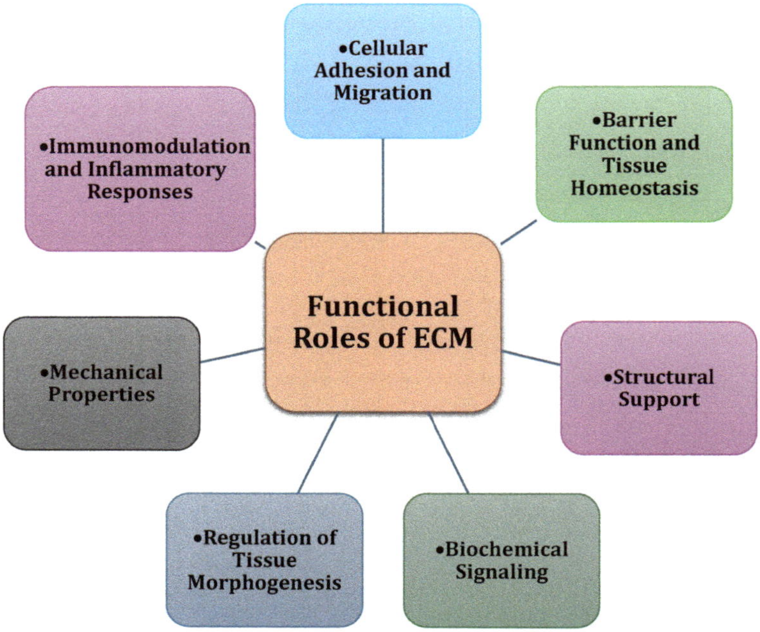

Fig. 5.2 Different functional roles of ECM

- **Cellular Adhesion and Migration**

The ECM plays a pivotal role in regulating cellular behavior through its ability to support cell adhesion, migration, and spatial organization. Cell surface receptors, including integrins and proteoglycans, interact with specific ECM ligands to anchor cells within tissues and facilitate their movement. Integrins, for instance, mediate cell-ECM adhesion and transmit mechanical signals that influence cytoskeletal dynamics and cellular responses. This interaction is critical during processes such as embryonic development, wound healing, and immune response, where cell migration guided by ECM gradients directs tissue morphogenesis and repair.

- **Biochemical Signaling**

Biochemical cues embedded within the ECM regulate diverse cellular functions by modulating signal transduction pathways and gene expression. Growth factors, cytokines, and other bioactive molecules are sequestered within the ECM matrix, where they bind to specific ECM components and receptors to regulate cell behavior. For example, fibroblast growth factors (FGFs) bound to heparan sulfate proteoglycans in the ECM promote cell proliferation and angiogenesis during tissue repair. Similarly, transforming growth factor-beta (TGF-β) complexes with ECM proteins regulate processes such as cell differentiation and matrix deposition, influencing tissue development and homeostasis.

- **Regulation of Tissue Morphogenesis**

During development and tissue regeneration, the ECM acts as a dynamic scaffold that guides tissue morphogenesis and organogenesis. Spatial and temporal gradients of ECM components and signaling molecules direct cell fate determination, tissue patterning, and organ formation [12]. For instance, the basement membrane ECM in epithelial tissues provides a structural template for cell polarization and differentiation, essential for the formation of functional barriers and glandular structures. Moreover, ECM-derived cues orchestrate stem cell niches in adult tissues, maintaining stem cell quiescence or promoting their activation and differentiation in response to injury or physiological demands.

- **Barrier Function and Tissue Homeostasis**

In epithelial and endothelial tissues, the ECM serves as a protective barrier that regulates molecular transport, fluid balance, and interactions with the external environment. The basement membrane, a specialized ECM structure underlying epithelial and endothelial layers, acts as a selective filter that regulates nutrient uptake, waste removal, and cell–cell communication. Disruptions in ECM barrier function, such as in diseases like cancer or inflammatory disorders, compromise tissue homeostasis and contribute to disease progression by altering cell-ECM interactions and signaling pathways.

- **Immunomodulation and Inflammatory Responses**

The ECM plays a critical role in modulating immune responses and inflammatory processes through interactions with immune cells and cytokine signaling. ECM components and fragments can act as damage-associated molecular patterns (DAMPs) that activate immune cells and initiate inflammatory responses. Conversely, ECM remodeling and degradation products can suppress inflammation and promote tissue repair by modulating immune cell recruitment, phenotype switching, and cytokine production [13]. Dysregulation of ECM-immune interactions contributes to chronic inflammation, fibrosis, and autoimmune diseases, highlighting the ECM's role in immune surveillance and tissue homeostasis.

- **Clinical Implications and Therapeutic Potential**

Understanding the ECM's functional roles in health and disease has profound clinical implications for tissue engineering, regenerative medicine, and therapeutic interventions. ECM-derived biomaterials and scaffolds are utilized in tissue engineering approaches to mimic native tissue microenvironments, promoting cell adhesion, proliferation, and differentiation for tissue repair and regeneration. Targeting ECM signaling pathways and interactions holds promise for developing novel therapies that modulate tissue remodeling, enhance wound healing, and mitigate pathological processes associated with aging, injury, or disease. By harnessing the ECM's inherent properties and bioactivity, researchers and clinicians are advancing towards personalized medicine approaches that optimize treatment outcomes and improve patient quality of life.

The ECM's multifaceted functional roles in structural support, mechanical properties, cellular signaling, and tissue morphogenesis underscore its importance in maintaining tissue integrity, regulating physiological processes, and supporting therapeutic interventions across diverse biomedical applications.

5.4 Biomimetic Approaches in ECM Design

Biomimetic approaches in ECM (Extracellular Matrix) design represent a transformative strategy in biomedical engineering aimed at replicating and harnessing the intricate properties and functions of natural ECM for therapeutic and regenerative applications. By drawing inspiration from biological systems, biomimetic ECM design seeks to develop synthetic materials and scaffolds that mimic the structural organization, biochemical composition, and dynamic interactions found in native tissues. These approaches integrate advanced materials science, bioengineering principles, and molecular biology to create biomaterials capable of promoting cell adhesion, proliferation, differentiation, and tissue regeneration in a controlled and biocompatible manner [14]. Through biomimetic ECM design, researchers aim to bridge the gap between synthetic biomaterials and natural tissues, offering innovative solutions to address complex medical challenges such as tissue repair, organ transplantation, and personalized regenerative therapies (Fig. 5.3).

5.4.1 Principles of Biomimicry

Biomimicry, as applied to Extracellular Matrix (ECM) design, encompasses a set of principles that emulate natural biological systems to create synthetic materials and scaffolds capable of replicating ECM architecture, organization, and biological signaling gradients. These principles leverage insights from how ECM components interact within tissues to guide cellular behavior, tissue morphogenesis, and homeostasis. By mimicking these natural frameworks, biomimetic ECM designs aim to enhance tissue engineering, regenerative medicine, and therapeutic interventions.

- **Understanding ECM Architecture and Organization**

Central to biomimetic ECM design is the replication of ECM architecture observed in native tissues. The ECM's intricate organization of collagen fibers, elastin networks, and glycosaminoglycan matrices varies across tissues, imparting specific mechanical

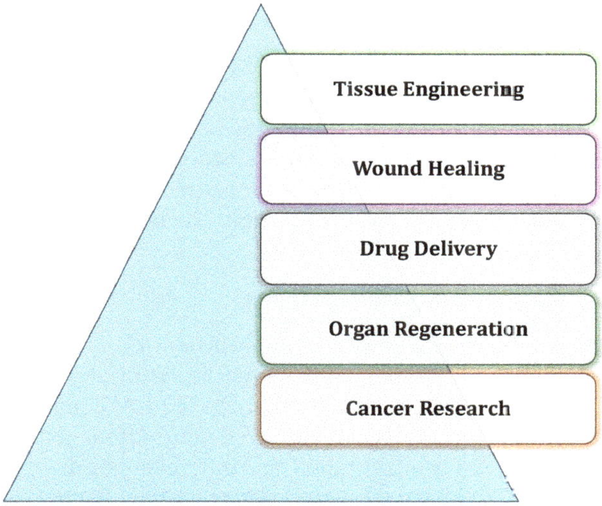

Fig. 5.3 Applications of biomimetic approaches in ECM design

properties and structural support [15]. Biomimetic approaches strive to recreate these hierarchical structures using advanced biomaterials and fabrication techniques. For example, electrospinning and 3D printing technologies enable the creation of fibrous scaffolds that mimic the fibrous architecture of ECM, facilitating cell adhesion and tissue integration.

- **Integrating Biological Signals**

Biomimetic ECM design involves integrating biological signals embedded within the ECM to regulate cellular responses and tissue regeneration. ECM components, such as growth factors, cytokines, and extracellular vesicles, are sequestered within the matrix and released in a spatiotemporal manner to modulate cell behavior [16]. Biomaterials can be engineered to incorporate these signaling molecules or mimic their presentation to cells, promoting specific cellular activities such as migration, proliferation, and differentiation. This integration of biological signals enhances the biomimetic scaffold's ability to guide tissue development and repair processes.

- **Replicating ECM Dynamics and Remodeling**

Natural ECM undergoes continuous remodeling in response to physiological cues and mechanical stimuli, crucial for tissue homeostasis and adaptation. Biomimetic ECM designs aim to replicate this dynamic remodeling by developing materials that degrade, remodel, and interact with cells in a manner similar to native ECM. Degradable polymers, hydrogels, and nanomaterials designed to mimic ECM turnover allow for controlled release of bioactive molecules and facilitate tissue regeneration without adverse inflammatory responses.

- **Biofunctionalization for Cell Interaction**

To facilitate cell-ECM interactions, biomimetic ECM designs incorporate biofunctionalization strategies that enhance cell adhesion, migration, and differentiation. Surface modifications with cell-adhesive peptides (e.g., RGD motifs) or growth factor-binding domains mimic ECM protein interactions, promoting specific cellular responses. These biofunctionalized surfaces or scaffolds create microenvironments conducive to cell attachment and tissue formation, crucial for successful integration and function of engineered tissues or implants.

- **Creating Spatial and Temporal Gradients**

Natural ECM gradients of biochemical cues play a pivotal role in guiding tissue development and morphogenesis. Biomimetic approaches replicate these gradients by designing scaffolds with spatially defined variations in ECM composition, stiffness, or bioactive molecule distribution. Gradient-based scaffolds influence cell behavior, directing cell migration, tissue organization, and differentiation along specific paths. This spatial control over cellular responses enhances the biomimetic scaffold's ability to recapitulate complex tissue architectures and functions.

- **Regulation of Mechanical Properties**

Biomimetic ECM design also focuses on regulating mechanical properties to match those of native tissues, ensuring optimal function and durability of engineered constructs. By mimicking the elastic modulus, tensile strength, and viscoelastic behavior of ECM, biomaterials can withstand physiological forces and mechanical stresses encountered in vivo. Tunable biomaterials and composite structures allow for customization of mechanical properties to meet specific tissue requirements, enhancing the long-term stability and performance of engineered tissues.

- **Application in Disease Modeling and Drug Screening**

Beyond tissue engineering, biomimetic ECM designs find applications in disease modeling and drug screening platforms. Engineered ECM scaffolds can recreate disease-specific microenvironments to study pathological mechanisms, drug responses, and personalized treatment strategies. By incorporating patient-derived cells and disease biomarkers into biomimetic models, researchers can simulate complex disease states and evaluate therapeutic interventions in a physiologically relevant context.

Advancing biomimetic ECM design involves overcoming challenges such as scalability, reproducibility, and regulatory approval for clinical translation. Innovations in biomaterial synthesis, bio fabrication techniques, and biomanufacturing are critical to developing biomimetic scaffolds with enhanced biomimicry and therapeutic efficacy [17]. Moreover, interdisciplinary collaborations between biomaterial scientists, engineers, biologists, and clinicians are essential to harnessing the full potential of biomimetic ECM designs in addressing unmet medical needs and advancing personalized regenerative therapies. Principles of biomimicry in ECM design offer innovative strategies to replicate complex architecture, biological signals, and gradients found in natural ECM. By integrating these principles into biomaterials and scaffolds, researchers are paving the way for advancements in tissue engineering, regenerative medicine, and disease modeling, with the potential to revolutionize healthcare and improve patient outcomes.

Synthetic ECM materials and scaffolds represent a pivotal advancement in biomedical engineering, aiming to mimic and enhance the functionalities of natural Extracellular Matrix (ECM) for various applications in tissue engineering, regenerative medicine, and biomedical research. These synthetic materials are designed to replicate the structural, mechanical, and biochemical properties of native ECM while offering tunable characteristics and controlled release of bioactive molecules. Synthetic ECM materials are typically composed of biocompatible polymers such as polyethylene glycol (PEG), polylactic acid (PLA), poly (lactic-co-glycolic acid) (PLGA), and hydrogels derived from natural polymers like gelatin or alginate. These materials can be engineered with precise control over their physical and chemical properties, including degradation kinetics, mechanical stiffness, and surface topography, to mimic specific aspects of native ECM necessary for cellular interactions and tissue regeneration. One of the key features of synthetic ECM materials is their ability to replicate the mechanical properties of different tissues. Through modifications in polymer composition and crosslinking density, synthetic scaffolds can match the elastic modulus and tensile strength required for specific tissue applications. For example, hydrogels can be tailored to mimic the softness of brain tissue or the stiffness of bone, providing optimal mechanical support for cells to adhere, proliferate, and differentiate within engineered constructs.

Synthetic ECM materials are functionalized with bioactive molecules such as cell-adhesive peptides (e.g., RGD motifs), growth factors, and cytokines to enhance cellular interactions and promote tissue regeneration. These molecules are either incorporated directly into the polymer matrix or tethered onto scaffold surfaces to mimic ECM signaling cues that regulate cell behavior, including adhesion, migration, and differentiation [18]. This biofunctionalization enables precise control over cellular responses within synthetic ECM scaffolds, facilitating tissue-specific regeneration and repair. Synthetic ECM materials are designed to degrade and remodel in a controlled manner, mimicking the turnover and remodeling processes observed in natural ECM. Degradable polymers such as PLGA or PEG-based hydrogels undergo enzymatic or hydrolytic degradation, allowing for gradual release of encapsulated bioactive molecules and facilitating integration with host tissues. This dynamic degradation profile supports tissue remodeling and regeneration while minimizing inflammatory responses and promoting long-term functional outcomes. Synthetic ECM materials can be fabricated using a variety of techniques, including electrospinning, 3D bioprinting, and micro molding, to create scaffolds with precise architecture and spatial organization. Electrospinning produces nanofibrous structures that mimic the fibrous architecture of ECM, promoting cell adhesion and tissue ingrowth. 3D bioprinting allows for layer-by-layer deposition of biomaterials and cells to create complex, patient-specific constructs with controlled geometry and cell distribution. These fabrication methods enable the customization of synthetic ECM scaffolds for specific tissue types and clinical applications.

Synthetic ECM materials hold promise for a wide range of clinical applications, including wound healing, orthopedic repairs, cardiovascular tissue engineering, and drug delivery systems. In wound healing, synthetic scaffolds promote rapid tissue regeneration and wound closure by providing a supportive microenvironment for cell migration and proliferation. In orthopedics, biodegradable scaffolds facilitate bone regeneration and repair by mimicking the mineralized matrix of natural bone tissue. Furthermore, synthetic ECM-based drug delivery systems enable controlled release of therapeutic agents, enhancing treatment efficacy and minimizing systemic side effects in targeted therapies.

Despite their significant potential, synthetic ECM materials face challenges such as scalability, reproducibility, and long-term biocompatibility in clinical settings. Innovations in biomaterial design, fabrication techniques, and integration of advanced imaging and computational modeling are needed to optimize the performance and safety of synthetic ECM scaffolds. Future research directions include enhancing scaffold bio functionality, improving host tissue integration, and developing personalized regenerative therapies that harness the full potential of synthetic ECM materials to address complex medical needs and improve patient outcomes. Synthetic ECM materials represent a versatile platform in biomedical engineering, offering tailored solutions for tissue regeneration, drug delivery, and disease modeling. By replicating key features of natural ECM while providing flexibility in design and functionality, these materials pave the way for innovative therapies that promote healing, restore tissue function, and advance personalized medicine approaches.

5.5 Applications in Regenerative Medicine and Medical Devices

Applications in regenerative medicine and medical devices harness advanced biomaterials and scaffolds, inspired by the principles of tissue engineering and biomimicry, to address complex medical challenges and improve patient outcomes. These applications encompass a broad spectrum of therapeutic strategies aimed at restoring tissue function, promoting regeneration, and enhancing the integration of biomedical implants within the body (Fig. 5.4) [19]. From engineered tissues and organs to innovative drug delivery systems and diagnostic platforms, these technologies represent a frontier in personalized medicine, offering tailored solutions that mimic the natural environment of tissues and organs to support healing and long-term health.

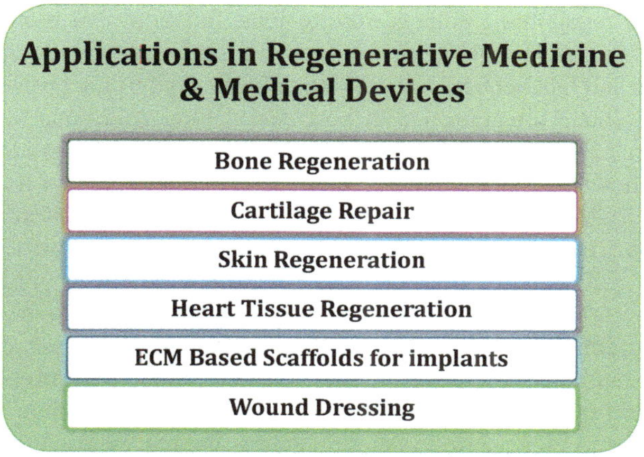

Fig. 5.4 Applications in regenerative medicine and medical devices

5.5.1 Tissue Engineering with ECM Mimics

Tissue engineering with ECM mimics represents a transformative approach in biomedical research aimed at creating functional tissues and organs for regenerative medicine applications. By replicating the intricate architecture, biochemical composition, and dynamic properties of natural Extracellular Matrix (ECM), these biomimetic scaffolds provide a supportive microenvironment that promotes cell adhesion, proliferation, differentiation, and tissue morphogenesis. ECM mimics in tissue engineering are designed to mimic the structural and functional characteristics of native ECM components. This includes using biocompatible polymers such as hydrogels, scaffolds, and nanofibers that replicate the mechanical properties and spatial organization observed in tissues like bone, cartilage, skin, and blood vessels. For example, collagen-based scaffolds can emulate the fibrous architecture of ECM in skin, facilitating cell attachment and promoting wound healing. Similarly, decellularized ECM matrices retain native tissue-specific biochemical cues and provide a natural substrate for cell colonization and tissue regeneration.

ECM mimics integrates biological signaling cues, including growth factors, cytokines, and cell-adhesive peptides, to regulate cellular responses and tissue development. These bioactive molecules are incorporated into scaffolds or delivered in a controlled manner to mimic the spatiotemporal gradients found in natural ECM. For instance, scaffolds functionalized with bone morphogenetic proteins (BMPs) promote osteogenic differentiation of mesenchymal stem cells, facilitating bone tissue regeneration. By mimicking ECM signaling pathways, biomimetic scaffolds direct cell behavior and support tissue-specific functions critical for successful integration and functionality of engineered tissues [20]. ECM mimics are applied in the regeneration of complex tissues and organs that require precise structural integrity and functional performance. Engineered ECM scaffolds are utilized in orthopedic applications to repair bone defects, restore joint function, and promote cartilage regeneration. In cardiovascular tissue engineering, scaffolds mimicking the mechanical properties and biochemical cues of native ECM support the growth of vascular tissues and facilitate integration with host vasculature. Additionally, ECM-based approaches are explored in liver, kidney, and neural tissue regeneration, aiming to restore organ function and mitigate the shortage of donor organs for transplantation.

The use of ECM mimics in tissue engineering offers several advantages for clinical translation. These biomaterials provide a biocompatible and bioactive platform that enhances cell viability, proliferation, and differentiation in vitro and in vivo [21]. By promoting tissue-specific functions and supporting angiogenesis and innervation, ECM mimics facilitate the formation of functional tissues with enhanced physiological relevance. Moreover, biomimetic scaffolds reduce immune rejection and improve bio integration when implanted in patients, minimizing adverse reactions and supporting long-term tissue remodeling and regeneration. Despite significant progress, challenges remain in the development and application of ECM mimics in tissue engineering. These include achieving optimal biomaterial properties, scalability of fabrication techniques, and regulatory considerations for clinical approval. Innovations in biomaterial design, such as bio fabrication technologies (e.g., 3D bioprinting), advanced imaging for scaffold characterization, and personalized medicine approaches, are advancing the field towards tailored therapies that address individual patient needs and specific tissue requirements.

Future research directions in tissue engineering with ECM mimics focus on enhancing scaffold bio functionality, improving vascularization of engineered tissues, and integrating bio responsive materials for controlled release of therapeutic agents. Advances in stem cell biology and genetic engineering also hold promise for developing personalized regenerative therapies that harness the full potential of ECM mimics to repair and regenerate damaged tissues. Moreover, interdisciplinary collaborations between scientists, engineers, clinicians, and regulatory bodies are essential for translating ECM-based technologies from bench to bedside, paving the way for transformative treatments in regenerative medicine and personalized healthcare.

Tissue engineering with ECM mimics represents a promising strategy for developing biomimetic scaffolds that recapitulate the complexity and functionality of native ECM. By leveraging biological cues and mimicking tissue-specific microenvironments, these advanced biomaterials facilitate the regeneration of tissues and organs, offering potential solutions for treating injuries, degenerative diseases, and congenital defects. As research progresses and technology advances, ECM mimics the key to revolutionizing regenerative medicine and enhancing quality of life for patients worldwide.

5.5.2 ECM Based Therapies

Extracellular Matrix (ECM)-based therapies represent a promising frontier in biomedical research, offering innovative approaches for wound healing, tissue repair, and organ regeneration. The ECM, a complex network of proteins, glycosaminoglycans, and bioactive molecules surrounding cells in tissues, plays a critical role in regulating cellular behavior and maintaining tissue homeostasis. Harnessing the regenerative potential of ECM components and mimicking their biochemical cues has led to the development of therapeutic strategies that enhance tissue repair and promote functional recovery in various medical conditions.

5.5.2.1 Role of ECM in Wound Healing

In wound healing, ECM-based therapies aim to create a supportive microenvironment that facilitates the natural healing process. ECM scaffolds, such as decellularized dermal matrices or synthetic hydrogels, provide a structural framework for cell migration, proliferation, and tissue remodeling at the wound site [22]. These scaffolds mimic the ECM's native architecture and biochemical composition, promoting angiogenesis, collagen deposition, and epithelialization to accelerate wound closure. Bioactive molecules embedded within ECM scaffolds, such as growth factors (e.g., PDGF, VEGF) and cytokines, further stimulate cellular responses and modulate inflammatory pathways to promote healing without excessive scar formation.

5.5.2.2 Applications in Bone and Cartilage Regeneration

In orthopedic and musculoskeletal medicine, ECM-based therapies are employed to regenerate bone and cartilage tissues damaged by trauma, disease, or aging. Scaffolds derived from ECM components like collagen and hyaluronic acid provide a conducive environment for osteogenic and chondrogenic differentiation of mesenchymal stem cells (MSCs) or progenitor cells. These scaffolds support tissue ingrowth and vascularization while promoting matrix deposition and mineralization, essential for restoring structural integrity and function in bone defects or cartilage lesions [23]. ECM-based approaches also enhance the integration of implants and grafts into native tissues, improving long-term outcomes for patients undergoing joint reconstruction or repair procedures.

5.5.2.3 Cardiovascular Applications

In cardiovascular medicine, ECM-based therapies hold promise for repairing damaged myocardium following myocardial infarction (heart attack) or restoring vascular function in ischemic conditions. ECM scaffolds are engineered to mimic the mechanical properties and biochemical cues of cardiac or vascular tissues promote cardiomyocyte survival, neovascularization, and functional recovery [24]. These scaffolds may be seeded with cardiac progenitor cells or stem cells to enhance tissue regeneration and improve heart function. Moreover, ECM-derived bioactive molecules can modulate inflammatory responses and promote tissue remodeling, supporting the repair of injured cardiovascular tissues and reducing the risk of heart failure.

5.5.2.4 Liver and Kidney Regeneration

Liver and kidney diseases present significant challenges due to limited organ availability for transplantation and the complexity of tissue regeneration. ECM-based therapies offer alternative approaches to stimulate hepatic or renal tissue repair and regeneration. ECM scaffolds seeded with hepatocytes or renal progenitor cells provide a supportive niche for cell attachment, proliferation, and metabolic function, facilitating liver or kidney regeneration in acute or chronic disease settings [25]. Bioengineered ECM matrices also modulate immune responses, promote angiogenesis, and integrate with host tissues to restore organ function and improve patient outcomes without the need for whole organ transplantation.

5.5.2.5 Neural Repair and Regeneration

In neurological disorders and spinal cord injuries, ECM-based therapies are explored to promote neural repair and functional recovery. ECM scaffolds are designed to mimic the neurogenic niche support neuronal survival, axonal growth, and synaptic connectivity in damaged or degenerative brain and spinal cord tissues. These scaffolds may incorporate neurotrophic factors, extracellular vesicles, or neural stem cells to enhance neuro regeneration and repair circuitry disrupted by injury or disease. ECM-derived molecules also modulate glial scar formation and inflammatory responses, promoting neural tissue remodeling and functional restoration in neurodegenerative conditions [26].

Despite promising advances, ECM-based therapies face challenges such as optimizing scaffold properties, enhancing host tissue integration, and ensuring long-term safety and efficacy in clinical settings. Innovations in biomaterial design, bio fabrication techniques, and personalized medicine approaches are critical for overcoming these challenges and advancing ECM-based therapies from preclinical research to clinical applications. Future research directions include enhancing scaffold bio functionality, improving vascularization of engineered tissues, and integrating bio responsive materials for controlled release of therapeutic agents tailored to individual patient needs. ECM-based therapies represent a versatile and promising approach for tissue engineering, wound healing, and organ regeneration across various medical disciplines. By harnessing the regenerative potential of ECM components and mimicking their biological cues, these therapies offer innovative solutions to address complex medical challenges and improve patient outcomes.

5.6 Biophysical and Biochemical Properties of Biomimetic ECM

Biomimetic ECM (Extracellular Matrix) refers to engineered materials designed to replicate the intricate biophysical and biochemical properties (Fig. 5.5) of natural ECM found in tissues. These biomaterials are crafted to mimic the structural organization, mechanical characteristics, and biochemical signaling cues essential for regulating cellular behaviors and tissue functions. Biophysically, biomimetic ECMs are tailored to resemble the physical architecture and mechanical properties of native ECM, such as elasticity, stiffness, and porosity, which vary across different tissue types like bone, skin, and cartilage. By integrating these biophysical features, biomimetic ECMs provide a supportive microenvironment that promotes cell adhesion, migration, proliferation, and differentiation akin to natural tissue matrices. Biochemically, these materials incorporate bioactive molecules, growth factors, and cell-binding motifs that mimic ECM signaling pathways to modulate cellular responses and facilitate tissue regeneration processes [27]. Thus, biomimetic ECMs represent a paradigm shift in biomedical engineering, offering versatile platforms for tissue engineering, regenerative medicine, and therapeutic interventions aimed at mimicking and enhancing the biological functionality of native ECM in clinical applications.

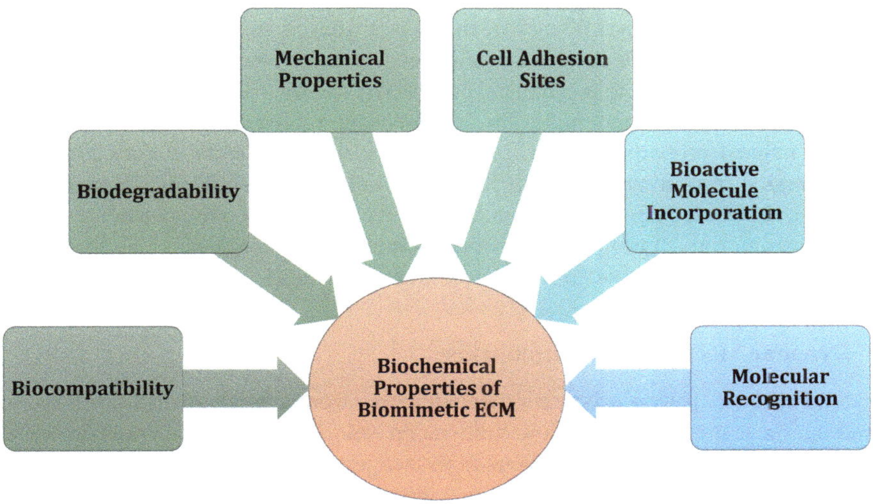

Fig. 5.5 Biochemical properties of biomimetic ECM

5.6.1 Mechanical Properties

Mechanical properties play a crucial role in tissue engineering, influencing cell behavior, tissue development, and the overall functionality of engineered constructs. These properties encompass a range of characteristics such as stiffness, elasticity, viscoelasticity, and strength, which are essential for mimicking the native tissue microenvironment and ensuring optimal performance of biomaterials in biomedical applications.

- **Characterization Methods**

Characterizing the mechanical properties of biomaterials and tissue constructs involves various techniques tailored to measure specific parameters. Common methods include tensile testing, where samples are subjected to controlled tension to measure their stress–strain relationship and determine properties like Young's modulus and ultimate tensile strength. Compression testing assesses material behavior under compressive forces, crucial for understanding load-bearing capabilities in applications such as cartilage or bone tissue engineering. Additionally, techniques like atomic force microscopy (AFM) and rheometry provide nanoscale and viscoelastic measurements, respectively, offering insights into material stiffness, adhesion forces, and response to dynamic loading conditions.

- **Significance in Tissue Engineering**

Mechanical properties are pivotal in tissue engineering as they dictate how cells interact with biomaterial scaffolds and influence tissue formation and maturation. Cells sense and respond to mechanical cues from their microenvironment, including substrate stiffness and topography, through mechano transduction pathways. Tuning the mechanical properties of scaffolds to match those of native tissues promotes cell adhesion, proliferation, and differentiation along tissue-specific lineages. For example, scaffolds designed for cardiac tissue regeneration require elasticity similar to myocardium to support cardiomyocyte contraction, while bone scaffolds necessitate stiffness comparable to cortical or trabecular bone for effective osteogenic differentiation.

- **Dynamic Mechanical Environments**

Natural tissues experience dynamic mechanical environments characterized by cyclic loading, shear stress, and fluid flow, influencing cellular behavior and tissue function. Mimicking these dynamic conditions in tissue engineering is crucial for enhancing biomaterial integration and promoting functional tissue regeneration. Bioreactors and mechanical stimulators apply controlled mechanical forces to engineered constructs, simulating physiological conditions and promoting tissue maturation. These dynamic environments enhance cell viability, matrix deposition, and vascularization within scaffolds, ensuring robust tissue formation and integration upon implantation.

- **Impact on Cell Mechanobiology**

Mechanical properties not only provide structural support but also regulate cellular mechanobiology, affecting cell morphology, cytoskeletal organization, and gene expression patterns. Cells sense substrate stiffness through integrin-mediated adhesions and adjust their behavior, accordingly, influencing processes such as proliferation, migration, and differentiation. Changes in mechanical cues can alter stem cell fate decisions, guiding lineage commitment towards specific tissue phenotypes. Understanding these mechano transduction pathways is essential for designing biomaterials that promote desired cellular responses and facilitate tissue regeneration in diverse physiological contexts.

- **Clinical Relevance and Biomaterial Design**

In clinical applications, matching the mechanical properties of biomaterial scaffolds to those of native tissues is critical for successful integration and long-term functionality. Deviations in scaffold stiffness or elasticity can lead to implant failure, inflammation, or impaired tissue regeneration. Biomaterials designed with tailored mechanical properties enhance biocompatibility, minimize host immune responses, and support tissue ingrowth and remodeling. For example, synthetic bone grafts with adjustable stiffness promote osteointegration and bone healing, while compliant vascular grafts mimic arterial elasticity to prevent thrombosis and promote vascular remodeling.

- **Challenges and Advances**

Challenges in tissue engineering include replicating the complex mechanical gradients and anisotropic properties found in native tissues, particularly in load-bearing structures like ligaments or tendons. Advances in biomaterial science and fabrication techniques, such as 3D printing and electrospinning, enable precise control over scaffold architecture and mechanical properties, facilitating the development of customized constructs for patient-specific applications. Moreover, bioinspired materials incorporating smart polymers and responsive hydrogels respond to physiological stimuli, adjusting their mechanical properties dynamically to support tissue regeneration and therapeutic outcomes.

- **Future Directions and Innovations**

Future research directions aim to refine characterization methods for biomaterial mechanical properties, enhancing accuracy and reproducibility in predicting in vivo performance. Innovations in biomimetic scaffolds and mechanobiology continue to drive advancements in tissue engineering, paving the way for personalized regenerative therapies that optimize scaffold mechanics to promote tissue-specific repair and regeneration. Integrating computational modeling and advanced imaging techniques further enhances our understanding of how mechanical cues influence cellular responses, guiding the design of next-generation biomaterials that mimic the complexity and functionality of native tissues.

Mechanical properties are fundamental in tissue engineering, influencing biomaterial design, cell mechanobiology, and tissue function in dynamic physiological environments. By harnessing these principles, researchers are advancing biomaterials that not only replicate the structural integrity of native tissues but also promote effective tissue regeneration and functional recovery in diverse clinical applications.

5.6.2 *Biochemical Signaling*

Biochemical signaling within ECM mimicry plays a pivotal role in regulating cellular behavior, differentiation, and tissue development, crucial for effective tissue engineering and regenerative medicine applications. These signaling mechanisms involve a complex interplay of growth factors, cytokines, and other bioactive molecules embedded within biomaterial scaffolds to replicate the dynamic biochemical cues present in natural Extracellular Matrix (ECM).

- **Role of Growth Factors**

Growth factors are potent signaling molecules that orchestrate cellular responses by binding to specific receptors on cell surfaces, triggering downstream signaling pathways that regulate processes such as cell proliferation, migration, and differentiation. In ECM mimicry, growth factors like transforming growth factor-beta (TGF-β), fibroblast growth factors (FGFs), and platelet-derived growth factors (PDGFs) are incorporated into scaffolds to promote tissue-specific differentiation of stem cells or progenitor cells. For instance, TGF-β stimulates chondrogenic differentiation in cartilage tissue engineering, while PDGF enhances angiogenesis and wound healing responses in vascular and dermal applications.

- **Cytokines and Chemokines**

Cytokines and chemokines within ECM mimics serve as mediators of immune responses, inflammation, and tissue repair processes [28]. These small proteins regulate cell-to-cell communication and immune cell activation, influencing tissue remodeling and regeneration. In biomaterial scaffolds, cytokines such as interleukins (ILs) and tumor necrosis factors (TNFs) modulate inflammatory pathways and promote tissue healing by recruiting immune cells and facilitating tissue remodeling. Chemokines play a crucial role in guiding cell migration and homing to specific sites within engineered tissues, aiding in the integration and functionality of implanted constructs.

- **Bioactive Peptides and ECM Proteins**

Bioactive peptides derived from ECM proteins like collagen, fibronectin, and laminin mimic cell-binding motifs and signaling domains critical for cell adhesion and differentiation. These peptides are integrated into biomaterial scaffolds to enhance cellular interactions, promote matrix deposition, and facilitate tissue-specific regeneration. For example, RGD (arginine-glycine-aspartate) peptides mimic ECM protein interactions with integrin receptors, promoting cell adhesion and spreading necessary for tissue formation in bone, skin, and cardiovascular applications.

- **Temporal and Spatial Control**

The spatiotemporal presentation of biochemical signals within ECM mimics is critical for orchestrating cellular behaviors and tissue morphogenesis. Biomaterials are engineered to release growth factors and bioactive molecules in a controlled manner, mimicking the natural gradients and concentrations observed in native ECM. This temporal control ensures that cells receive signals in a sequence and dosage suitable for guiding their proliferation, differentiation, and tissue-specific functions. Techniques such as microencapsulation, nanoparticle delivery systems, and scaffold functionalization enable precise modulation of biochemical signaling dynamics within engineered tissues.

- **Regulation of Cellular Differentiation**

Biochemical signaling in ECM mimicry regulates cellular differentiation by influencing gene expression profiles and lineage commitment of stem cells or progenitor cells seeded within biomaterial scaffolds. Growth factors and cytokines activate specific transcription factors and signaling cascades that drive cellular fate decisions towards desired phenotypes, such as osteoblasts in bone tissue engineering or cardiomyocytes in cardiac regeneration. The integration of multiple signaling cues within biomaterials enhances the efficiency and specificity of differentiation protocols, ensuring the generation of functional tissues with physiological relevance and therapeutic efficacy.

- **Clinical Applications and Therapeutic Strategies**

In clinical applications, biochemical signaling strategies within ECM mimics are tailored to address specific tissue repair and regeneration challenges. Scaffolds designed with growth factor gradients promote vascularization and tissue integration in orthopedic implants or wound healing dressings. Bioactive peptide-functionalized matrices enhance neuronal differentiation and neurite outgrowth in neural tissue engineering [29]. Moreover, personalized medicine approaches utilize patient-derived cells and tailored biochemical cues to optimize therapeutic outcomes and promote tissue-specific repair in personalized regenerative therapies.

- **Challenges and Future Directions**

Challenges in utilizing biochemical signaling within ECM mimics include maintaining bioactivity during scaffold fabrication, achieving sustained release profiles, and optimizing signal stability in complex physiological environments. Advances in biomaterial science, nanotechnology, and controlled-release strategies are addressing these challenges by enhancing the bioavailability and efficacy of growth factors and cytokines within engineered tissues. Future research directions aim to further refine biochemical signaling approaches, integrate multi-functional biomaterial platforms, and elucidate novel signaling pathways to advance the field of tissue engineering and regenerative medicine. Biochemical signaling within ECM mimics represents a sophisticated approach to mimic and harness the regulatory mechanisms of natural ECM in tissue engineering and regenerative medicine. By leveraging growth factors, cytokines, and bioactive peptides, researchers are advancing biomaterials that orchestrate cellular behaviors, promote tissue-specific differentiation, and facilitate functional tissue regeneration in diverse clinical applications.

Exercise 3

Multiple Choice Questions

1. What is the primary role of the extracellular matrix (ECM) in tissue engineering?
a) Structural support
b) DNA replication
c) Immune defense
d) Oxygen transport

2. Which of the following is a major component of natural ECM used in medical applications?
a) Collagen
b) Hemoglobin
c) Insulin
d) Glucose

3. How does biomimicry of ECM contribute to regenerative medicine?
a) By promoting tissue destruction
b) By replicating natural healing environments
c) By preventing cell migration
d) By blocking growth factor signaling

4. In what way do synthetic ECM biomaterials promote cell adhesion?
a) By increasing immune response
b) By mimicking biochemical cues of natural ECM
c) By decreasing surface area
d) By blocking receptor-ligand interactions

5. Which property of the ECM is crucial for mimicking it in artificial scaffolds?
a) Electrical conductivity
b) Porosity and mechanical strength
c) Color and texture
d) Magnetic properties

6. What is a common challenge in replicating the ECM for biomedical purposes?
a) Replicating the exact genetic code
b) Achieving the right biomechanical and biochemical properties
c) Preventing bacterial growth
d) Increasing thermal conductivity

7. How do hydrogels mimic ECM in tissue regeneration?
a) By being a liquid
b) By providing a 3D network for cell growth
c) By dissolving in the body
d) By blocking nutrient flow

8. Which type of cells are typically influenced by ECM biomimicry in wound healing?
a) Erythrocytes
b) Fibroblasts
c) Neurons
d) Adipocytes

9. Why is the mechanical stiffness of ECM biomaterials important in tissue repair?
a) It regulates cell differentiation and proliferation
b) It increases the speed of blood clotting
c) It enhances immune system response
d) It minimizes inflammation

5.7 Current Limitations and Challenges

Addressing the challenges and advancing the future directions of ECM-based therapies and biomimetic scaffolds in tissue engineering and regenerative medicine involves navigating several critical areas that impact their clinical applicability and efficacy:

- **Immunogenicity and Biocompatibility**: One of the primary challenges is minimizing immune responses and ensuring long-term biocompatibility of ECM-derived materials. Variability in immune reactions to ECM components, especially in decellularized matrices or biomaterial implants, requires thorough characterization and modification to mitigate adverse immune responses and promote tissue integration.
- **Scalability of Production**: Scaling up the production of biomimetic ECM scaffolds from laboratory-scale to clinical applications poses significant challenges. Standardizing manufacturing processes while maintaining scaffold quality and consistency is essential for widespread adoption and commercial viability. Innovations in bioprocessing techniques and automation are crucial for achieving scalable production without compromising scaffold performance.
- **Clinical Translation and Regulatory Approval**: Moving from preclinical studies to clinical trials involves navigating regulatory pathways and demonstrating safety, efficacy, and reproducibility of ECM-based therapies. Meeting stringent regulatory requirements and conducting robust clinical trials are necessary steps to gain approval for clinical use and ensure patient safety.
- **Mechanical and Structural Integrity**: Achieving mechanical properties that match those of native tissues remains a challenge in tissue engineering. Developing biomaterial scaffolds with tunable stiffness, elasticity, and durability is essential for supporting tissue-specific functions and withstanding physiological loads over extended periods.
- **Biofunctionalization and Controlled Release**: Enhancing the bioactivity and controlled release of growth factors, cytokines, and other bioactive molecules within ECM scaffolds is crucial for guiding cellular responses and promoting tissue regeneration. Developing strategies for sustained release and spatial control of biochemical signals can optimize therapeutic outcomes and tissue integration.
- **Integration with Host Tissues**: Promoting seamless integration between biomimetic scaffolds and host tissues is critical for achieving functional tissue regeneration. Strategies to enhance vascularization, minimize fibrotic responses, and support tissue remodeling are essential for improving long-term scaffold performance and patient outcomes.
- **Cost-effectiveness and Accessibility**: Addressing the cost-effectiveness and accessibility of ECM-based therapies is vital for their widespread clinical adoption. Innovations in biomaterial design, manufacturing efficiency, and therapeutic efficacy are needed to reduce production costs and increase accessibility to patients in diverse healthcare settings.

- **Personalized Medicine and Biomarker Development**: Advancing personalized medicine approaches involves tailoring ECM-based therapies to individual patient needs based on genetic, physiological, and disease-specific factors. Biomarker identification and validation are essential for predicting patient responses to treatment and optimizing therapeutic outcomes in personalized regenerative medicine strategies.

In summary, overcoming these challenges and advancing future directions in ECM-based therapies requires interdisciplinary collaboration, technological innovation, and rigorous translational research efforts. Addressing immunogenicity concerns, enhancing scalability, navigating regulatory pathways, optimizing scaffold properties, and advancing personalized medicine approaches are critical steps toward realizing the full potential of biomimetic ECM scaffolds in clinical applications for tissue engineering and regenerative medicine.

5.8 Summary

The exploration of biomimicry in Extracellular Matrix (ECM) for medical applications reveals a promising frontier in biomedical research and regenerative medicine. Throughout this chapter, we have delved into the structural, biochemical, and mechanical intricacies of ECM and how mimicking these characteristics can profoundly impact tissue engineering and therapeutic strategies. Firstly, understanding the complex composition and organization of natural ECM has underscored the importance of biomimicry in designing scaffolds and biomaterials that emulate the native tissue microenvironment. By incorporating ECM-derived components such as collagen, glycosaminoglycans, and bioactive molecules like growth factors and cytokines, researchers can create scaffolds that not only provide structural support but also regulate cellular behaviors crucial for tissue regeneration and repair. Secondly, the role of biochemical signaling in ECM mimicry has been pivotal. By mimicking the spatial and temporal presentation of growth factors and cytokines within native ECM, biomimetic scaffolds can guide cell adhesion, proliferation, differentiation, and tissue-specific function. This biochemical mimicry is essential for enhancing therapeutic outcomes in wound healing, organ regeneration, and orthopedic interventions, where precise control over cellular responses is paramount.

Thirdly, the mechanical properties of biomimetic ECM have emerged as a critical determinant of scaffold functionality and integration with host tissues. Mimicking the elasticity, stiffness, and dynamic mechanical environments of natural ECM is essential for promoting cell mechano transduction, supporting tissue maturation, and ensuring long-term biomechanical stability of engineered constructs. Addressing the challenges and future directions in ECM-based therapies—such as immunogenicity, biocompatibility concerns, scalability, and clinical translation—highlights ongoing efforts to optimize scaffold design, manufacturing processes, and regulatory pathways. Innovations in bioprocessing, controlled release systems, and personalized medicine approaches are paving the way for more effective and accessible treatments tailored to individual patient needs. Biomimetic ECM holds immense promise for advancing regenerative medicine, personalized therapies, and biomedical innovations. By harnessing the biological insights and engineering principles of natural ECM, researchers are poised to revolutionize medical treatments, offering new hope for patients suffering from tissue damage, degenerative diseases, and congenital defects.

Points to Remember

- **ECM mimics in tissue engineering are inspired by nature's intricate design, aiming to replicate the structural and functional properties of natural ECM found in tissues like bone, cartilage, and skin.**
- **Biomimetic ECM scaffolds can dynamically adapt to changes in their environment, responding to biochemical and mechanical cues to promote tissue regeneration and repair.**
- **Incorporating growth factors and cytokines into ECM mimics enhances their bioactivity, guiding cellular behaviors crucial for tissue-specific differentiation and functional integration.**
- **Advances in biomimetic ECM technologies are driving personalized medicine approaches, tailoring treatments to individual patient needs based on genetic, physiological, and disease-specific factors.**
- **ECM-based therapies offer regenerative potential across diverse medical fields, from wound healing and orthopedics to cardiovascular and neural regeneration, addressing critical healthcare challenges.**
- **Innovations in biomaterial fabrication techniques, such as 3D bioprinting and electrospinning, enable precise control over scaffold architecture and mechanical properties, enhancing therapeutic efficacy.**
- **Bridging the gap between laboratory research and clinical application, biomimetic ECM technologies are progressing through rigorous testing and regulatory processes to reach patients worldwide, offering new hope in the treatment of complex medical conditions.**

References

1. Yue B (2014) Biology of the extracellular matrix: an overview. J Glaucoma 23(8 Suppl 1), S20–S23. https://doi.org/10.1097/IJG.0000000000000108
2. Satir P (2005) Tour of organelles through the electron microscope: a reprinting of Keith R. Porter's classic Harvey Lecture with a new introduction. Anatomical Record Part A Discov Mol Cell Evolution Biol 287A(2):1184–1204. https://doi.org/10.1002/ar.a.20222
3. Naba A (2023) Ten years of extracellular matrix proteomics: accomplishments, challenges, and future perspectives. Mol Cell Proteomics 22(4):100528. https://doi.org/10.1016/j.mcpro.2023.100528
4. Muse S (2005) Genomics and bioinformatics. In: Introduction to biomedical engineering, pp 799–831. https://doi.org/10.1016/B978-0-12-238662-6.50015-X

5. Kim Y, Ko H, Kwon IK, Shin K (2016) Extracellular matrix revisited: roles in tissue engineering. Int Neurourol J 20(Suppl 1):S23–S29. https://doi.org/10.5213/inj.1632600.318
6. Rosso F, Giordano A, Barbarisi M, Barbarisi A (2004) From cell-ECM interactions to tissue engineering. J Cell Physiol 199(2):174–180. https://doi.org/10.1002/jcp.10471
7. Karamanos NK (2019) Extracellular matrix: key structural and functional meshwork in health and disease. FEBS J 286(15):2826–2829. https://doi.org/10.1111/febs.14992
8. Karamanos NK, Theocharis AD, Piperigkou Z, Manou D, Passi A, Skandalis SS, Vynios DH, Orian-Rousseau V, Ricard-Blum S, Schmelzer CEH, Duca L, Durbeej M, Afratis NA, Troeberg L, Franchi M, Masola V, Onisto M (2021) A guide to the composition and functions of the extracellular matrix. FEBS J 288(24):6850–6912. https://doi.org/10.1111/febs.15776
9. Afshar K, Sanaei MJ, Ravari MS, Pourbagheri-Sigaroodi A, Bashash D (2023) An overview of extracellular matrix and its remodeling in the development of cancer and metastasis with a glance at therapeutic approaches. Cell Biochem Funct 41(8):930–952. https://doi.org/10.1002/cbf.3846
10. Khalili A, Ahmad M (2015) A review of cell adhesion studies for biomedical and biological applications. Int J Mol Sci 16(8):18149–18184. https://doi.org/10.3390/ijms160818149
11. Sanz-Fraile H, Otero J (2024) Mechanical properties of the extracellular matrix. In: Maia FRA, Oliveira JM, Reis RL (eds) Handbook of the extracellular matrix. Springer, Cham. https://doi.org/10.1007/978-3-030-92090-6_5-1
12. Pally D, Naba A (2024) Extracellular matrix dynamics: A key regulator of cell migration across length-scales and systems. Curr Opin Cell Biol 86:102309. https://doi.org/10.1016/j.ceb.2023.102309
13. Riley HJ, Bradshaw AD (2020) The influence of the extracellular matrix in inflammation: findings from the SPARC-Null Mouse. Anatom Record (Hoboken, N.J.: 2007) 303(6):1624–1629. https://doi.org/10.1002/ar.24133
14. Levett PA, Melchels FP, Schrobback K, Hutmacher DW, Malda J, Klein TJ (2013) A biomimetic extracellular matrix for cartilage tissue engineering centered on photocurable gelatin, hyaluronic acid and chondroitin sulfate. Acta Biomater 10(1):214–223. https://doi.org/10.1016/j.actbio.2013.10.005
15. Sá AAM, Viana DM (2023) Design and biomimicry: a review of interconnections and creative potentials. Biomimetics (Basel, Switzerland) 8(1):61. https://doi.org/10.3390/biomimetics8010061
16. Pellowe AS, Gonzalez AL (2015) Extracellular matrix biomimicry for the creation of investigational and therapeutic devices. Wiley Interdiscip Rev Nanomed Nanobiotechnol 8(1):5–22. https://doi.org/10.1002/wnan.1349
17. Jung YH, Park K, Kim M et al (2023) Development of an extracellular matrix plate for drug screening using patient-derived tumor organoids. BioChip J 17:284–292. https://doi.org/10.1007/s13206-023-00099-y
18. Ligorio C, Mata A (2023) Synthetic extracellular matrices with function-encoding peptides. Nat Rev Bioeng 1:518–536. https://doi.org/10.1038/s44222-023-00055-3
19. Turner NJ (2021) Special issue: application of extracellular matrix in regenerative medicine. Appl Sci 11(7):3262. https://doi.org/10.3390/app11073262
20. Wang Y, Du M, Wu T, Su T, Ai L, Jiang D (2023) The application of ECM-derived biomaterials in cartilage tissue engineering. Mechanobiol Med 1(1):100007. https://doi.org/10.1016/j.mbm.2023.100007
21. Nicolas J, Magli S, Rabbachin L, Sampaolesi S, Nicotra F, Russo L (2020) 3D Extracellular matrix mimics: fundamental concepts and role of materials chemistry to influence stem cell fate. Biomacromol 21(6):1968–1994. https://doi.org/10.1021/acs.biomac.0c00045
22. Agren MS, Werthén M (2007) The extracellular matrix in wound healing: a closer look at therapeutics for chronic wounds. Int J Low Extrem Wounds 6(2):82–97. https://doi.org/10.1177/1534734607301394
23. Lin X, Patil S, Gao YG, Qian A (2020) The bone extracellular matrix in bone formation and regeneration. Front Pharmacol 11:757. https://doi.org/10.3389/fphar.2020.00757

24. Bloksgaard M, Lindsey M, Martinez-Lemus LA (2018) Extracellular matrix in cardiovascular pathophysiology. Am J Physiol Heart Circ Physiol 315(6):H1687–H1690. https://doi.org/10.1152/ajpheart.00631.2018
25. McQuitty CE, Williams R, Chokshi S, Urbani L (2020) Immunomodulatory role of the extracellular matrix within the liver disease microenvironment. Front Immunol 11:574276. https://doi.org/10.3389/fimmu.2020.574276
26. Chelyshev YA, Kabdesh IM, Mukhamedshina YO (2022) Extracellular matrix in neural plasticity and regeneration. Cell Mol Neurobiol 42(3):647–664. https://doi.org/10.1007/s10571-020-00986-0
27. Tian H, Shi H, Yu J, Ge S, Ruan J (2022) Biophysics role and biomimetic culture systems of ECM stiffness in cancer EMT. Global Challenges 6(6). https://doi.org/10.1002/gch2.202100094
28. da Silva PHR, Borges BC, Uehara IA, Soldi LR, de Araújo RA, Silva MJB (2021) Chemokines and the extracellular matrix: Set of targets for tumor development and treatment. Cytokine 144:155548. https://doi.org/10.1016/j.cyto.2021.155548
29. Capella-Monsonís H, Crum RJ, Hussey GS, Badylak SF (2024) Advances, challenges, and future directions in the clinical translation of ECM biomaterials for regenerative medicine applications. Adv Drug Deliv Rev 211:115347. https://doi.org/10.1016/j.addr.2024.115347

Chapter 6
ECM in Diagnostic Innovations

6.1 Introduction

The extracellular matrix (ECM) is a complex and dynamic network of proteins, glycoproteins, proteoglycans, and polysaccharides that provides structural and biochemical support to surrounding cells. Beyond its traditional role as a mere scaffold, the ECM actively regulates cellular functions, including adhesion, migration, proliferation, and differentiation. The ECM's composition and structure vary across tissues, reflecting its specialized roles in different organs [1]. The highly organized nature of the ECM contributes to the maintenance of tissue homeostasis, while its dysfunction is often associated with various pathological conditions, including cancer, fibrosis, and cardiovascular diseases. Understanding the ECM's intricacies offers valuable insights into its pivotal role in tissue function and disease progression, setting the stage for its use in diagnostic applications. The ECM's structure is primarily composed of fibrous proteins like collagen and elastin, glycoproteins such as fibronectin and laminin, and proteoglycans that provide resilience and elasticity. These components form a highly organized network that interacts with cell surface receptors, particularly integrins, to influence cellular behavior. The ECM is not static; it undergoes continuous remodeling through the balanced activity of matrix metalloproteinases (MMPs) and their inhibitors, which regulate the degradation and synthesis of ECM components. This dynamic remodeling is crucial in wound healing and tissue repair but can also contribute to disease when dysregulated. For example, abnormal ECM remodeling is a hallmark of cancer, where it facilitates tumor progression and metastasis, making it a valuable target for diagnostic exploration.

In diagnostics, the ECM's role extends beyond its structural function, as alterations in ECM composition and organization are early indicators of disease. The unique molecular signatures within the ECM, such as modified collagen fibers, increased proteoglycan levels, or altered glycosylation patterns, serve as potential biomarkers for various diseases. In cancer, ECM components like tenascin-C and periostin are

D. Suhag, *Extracellular Matrix biomimicking for Theranostics*, Biomedical Materials for Multi-functional Applications, https://doi.org/10.1007/978-981-95-7900-6_6

upregulated, contributing to an invasive microenvironment. Detecting these changes can aid in early diagnosis and monitoring of disease progression, offering a non-invasive approach to understanding the underlying pathology [2]. As a result, ECM-based biomarkers are increasingly recognized for their potential to improve disease detection and personalized medicine. The importance of the ECM in diagnostic innovations lies in its direct involvement in disease pathophysiology. The ECM acts as a reservoir for growth factors and cytokines, modulating their availability and activity in the tissue microenvironment. This capability allows the ECM to influence cell signaling pathways that drive disease processes. For instance, in fibrotic diseases, the ECM becomes excessively stiff due to the overproduction of collagen, which can be detected through advanced imaging techniques like magnetic resonance elastography [3]. By quantifying ECM stiffness, clinicians can non-invasively assess the severity of fibrosis, guiding treatment decisions and predicting patient outcomes.

Imaging technologies have advanced our ability to visualize ECM components and their alterations in diseased tissues. Techniques such as multiphoton microscopy, optical coherence tomography, and super-resolution microscopy enable detailed visualization of the ECM's architecture at the molecular level. These imaging modalities are particularly valuable in diagnosing cancer, where ECM alterations are often more pronounced than changes in the cells themselves. By providing real-time, high-resolution images of ECM changes, these technologies enhance our understanding of disease mechanisms and offer new avenues for early detection. Microfluidic devices also play a significant role in ECM diagnostics, allowing for the creation of ECM-on-a-chip models that mimic the native tissue environment. These devices facilitate the study of ECM-cell interactions under controlled conditions, providing insights into how ECM alterations drive disease progression [4]. By integrating ECM components with microfluidic platforms, researchers can develop more accurate disease models for drug testing and biomarker discovery. This approach enhances our ability to predict how changes in the ECM contribute to disease, leading to more effective diagnostic and therapeutic strategies. ECM's structural complexity and functional diversity make it an invaluable component in the field of diagnostics. Its ability to reflect pathological changes at an early stage provides a unique advantage in disease detection and monitoring. As research into ECM dynamics continues to evolve, the potential for ECM-based diagnostic innovations will likely expand, offering new tools for clinicians and researchers to tackle complex diseases. The integration of ECM analysis with advanced imaging, biomarker discovery, and microfluidic technologies heralds a new era in precision diagnostics, where the ECM plays a central role in unraveling the complexities of disease.

6.2 Role of ECM in Disease Pathophysiology

The extracellular matrix (ECM) plays a crucial role in maintaining tissue architecture and regulating cellular functions. However, alterations in the ECM can significantly contribute to the development and progression of various diseases. In cancer, for example, the ECM undergoes extensive remodeling, characterized by increased stiffness, altered composition, and enhanced degradation. These changes create a permissive environment for tumor growth and metastasis by disrupting normal cell-ECM interactions and facilitating malignant cell behavior. Proteins such as collagen, fibronectin, and laminin are often overexpressed or structurally modified in cancerous tissues, contributing to an altered mechanical landscape that promotes invasive cell migration and resistance to apoptosis [5]. The remodeled ECM not only supports tumor cells but also actively communicates with them, influencing their behavior and driving disease progression. In fibrotic diseases, such as liver cirrhosis, pulmonary fibrosis, and systemic sclerosis, the ECM plays a central role in disease pathophysiology through excessive accumulation of collagen and other matrix proteins. This pathological remodeling leads to increased tissue stiffness, disrupting normal organ function. For instance, in liver fibrosis, the deposition of excessive ECM disrupts the normal architecture of the liver, impairing blood flow and hepatocyte function, eventually leading to liver failure. The fibrotic ECM acts as a continuous source of mechanical stress on resident cells, perpetuating a cycle of inflammation, cellular activation, and further ECM deposition. The dysregulated balance between ECM synthesis and degradation is often driven by chronic inflammation and persistent activation of myofibroblasts, key cells responsible for ECM production in fibrosis.

Cardiovascular diseases also involve significant ECM alterations that contribute to pathological remodeling of the heart and blood vessels [6]. In conditions like atherosclerosis, the ECM of the arterial wall becomes thickened and rigid due to the deposition of lipids, inflammatory cells, and ECM proteins. This altered ECM environment not only impairs vascular elasticity but also promotes the formation of plaques that can lead to vascular occlusion, heart attacks, or strokes. In heart failure, ECM remodeling contributes to myocardial stiffness and impaired contractility, often resulting in detrimental changes to heart function. Matrix metalloproteinases (MMPs) play a crucial role in this process by degrading ECM components, which, when dysregulated, can lead to the breakdown of normal tissue architecture and further exacerbate disease progression. ECM is also central to the pathogenesis of neurodegenerative diseases, where its components influence neural cell survival, plasticity, and synaptic function. In conditions like Alzheimer's disease, ECM components such as proteoglycans and tenascin are altered, contributing to the formation of amyloid plaques and neurofibrillary tangles. ECM in the central nervous system also regulates the permeability of the blood–brain barrier, which can be compromised in neurodegenerative diseases, leading to increased inflammation and neuronal damage [7]. Additionally, the ECM's role in synaptic plasticity and neural repair is impaired in these conditions, contributing to the progressive loss of cognitive function and other neurological deficits.

ECM is not just a passive structural element but an active participant in disease pathophysiology across a wide range of conditions. Its ability to regulate cellular behavior, signal transduction, and tissue mechanics makes it a critical factor in the progression of diseases such as cancer, fibrosis, cardiovascular disorders, and neurodegeneration. Understanding the dynamic interactions between the ECM and cells provides valuable insights into disease mechanisms and highlights the ECM as a potential target for therapeutic intervention. By addressing the pathological alterations of the ECM, novel diagnostic and therapeutic strategies can be developed to improve disease outcomes and patient care.

Exercise 1

Fill Ups
(Hyaluronic acid, Collagen, Cancer, Elastin, Fibronectin, Matrix metalloproteinases (MMPs), Fibrosis)

1. The extracellular matrix (ECM) provides structural and biochemical support to surrounding cells and consists of proteins such as ______ and ______.

2. Alterations in ECM composition are often associated with diseases such as ______ and ______, making ECM components potential biomarkers.

3. One of the key ECM proteins that can serve as a biomarker in cancer detection is ______, which is involved in cell adhesion and migration.

4. Degradation of ECM components by enzymes like ______ can indicate the presence of tissue remodeling during disease progression.

5. Biomarkers from ECM turnover, such as ______, are used to monitor the progression of fibrotic diseases and tissue repair.

6.3 ECM-Based Biomarkers for Disease Detection

ECM-based biomarkers are emerging as powerful tools for disease detection due to their ability to reflect the molecular and structural changes that occur in tissues during pathological conditions (Fig. 6.1). The extracellular matrix (ECM) is a dynamic network of proteins, glycoproteins, and proteoglycans that not only provides structural support to tissues but also regulates critical cellular processes such as signaling, migration, and differentiation. During disease progression, the ECM undergoes significant remodeling, including changes in its composition, stiffness, and organization. These alterations often precede visible clinical symptoms, making ECM components promising early biomarkers for detecting diseases like cancer, fibrosis, and cardiovascular disorders. By targeting specific ECM molecules, such as modified collagen, laminin fragments, or proteoglycans, diagnostic assays can identify disease presence and stage with high sensitivity and specificity [8]. The clinical potential of ECM-based biomarkers lies in their unique ability to capture disease-specific signatures that are not detectable through traditional cellular biomarkers. For example, in cancer, the breakdown products of ECM proteins, such as collagen and elastin, are released into the bloodstream and can serve as indicators of tumor growth and metastasis. Similarly, in fibrotic diseases, elevated levels of ECM components like hyaluronic acid and specific collagen isoforms correlate with disease severity and progression. These biomarkers can be detected using various techniques, including blood tests, imaging, and tissue biopsies, providing minimally invasive and real-time insights into the disease state.

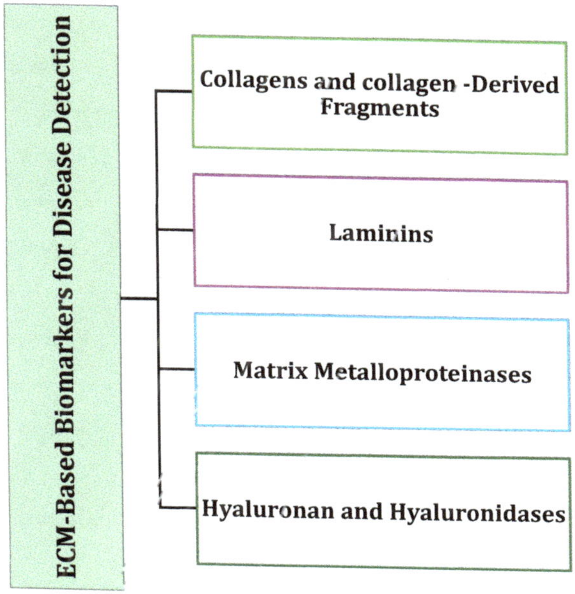

Fig. 6.1 ECM-based biomarkers for disease detection

6.4 Identification of ECM Components as Biomarkers

The identification of ECM components as biomarkers has gained significant attention due to the ECM's dynamic involvement in disease processes and its ability to reflect early pathological changes. Unlike traditional cellular biomarkers, ECM components provide insights into the extracellular environment, where many critical disease-related alterations occur. In cancer, for instance, the overexpression and structural modification of ECM proteins such as collagen, fibronectin, and laminin are associated with tumor growth and metastasis. Specific fragments of these proteins, released during ECM remodeling, can be detected in body fluids, serving as non-invasive biomarkers for early cancer diagnosis and monitoring (Fig. 6.2) [9]. These ECM-derived biomarkers are particularly valuable because they capture the biochemical and mechanical changes in the tumor microenvironment that are not evident through conventional diagnostic methods. Proteolytic enzymes like matrix metalloproteinases (MMPs), which degrade ECM components, also serve as important biomarkers. Elevated levels of MMPs and their specific breakdown products are frequently associated with cancer, cardiovascular diseases, and fibrotic conditions. For example, increased MMP activity is linked to the degradation of basement membrane components, facilitating cancer cell invasion and metastasis. Detecting these enzyme levels or their degradation products in blood or tissue samples can provide early indications of disease onset and progression. In cardiovascular diseases, MMPs contribute to the remodeling of vascular ECM, leading to conditions such as atherosclerosis and heart failure, where specific ECM fragments can be used to

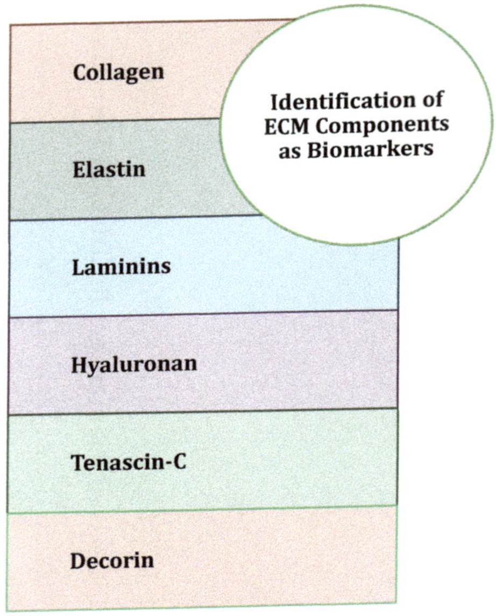

Fig. 6.2 Identification of ECM components as biomarkers

assess disease severity and predict outcomes. Glycosaminoglycans (GAGs), such as hyaluronic acid and chondroitin sulfate, are another group of ECM components with significant biomarker potential. These molecules are involved in regulating tissue hydration, elasticity, and cell signaling. Changes in GAG composition and structure are associated with various pathological conditions, including cancer, arthritis, and fibrosis [10]. Elevated serum levels of hyaluronic acid, for instance, are indicative of liver fibrosis and can be used as a non-invasive marker to assess liver health. Similarly, altered patterns of sulfation in GAGs are observed in cancers, influencing cell proliferation and metastasis. The unique modifications in these ECM components provide disease-specific signatures that can be harnessed for diagnostic purposes.

Another promising area in ECM biomarker identification is the use of specific collagen isoforms and their cross-linked fragments. Collagen, being the most abundant protein in the ECM, undergoes extensive remodeling in diseases such as fibrosis and cancer. Specific collagen types, like collagen I and III, are often upregulated in fibrotic tissues, while abnormal cross-linking of collagen fibers contributes to increased tissue stiffness. These structural changes can be detected through biomarkers like pro-peptides of collagen synthesis or specific collagen degradation fragments in blood samples. For example, the presence of collagen degradation products in the blood can indicate ongoing fibrotic activity in organs such as the liver, lungs, or kidneys, allowing for early diagnosis and monitoring of disease progression.

The identification of ECM components as biomarkers is further enhanced by advanced analytical techniques, including mass spectrometry, proteomics, and imaging technologies. Mass spectrometry allows for the precise quantification and characterization of ECM proteins and their modifications, facilitating the discovery of novel biomarkers with high diagnostic value. Proteomics provides a broader view of the ECM's molecular landscape, enabling the identification of disease-specific patterns of protein expression and post-translational modifications. These technologies, combined with bioinformatics, allow for the integration of ECM biomarker data into clinical practice, paving the way for personalized diagnostics.

6.4.1 ECM Remodeling as a Diagnostic Tool

ECM remodeling is a fundamental process that involves the continuous breakdown and rebuilding of the extracellular matrix, driven by a variety of enzymes, growth factors, and cellular interactions. In healthy tissues, ECM remodeling maintains tissue structure and function, playing a critical role in processes like wound healing and tissue regeneration. However, in disease states, this remodeling process becomes dysregulated, leading to abnormal ECM composition, structure, and mechanical properties. The altered ECM environment not only reflects the presence of disease but also actively contributes to its progression. These changes can be detected and quantified, making ECM remodeling a valuable diagnostic tool for various conditions, including cancer, fibrosis, and cardiovascular diseases. In cancer, ECM remodeling is a hallmark of tumor progression and metastasis, as the tumor microenvironment undergoes extensive alterations that support malignant behavior. The abnormal ECM in tumors is often characterized by increased collagen deposition, altered cross-linking, and enhanced proteolytic activity, creating a stiff and dense matrix that facilitates cancer cell invasion [11]. Diagnostic approaches that target these remodeling changes include the measurement of ECM degradation products, such as collagen fragments, in blood or tissue samples. Elevated levels of these fragments have been associated with various cancers, including breast, prostate, and lung cancer, serving as potential biomarkers for early detection and monitoring. Advanced imaging techniques, such as multiphoton microscopy and second harmonic generation, can also visualize ECM remodeling in vivo, providing real-time insights into tumor development. In fibrotic diseases, ECM remodeling is a key driver of pathology, characterized by excessive deposition of ECM components like collagen and fibronectin, leading to tissue stiffening and scarring. This pathological remodeling disrupts normal organ function, as seen in conditions like liver fibrosis, pulmonary fibrosis, and renal fibrosis. Diagnostic tools that assess ECM remodeling in fibrosis focus on measuring biomarkers of collagen turnover, such as specific collagen pro-peptides and degradation products. Non-invasive imaging techniques, like magnetic resonance elastography (MRE), can also detect changes in tissue stiffness associated with fibrosis, providing a quantitative assessment of disease severity. By evaluating ECM remodeling, clinicians can diagnose fibrosis at earlier stages and track disease progression more effectively than with conventional diagnostic methods.

Cardiovascular diseases, including atherosclerosis and heart failure, are also characterized by significant ECM remodeling that contributes to vascular and myocardial dysfunction. In atherosclerosis, the ECM of the arterial wall is remodeled due to inflammation, lipid accumulation, and the activity of proteolytic enzymes, leading to plaque formation and vessel stiffening. Diagnostic approaches that assess ECM remodeling in cardiovascular diseases often measure circulating levels of ECM degradation products, such as elastin and collagen fragments, which are indicative of vascular remodeling and plaque instability. Imaging modalities like ultrasound elastography and cardiac magnetic resonance imaging (MRI) can further visualize ECM changes in the heart and blood vessels, providing valuable diagnostic information on the extent and impact of remodeling.

ECM remodeling is also involved in neurodegenerative diseases, where it affects the integrity of the blood–brain barrier and neural tissue architecture. In conditions like Alzheimer's disease, ECM remodeling enzymes, including MMPs, are upregulated, leading to the degradation of key ECM components and contributing to neuronal damage [5]. Biomarkers of ECM remodeling, such as elevated MMP levels and altered proteoglycan patterns, can be detected in cerebrospinal fluid and blood, offering potential diagnostic tools for early detection and disease monitoring. Additionally, imaging techniques that target ECM changes in the brain, such as diffusion tensor imaging (DTI), can provide insights into the impact of ECM remodeling on neural connectivity and cognitive function. The use of ECM remodeling as a diagnostic tool is further enhanced by advancements in technology, including high-throughput proteomics, mass spectrometry, and bioinformatics. These technologies enable the detailed characterization of ECM components and their modifications, facilitating the discovery of novel remodeling biomarkers with high specificity and sensitivity. By integrating ECM biomarker data with clinical information, personalized diagnostic approaches can be developed, tailoring detection and monitoring strategies to individual patients. The ability to detect ECM remodeling early in disease progression allows for timely intervention, potentially improving treatment outcomes and patient care.

ECM remodeling offers a unique window into the pathological processes underlying many diseases, making it an invaluable diagnostic tool. By focusing on the dynamic changes in the ECM, clinicians and researchers can gain a deeper understanding of disease mechanisms, identify early indicators of disease, and develop more precise diagnostic and therapeutic strategies.

Exercise 2

Fill Ups
(Biochemical, Fluids, Fast, Sensors, Molecules, 3D, Cost-effective)

1. Microfluidic devices allow for the precise control of ______ and ______ at the microscale for ECM analysis.

2. One advantage of using microfluidic devices in ECM studies is the ability to mimic the ______ environment of tissues.

3. In ECM analysis, microfluidic devices can be used to study cell behavior in response to ______ gradients.

4. Microfluidic platforms are valuable in drug testing because they provide a ______ and ______ method to assess ECM changes.

5. The integration of ______ into microfluidic devices allows real-time monitoring of ECM component interactions.

6.4.2 *Biomarkers for Cancer, Fibrosis, and Cardiovascular Diseases*

- **ECM-Based Biomarkers for Cancer**

In cancer, the extracellular matrix (ECM) undergoes extensive remodeling, making ECM-based biomarkers valuable tools for disease detection and monitoring. Tumor progression is often associated with increased deposition and cross-linking of ECM components such as collagen, fibronectin, and laminin, which create a stiff and supportive environment for cancer cells. These alterations can be identified through specific biomarkers that reflect changes in the ECM. For instance, collagen degradation products, such as type I and type III collagen fragments, are frequently elevated in the blood of cancer patients and can serve as indicators of tumor invasion and metastasis. Certain ECM proteins, such as tenascin-C and periostin, are overexpressed in the tumor microenvironment, correlating with aggressive cancer phenotypes. The detection of these ECM-derived biomarkers provides a non-invasive approach to identifying cancer presence, assessing disease progression, and predicting patient outcomes [12]. Matrix metalloproteinases (MMPs) play a critical role in cancer by degrading ECM components and facilitating tumor cell migration. Elevated levels of MMPs, particularly MMP-2 and MMP-9, have been detected in various cancers, including breast, prostate, and colorectal cancer, and are associated with poor prognosis. The activity of these enzymes releases ECM fragments into the bloodstream, which can be detected and quantified as biomarkers of cancer progression. Other ECM remodeling enzymes like lysyl oxidase (LOX), which contribute to ECM cross-linking and stiffness, are also elevated in cancer and have emerged as potential diagnostic markers. These ECM-based biomarkers not only aid in early cancer detection but also offer insights into the tumor microenvironment, enhancing our understanding of cancer biology and improving patient management strategies.

- **ECM-Based Biomarkers for Fibrosis**

Fibrotic diseases, such as liver cirrhosis, pulmonary fibrosis, and systemic sclerosis, are characterized by excessive ECM deposition and tissue stiffening due to the overproduction of collagen, fibronectin, and other matrix components. The identification of ECM-based biomarkers in fibrosis focuses on detecting these excessive matrix components or their breakdown products, which are indicative of ongoing fibrotic activity. For example, serum biomarkers such as the N-terminal pro-peptide of type III collagen (PIIINP) and type IV collagen fragments are elevated in liver fibrosis and serve as non-invasive indicators of liver damage. Similarly, levels of hyaluronic acid, a glycosaminoglycan involved in ECM remodeling, are increased in fibrotic conditions and can be used to assess disease severity and progression. Fibrosis is driven by activated myofibroblasts, which excessively synthesize ECM components while reducing their degradation, leading to tissue scarring. Biomarkers such as tissue inhibitor of metalloproteinases (TIMPs), which inhibit ECM degradation enzymes, are often elevated in fibrosis and serve as indicators of disease activity

[13]. In pulmonary fibrosis, the detection of ECM degradation products, such as matrix metalloproteinase-derived collagen fragments, provides valuable information about disease stage and lung function. Advanced imaging techniques like magnetic resonance elastography (MRE) can also visualize ECM stiffness in fibrotic tissues, complementing biomarker data and offering a comprehensive assessment of the fibrotic burden. These ECM-based biomarkers not only aid in diagnosing fibrosis but also enable the monitoring of therapeutic responses, providing a more dynamic view of disease progression.

- **ECM-Based Biomarkers for Cardiovascular Diseases**

Cardiovascular diseases, including atherosclerosis, heart failure, and hypertension, involve significant ECM remodeling that contributes to vascular and myocardial dysfunction. In these conditions, the ECM of the heart and blood vessels is continuously remodeled, often leading to increased stiffness and impaired function. ECM-based biomarkers, such as circulating collagen degradation products and elastin fragments, are valuable for assessing the extent of vascular damage and remodeling. For example, elevated levels of elastin-derived peptides are associated with atherosclerosis and reflect the breakdown of the arterial ECM, which contributes to plaque formation and vessel stiffening. The detection of these biomarkers allows for early identification of cardiovascular risk and the assessment of disease progression. Matrix metalloproteinases (MMPs), particularly MMP-1, MMP-2, and MMP-9, play a significant role in cardiovascular ECM remodeling by degrading collagen and other ECM components. Elevated levels of these enzymes and their inhibitors, such as TIMP-1, are often detected in patients with heart failure and myocardial infarction, indicating ongoing ECM degradation and remodeling. These biomarkers can be measured in blood samples and provide insights into the severity of cardiac remodeling and the risk of adverse outcomes [14]. In addition to enzyme activity, specific ECM proteins like osteopontin, which are involved in ECM mineralization and inflammation, are also elevated in cardiovascular diseases and serve as potential biomarkers for diagnosis and prognosis.

- **Integrating ECM-Based Biomarkers in Clinical Diagnostics**

The integration of ECM-based biomarkers into clinical diagnostics offers a promising approach for the early detection and management of cancer, fibrosis, and cardiovascular diseases. These biomarkers provide a unique advantage by reflecting the dynamic changes in the ECM that are often early indicators of disease, allowing for timely intervention. Combining ECM biomarkers with advanced imaging techniques enhances diagnostic accuracy, as imaging can visualize ECM alterations in tissues, while biomarkers offer quantifiable data that reflect disease activity. For example, in cancer, combining ECM biomarkers with imaging modalities like MRI or PET scans improves the ability to detect tumors and assess their aggressiveness, leading to more personalized treatment strategies.

ECM-based biomarkers enable the monitoring of therapeutic responses, as changes in ECM remodeling can indicate the effectiveness of treatment. For instance, in fibrotic diseases, a reduction in circulating collagen fragments following antifibrotic therapy suggests a positive treatment response, allowing clinicians to adjust treatment plans accordingly. In cardiovascular diseases, the normalization of ECM biomarkers post-treatment reflects improved vascular or cardiac remodeling, guiding long-term patient management. The ability to track disease dynamics through ECM-based biomarkers not only improves diagnostic precision but also enhances patient outcomes by enabling a more responsive approach to treatment.

- **Future Directions and Challenges**

While ECM-based biomarkers hold great potential in diagnostics, their clinical translation faces challenges, including the need for standardization, validation, and large-scale studies to establish their efficacy and reliability. The complexity of ECM remodeling, influenced by factors such as age, gender, and comorbidities, requires comprehensive profiling of ECM biomarkers to ensure accurate disease diagnosis. The development of sensitive and specific assays for detecting ECM biomarkers in body fluids remains a critical area of research. Advances in proteomics, bioinformatics, and high-throughput technologies will likely drive the discovery of new ECM biomarkers and improve their integration into clinical practice.

ECM-based biomarkers represent a promising frontier in the early detection and management of cancer, fibrosis, and cardiovascular diseases. By capturing the biochemical and structural changes in the ECM, these biomarkers provide a unique perspective on disease processes that complement traditional cellular markers.

6.5 Imaging Techniques for ECM Visualization

Imaging techniques for ECM visualization have become crucial in understanding the role of the extracellular matrix (ECM) in health and disease (Fig. 6.3). The ECM, a complex network of proteins, glycoproteins, and polysaccharides, provides structural support to tissues and regulates critical cellular functions. However, alterations in the ECM's composition and structure often accompany diseases such as cancer, fibrosis, and cardiovascular disorders. Visualizing these changes in the ECM can provide valuable insights into disease mechanisms and progression. Advanced imaging techniques, including multiphoton microscopy, second harmonic generation, and magnetic resonance imaging (MRI), allow for high-resolution, real-time visualization of ECM components in their native environment [15]. These techniques enable researchers and clinicians to observe ECM remodeling, identify disease-specific patterns, and assess the impact of therapeutic interventions. The ability to visualize ECM in vivo offers a significant advantage over traditional histological methods, which often require invasive biopsies and cannot capture dynamic changes in the ECM. Non-invasive imaging approaches provide detailed structural and functional information about the ECM, enhancing our ability to diagnose and monitor diseases at various stages. For example, second harmonic generation microscopy can specifically highlight collagen fibers without the need for exogenous dyes, revealing

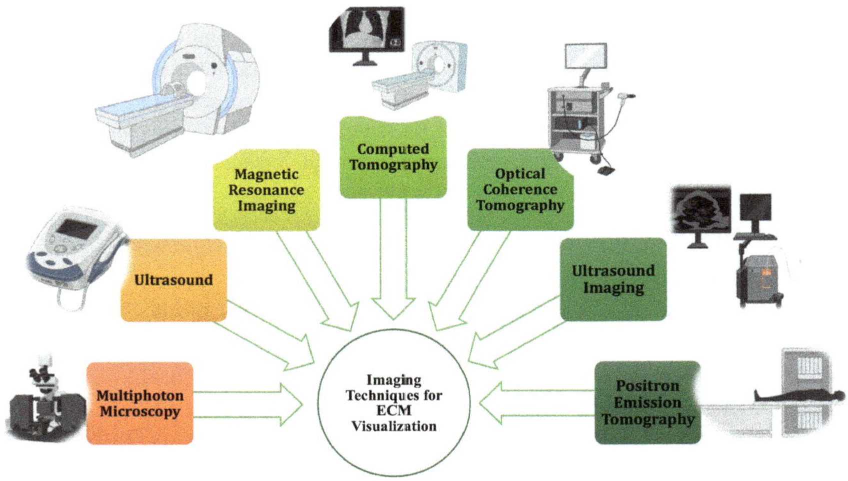

Fig. 6.3 Various imaging techniques for ECM visualization

structural abnormalities associated with tumor progression or tissue fibrosis. Similarly, MRI techniques, such as diffusion tensor imaging, can map ECM alterations in neural tissues, providing critical information in neurodegenerative diseases.

6.5.1 Traditional Imaging Techniques (MRI, CT, Ultrasound)

Magnetic Resonance Imaging (MRI), Computed Tomography (CT), and Ultrasound are traditional imaging techniques widely used in clinical practice to visualize internal structures, including tissues, organs, and pathological changes. Each technique offers unique capabilities, making them indispensable tools in modern diagnostics. MRI utilizes powerful magnets and radio waves to generate detailed images of soft tissues, providing high-resolution views of anatomical structures and detecting abnormalities in the extracellular matrix (ECM) without exposure to ionizing radiation. MRI's ability to differentiate between various tissue types and its superior contrast resolution make it particularly valuable for assessing soft tissue structures, such as the brain, muscles, and cartilage, where ECM alterations are critical in disease pathophysiology [16]. For instance, MRI can detect fibrosis, edema, and tissue stiffening, reflecting underlying ECM remodeling processes in cardiovascular and musculoskeletal disorders.

CT imaging, on the other hand, uses X-rays to produce cross-sectional images of the body, offering a detailed view of both bone and soft tissue structures. CT scans are widely used for diagnosing a range of conditions, including trauma, infections, and cancers, providing valuable insights into ECM alterations in solid organs and tissues. With the ability to generate three-dimensional reconstructions, CT scans are particularly useful in visualizing the extent of tumors, vascular abnormalities, and calcifications [17]. Despite its lower soft tissue contrast compared to MRI, CT remains a critical tool in evaluating ECM changes in dense tissues such as bones and lungs. CT's rapid imaging capability and high spatial resolution make it a preferred choice for emergency diagnostics, guiding surgical planning, and monitoring disease progression in real-time. Ultrasound imaging employs high-frequency sound waves to create real-time images of internal body structures, making it a versatile and non-invasive diagnostic tool. Unlike MRI and CT, ultrasound does not involve ionizing radiation, making it safer for use in various patient populations, including pregnant women and pediatric patients. Ultrasound is particularly effective in visualizing soft tissues, fluid-filled structures, and superficial organs such as the liver, kidneys, and thyroid. It is also extensively used in assessing ECM components in musculoskeletal injuries, such as tendons, ligaments, and cartilage. For example, ultrasound can detect changes in tendon structure and echogenicity associated with fibrosis or inflammation, providing insights into ECM remodeling in sports injuries and degenerative conditions.

One of the key advantages of traditional imaging techniques like MRI, CT, and ultrasound is their accessibility and established role in clinical diagnostics. These imaging modalities are widely available in hospitals and clinics, with well-established protocols for a broad range of conditions. MRI, in particular, offers advanced techniques such as diffusion-weighted imaging (DWI) and magnetic resonance elastography (MRE), which can specifically target ECM properties like tissue stiffness and diffusion, enhancing the detection of fibrosis and other ECM-related abnormalities. CT's ability to provide rapid, high-resolution images makes it invaluable in emergency settings, such as detecting fractures or assessing vascular injuries, where ECM alterations are often implicated in the pathology.

Despite their strengths, traditional imaging techniques have limitations, particularly in visualizing detailed molecular and biochemical changes within the ECM. While MRI and CT provide excellent anatomical details, they often lack the specificity to directly detect subtle ECM alterations, such as changes in collagen cross-linking or proteoglycan content. Ultrasound, though excellent for assessing structural changes, is limited by operator dependency and difficulty visualizing deeper tissues or those obscured by bone or gas. These limitations highlight the need for complementary imaging techniques or molecular imaging approaches that can provide more specific insights into ECM composition and dynamics. Emerging technologies that combine traditional imaging with targeted contrast agents or molecular probes are being developed to bridge this gap, offering new possibilities for diagnosing and monitoring ECM-related diseases.

Traditional imaging techniques like MRI, CT, and ultrasound remain cornerstone tools in clinical diagnostics, providing critical insights into the structural changes associated with ECM remodeling in various diseases. While each modality has distinct advantages, their ability to capture detailed anatomical images plays a crucial role in detecting and monitoring conditions that involve ECM alterations, such as cancer, fibrosis, and cardiovascular diseases.

6.5.2 *Advanced Imaging Techniques: Optical Coherence Tomography, Multiphoton Microscopy*

Advanced imaging techniques, such as Optical Coherence Tomography (OCT) and Multiphoton Microscopy (MPM), have revolutionized our ability to visualize the extracellular matrix (ECM) with high resolution and specificity (Fig. 6.4). These techniques provide critical insights into the complex structure and function of the ECM, which plays a fundamental role in tissue integrity, cellular signaling, and disease progression. Unlike traditional imaging modalities like MRI and CT, advanced techniques like OCT and MPM offer the ability to capture detailed, real-time images of ECM components in their native environment without the need for extensive sample preparation or staining. This capacity to visualize ECM at the microscale level is essential for understanding the dynamic remodeling processes that occur in diseases such as cancer, fibrosis, and cardiovascular disorders.

- Optical Coherence Tomography (OCT) is a non-invasive imaging technique that uses low-coherence light to capture cross-sectional images of tissues with micrometer-scale resolution, similar to an "optical ultrasound." OCT is particularly effective in visualizing ECM structures in transparent and semi-transparent tissues, such as the retina, skin, and vascular tissues. Its ability to detect subtle changes in tissue architecture makes it invaluable in diagnosing conditions like age-related macular degeneration, diabetic retinopathy, and atherosclerosis, where ECM remodeling plays a crucial role [18]. In cardiovascular applications, OCT can provide detailed images of arterial plaques, allowing clinicians to assess plaque composition and stability, which are influenced by ECM components like collagen and elastin. This detailed visualization helps in the early detection of disease, guiding therapeutic interventions and improving patient outcomes.
- Multiphoton Microscopy (MPM) is another powerful imaging technique that uses two or more photons of lower energy to excite fluorescent molecules within tissues, providing deep-tissue penetration and reduced phototoxicity. MPM is highly effective in visualizing the ECM in living tissues, allowing researchers to

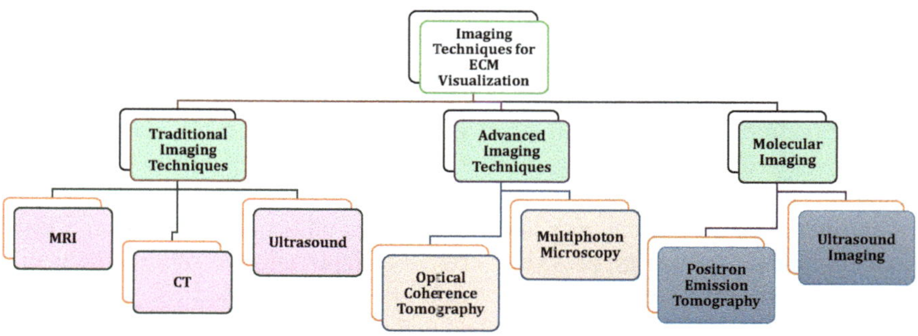

Fig. 6.4 Different types of imaging techniques for ECM visualization

observe cellular interactions with the ECM in real time. This technique is particularly valuable in cancer research, where it enables the study of tumor-ECM interactions, tumor cell migration, and ECM remodeling. MPM's ability to capture high-resolution images of ECM components, such as collagen fibers and elastin networks, provides detailed information about changes in ECM architecture that are associated with tumor growth and metastasis [19]. This capability is crucial for understanding the role of ECM in cancer progression and for developing targeted therapies that modify ECM properties to inhibit tumor spread.

In addition to their diagnostic applications, advanced imaging techniques like OCT and MPM are instrumental in studying ECM dynamics during tissue development, wound healing, and regenerative processes. By visualizing ECM remodeling in real time, these techniques provide insights into how cells remodel their surrounding matrix in response to injury or during tissue repair. For example, MPM can be used to monitor collagen deposition and degradation in wound healing, revealing how different cell types, such as fibroblasts and immune cells, contribute to ECM reconstruction. Similarly, OCT can track the formation of new ECM layers in vascular grafts or engineered tissues, helping to evaluate the success of regenerative therapies. This level of detail is not achievable with traditional imaging methods, highlighting the importance of advanced imaging in ECM research. One of the key advantages of OCT and MPM is their ability to provide label-free imaging of ECM components, minimizing the need for exogenous dyes or contrast agents. This feature allows for the natural visualization of ECM structures, preserving their native state and enabling more accurate analysis. For instance, OCT can detect variations in ECM optical properties, such as birefringence, which reflect changes in collagen organization and density—a key factor in diseases like fibrosis. MPM, on the other hand, uses endogenous fluorescence from ECM proteins, like collagen and elastin, to generate high-contrast images, making it a powerful tool for visualizing ECM without the interference of artificial labels. This capability is crucial for studying ECM in clinical settings, where maintaining tissue integrity and minimizing sample alteration are paramount.

The importance of advanced imaging techniques extends beyond diagnostics, as they also play a critical role in the evaluation of therapeutic efficacy (Fig. 6.4). By providing detailed images of ECM remodeling, OCT and MPM can monitor how treatments, such as antifibrotic drugs, cancer therapies, or regenerative medicine approaches, impact ECM structure and function. For example, in cancer treatment, MPM can be used to assess changes in ECM stiffness and alignment in response to therapy, helping to determine whether the treatment effectively disrupts the tumor microenvironment [20]. Similarly, OCT can evaluate the integrity of ECM in healing tissues, guiding decisions on wound management and reconstructive surgery. This ability to visualize treatment-induced changes in the ECM provides valuable feedback for clinicians and researchers, facilitating the development of more effective therapeutic strategies. The integration of OCT and MPM with other imaging modalities, such as confocal microscopy or photoacoustic imaging, enhances the ability to comprehensively assess ECM characteristics across different scales and depths. Such multimodal approaches combine structural, biochemical, and functional information, offering a holistic view of ECM alterations in disease and therapy. For instance, combining OCT with photoacoustic imaging can provide both high-resolution structural images and molecular insights, such as oxygenation status, within the ECM. This comprehensive imaging capability is particularly useful in complex diseases like cancer, where ECM changes are multifaceted and impact disease progression on multiple levels.

Advanced imaging techniques like OCT and MPM represent significant advancements in our ability to visualize the ECM with unprecedented clarity and detail. Their non-invasive, high-resolution imaging capabilities provide critical insights into ECM structure, function, and remodeling, enhancing our understanding of ECM's role in health and disease. By enabling the direct observation of ECM dynamics, these techniques not only improve diagnostic accuracy but also offer valuable tools for evaluating therapeutic interventions.

6.5.3 Molecular Imaging of ECM Components

Molecular imaging is an advanced technology that allows for the non-invasive visualization of biological processes at the molecular and cellular levels within living organisms. In the context of the extracellular matrix (ECM), molecular imaging plays a crucial role in understanding the complex architecture and dynamic interactions of ECM components, which include proteins like collagen, elastin, fibronectin, and glycosaminoglycans. These components provide structural support and regulate various cellular functions, such as adhesion, migration, and differentiation. By employing techniques such as magnetic resonance imaging (MRI) [21], positron emission tomography (PET), fluorescence imaging, and ultrasound, researchers can observe the real-time behavior of ECM molecules, assess their distribution, and identify pathological changes associated with diseases such as cancer, fibrosis, and cardiovascular disorders.

The molecular imaging of ECM components is instrumental in advancing our knowledge of disease progression and therapeutic response. For example, targeted imaging agents that bind specifically to ECM proteins can highlight areas of tissue remodeling, inflammation, or degradation, offering valuable insights into disease mechanisms. Additionally, imaging techniques enable the early detection of diseases by revealing subtle changes in ECM structure that might precede visible symptoms, thereby enhancing diagnostic accuracy. The ability to visualize ECM dynamics not only improves our understanding of tissue biology but also facilitates the development of targeted therapies and personalized medicine approaches, making molecular imaging a vital tool in modern biomedical research and clinical diagnostics.

6.5.4 Application of Imaging in Disease Diagnosis and Monitoring

- **Early Detection of Cancer**: Imaging techniques like PET and MRI can identify abnormal ECM remodeling and tumor microenvironment changes, allowing for early detection of cancers before they become clinically apparent. This early intervention is critical for improving patient outcomes [22].
- **Assessment of Fibrosis**: Imaging modalities, such as MRI with specific contrast agents, can visualize ECM components like collagen accumulation in organs affected by fibrosis (e.g., liver, lungs, and kidneys), helping to assess disease severity and progression.
- **Tracking Cardiovascular Disease**: Techniques like ultrasound and MRI can detect changes in ECM proteins such as elastin and collagen in the vascular walls, providing insights into atherosclerosis, aneurysms, and other cardiovascular conditions, allowing for timely intervention.
- **Monitoring Response to Therapy**: Molecular imaging can assess how ECM-targeted treatments (e.g., antifibrotic drugs or cancer therapies) affect the tumor microenvironment or fibrotic tissues, enabling clinicians to adjust treatment plans based on real-time feedback.
- **Diagnosis of Osteoarthritis and Cartilage Degradation**: Imaging techniques such as MRI can visualize ECM changes in cartilage, like the degradation of collagen and proteoglycans, providing critical information for the diagnosis and management of joint diseases.
- **Identifying Inflammatory Conditions**: Imaging methods, including fluorescence and PET, can detect ECM biomarkers associated with inflammation, aiding in diagnosing autoimmune diseases and monitoring inflammatory responses over time.
- **Detection of ECM-Related Genetic Disorders**: Techniques like MRI can reveal abnormal ECM structures linked to genetic disorders such as Marfan syndrome or Ehlers-Danlos syndrome, helping in early diagnosis and management of these conditions.
- **Guidance in Surgical Planning**: Imaging can map ECM abnormalities in tissues, guiding surgeons in planning procedures such as tumor resection or organ transplantation, ensuring more precise and effective interventions.

6.6 Microfluidic Devices for ECM Analysis

Microfluidic devices, also known as lab-on-a-chip systems, are innovative tools that enable the precise manipulation of small volumes of fluids within microscale channels. These devices have emerged as powerful platforms for analyzing the extracellular matrix (ECM), which plays a crucial role in regulating cell behavior, tissue structure, and disease progression. By recreating the intricate microenvironment of the

ECM, microfluidic devices allow researchers to study cell-ECM interactions, matrix composition, and mechanical properties in a controlled setting. This technology provides high-resolution data on how ECM components like collagen, fibronectin, and proteoglycans influence cellular functions, helping to bridge the gap between in vitro studies and in vivo conditions. The integration of microfluidic technology in ECM analysis offers several advantages, including the ability to mimic physiological conditions, apply mechanical forces, and create complex ECM gradients [23]. These devices are particularly valuable for studying disease models, such as cancer metastasis and fibrosis, where ECM remodeling plays a critical role. Microfluidic systems enable high-throughput screening of drug candidates targeting ECM components, making them essential for drug development and personalized medicine. The versatility and precision of microfluidic devices in simulating the ECM environment make them indispensable tools for advancing our understanding of tissue biology, disease mechanisms, and therapeutic responses.

6.6.1 ECM-on-a-Chip Models for Disease Detection

ECM-on-a-chip models are advanced microfluidic systems designed to replicate the complex environment of the extracellular matrix (ECM), providing a more realistic platform for studying tissue behavior and disease progression. These models integrate biomimetic ECM components within microfluidic channels to simulate the structural and biochemical cues of native tissues, allowing researchers to observe cell-ECM interactions in a controlled, miniaturized setting. By incorporating patient-derived cells and ECM proteins, ECM-on-a-chip devices can closely mimic the physiological conditions of healthy and diseased tissues, making them powerful tools for disease detection and analysis. These platforms enable real-time monitoring of cellular responses to ECM changes, providing crucial insights into the early stages of disease and potential biomarkers for diagnosis. One of the key applications of ECM-on-a-chip models is in cancer detection. Tumor growth and metastasis are heavily influenced by changes in the ECM, including alterations in matrix stiffness, composition, and degradation. By replicating the tumor microenvironment on a chip, researchers can study how cancer cells interact with the ECM, migrate, and invade surrounding tissues. These models can be used to identify specific ECM signatures associated with different types of cancer, aiding in early diagnosis and personalized treatment strategies [24]. For instance, ECM-on-a-chip platforms can detect abnormal ECM remodeling patterns, such as increased collagen crosslinking or altered glycoprotein expression, which are often precursors to cancer progression. ECM-on-a-chip devices are also valuable for detecting fibrotic diseases, such as liver cirrhosis, pulmonary fibrosis, and kidney fibrosis, where excessive ECM deposition leads to tissue scarring and organ dysfunction. These models can replicate the fibrotic microenvironment, allowing for the study of ECM stiffening and altered cell–matrix interactions that drive disease progression. By analyzing cellular responses to fibrotic ECM cues, these platforms help identify potential biomarkers and therapeutic targets for early intervention. Additionally, they provide a high-throughput method for screening antifibrotic drugs, enabling the rapid assessment of drug efficacy in reversing ECM changes.

In cardiovascular disease research, ECM-on-a-chip models are used to study atherosclerosis, myocardial fibrosis, and other conditions where ECM remodeling plays a critical role. These devices can mimic the mechanical forces, and biochemical gradients present in vascular tissues, providing a realistic platform for assessing how changes in ECM composition and stiffness contribute to disease [25]. By simulating the vascular ECM environment, these models enable the detection of early signs of vascular disease, such as increased collagen deposition or elastin degradation, which are often associated with plaque formation and vessel stiffening. ECM-on-a-chip platforms are also instrumental in studying neurodegenerative diseases, where ECM alterations in the brain's microenvironment can influence neuronal health and function. These models can recreate the brain ECM, allowing researchers to explore how changes in matrix components, such as laminin and hyaluronan, affect neural cell behavior and contribute to diseases like Alzheimer's and Parkinson's. By providing a closer look at the ECM's role in neuronal signaling and cell survival, ECM-on-a-chip devices can help identify early molecular changes that serve as biomarkers for neurodegenerative disorders. These microfluidic models are being used to study the role of ECM in inflammatory diseases, such as rheumatoid arthritis and inflammatory bowel disease (IBD). By replicating the inflamed ECM environment, ECM-on-a-chip devices allow researchers to examine how immune cells interact with the altered ECM and contribute to chronic inflammation. These insights are crucial for identifying early inflammatory markers and developing targeted therapies that can modulate the ECM to reduce inflammation and tissue damage.

ECM-on-a-chip models represent a significant advancement in disease detection and diagnostic innovation. By faithfully recreating the ECM environment of diseased tissues, these platforms enable a deeper understanding of the molecular changes that drive disease onset and progression. They provide a powerful tool for identifying disease-specific ECM alterations that can serve as early diagnostic markers, ultimately improving the ability to detect and treat diseases at their earliest stages.

6.6.2 Integration of Microfluidics with Biomarker Detection

The integration of microfluidics with biomarker detection has revolutionized the field of diagnostic technology by enabling rapid, sensitive, and precise analysis of biological samples. Microfluidic devices, often referred to as lab-on-a-chip systems, allow for the manipulation of tiny fluid volumes in channels that are often no wider than a human hair. This miniaturization offers significant advantages, including reduced sample and reagent usage, shorter processing times, and the ability to automate complex laboratory procedures on a single platform. When combined with biomarker detection, microfluidic devices provide a powerful tool for identifying disease-specific molecules such as proteins, nucleic acids, and metabolites, which are critical for early diagnosis and monitoring of various health conditions. One of the primary benefits of microfluidics in biomarker detection is the enhanced sensitivity and specificity it provides, which is crucial for detecting low-abundance biomarkers that traditional methods might lack [26]. Microfluidic devices can integrate multiple detection techniques, such as fluorescence, electrochemical sensors, and surface plasmon resonance, to capture and quantify biomarkers with high accuracy. This is particularly valuable in the early detection of diseases like cancer, where biomarkers may be present in very low concentrations. The ability to detect these subtle changes in biomarker levels can lead to earlier diagnosis, improving the chances of successful treatment. Microfluidics also facilitates the development of point-of-care (POC) diagnostic devices, which bring testing closer to patients and reduce the time between sample collection and result delivery. These portable devices can perform complex biomarker analyses in settings such as clinics, homes, or remote areas without the need for extensive laboratory infrastructure. For example, microfluidic POC devices have been developed for rapid detection of biomarkers in infectious diseases like COVID-19, where timely diagnosis is critical for patient management and controlling the spread of the virus. By providing immediate results, these devices empower healthcare providers to make faster and more informed decisions.

The versatility of microfluidics extends to the integration of multiplexed assays, allowing the simultaneous detection of multiple biomarkers from a single sample. This capability is particularly beneficial in diseases where a combination of biomarkers provides a more comprehensive view of the disease state, such as in cardiovascular diseases or cancer. Microfluidic platforms can separate and detect these biomarkers in parallel, significantly reducing the time and cost associated with traditional serial testing methods. This multiplexing ability enhances diagnostic accuracy and provides a broader understanding of the underlying disease mechanisms. Another significant application of microfluidic biomarker detection is in personalized medicine, where the goal is to tailor treatments based on the individual characteristics of each patient. Microfluidic devices can analyze biomarkers that reflect a patient's unique genetic, proteomic, and metabolic profile, enabling the selection of the most effective therapies. For instance, microfluidics can be used to monitor biomarkers associated with drug response, helping to identify which patients are most likely to benefit from specific treatments and minimizing the risk of adverse effects. This personalized approach optimizes therapeutic outcomes and improves overall patient care. In addition to diagnostics, microfluidics integrated with biomarker detection plays a crucial role in monitoring disease progression and treatment efficacy. By continuously or periodically measuring biomarker levels, these devices can provide real-time feedback on how a disease evolves or how a patient responds to therapy. This capability is especially valuable in chronic conditions like diabetes or cancer, where ongoing monitoring is necessary to adjust treatment strategies. Microfluidic devices can offer a minimally invasive way to track disease dynamics, enhancing the management of long-term health conditions.

The integration of microfluidics with biomarker detection is poised to drive future innovations in healthcare by enabling the development of more sophisticated, automated, and accessible diagnostic tools. As microfluidic technology continues to advance, with improvements in materials, fabrication techniques, and detection methods, it is expected that these devices will become even more powerful and widely adopted. The convergence of microfluidics and biomarker detection represents a transformative approach in the quest for earlier, more accurate, and more personalized disease detection and management, ultimately leading to better health outcomes and more efficient healthcare systems.

6.7 Technological Advances in ECM Analysis

Technological advances in extracellular matrix (ECM) analysis have revolutionized our understanding of tissue structure, function, and pathology. The ECM, a complex network of proteins, glycoproteins, and other biomolecules, plays a pivotal role in regulating cellular behavior and maintaining tissue integrity. Recent innovations in analytical techniques, such as advanced imaging, microfluidics, and molecular profiling, have provided unprecedented insights into the dynamic nature of the ECM. High-resolution imaging technologies like super-resolution microscopy, multiphoton microscopy, and atomic force microscopy allow researchers to visualize ECM components at the nanoscale, revealing intricate details of ECM architecture and its interactions with cells. These advancements are essential for uncovering the molecular mechanisms underlying various diseases, such as cancer, fibrosis, and cardiovascular disorders, where ECM remodeling is a key factor. Beyond imaging, breakthroughs in mass spectrometry, bioinformatics, and microfluidic platforms have significantly enhanced the ability to profile ECM composition and function. Mass spectrometry-based proteomics enables the comprehensive identification and quantification of ECM proteins, shedding light on how changes in ECM composition correlate with disease states [27]. Meanwhile, microfluidic technologies, such as ECM-on-a-chip models, provide a controlled environment to study cell-ECM interactions under physiologically relevant conditions. These platforms allow for high-throughput screening of drugs targeting ECM components, accelerating the development of new therapeutic strategies.

6.7.1 Nanotechnology in ECM Diagnostics

Nanotechnology is emerging as a transformative tool in the field of extracellular matrix (ECM) diagnostics, offering novel approaches to detect and analyze the intricate dynamics of the ECM in health and disease. The ECM plays a crucial role in maintaining tissue structure and regulating cellular functions, and its alterations are often associated with various diseases, including cancer, fibrosis, and cardiovascular disorders. Nanotechnology enables the development of highly sensitive diagnostic platforms that can detect subtle changes in ECM components at the molecular level, facilitating early disease detection and improving patient outcomes. By integrating nanoscale materials and devices, these advanced diagnostic techniques provide unparalleled precision in analyzing ECM structure, composition, and function. One of the key applications of nanotechnology in ECM diagnostics is the development of nano sensors that can detect ECM biomarkers with high sensitivity and specificity [28]. These sensors often utilize nanomaterials such as gold nanoparticles, carbon nanotubes, and quantum dots, which possess unique optical, electrical, and chemical properties. For example, gold nanoparticles can be functionalized with antibodies or peptides that specifically bind to ECM proteins like collagen or fibronectin, allowing for the rapid and precise detection of these molecules in biological samples. The high surface area-to-volume ratio of nanomaterials enhances the binding efficiency of target molecules, enabling the detection of ECM changes at ultra-low concentrations, which is critical for early disease diagnosis.

Nanotechnology also plays a significant role in imaging the ECM, enhancing the resolution and sensitivity of traditional imaging techniques. Nanoparticles can be engineered as contrast agents for imaging modalities such as magnetic resonance imaging (MRI), computed tomography (CT), and fluorescence microscopy. For instance, iron oxide nanoparticles are widely used as MRI contrast agents to visualize ECM alterations in fibrotic tissues or tumors, providing detailed information about ECM remodeling and disease progression. Similarly, fluorescent quantum dots can be used to label ECM components, enabling real-time tracking of ECM dynamics at the cellular level. These advanced imaging techniques allow for non-invasive monitoring of ECM changes, offering valuable insights into disease mechanisms and potential therapeutic targets. Another promising application of nanotechnology in ECM diagnostics is the use of nanocarriers for targeted delivery of diagnostic probes. Nanocarriers, such as liposomes, dendrimers, and polymeric nanoparticles, can be loaded with imaging agents, drugs, or nucleic acids and designed to specifically target ECM components associated with disease. This targeted approach enhances the accumulation of diagnostic probes in diseased tissues, improving the accuracy of detection and reducing off-target effects. For example, nanoparticles engineered to target matrix metalloproteinases (MMPs), enzymes that degrade ECM components in cancer and fibrosis, can provide precise information on the activity and localization of these enzymes, aiding in the diagnosis and monitoring of disease progression.

Nanotechnology is also advancing the field of liquid biopsy, a minimally invasive diagnostic method that analyzes circulating biomarkers in bodily fluids such as blood, urine, or saliva. Nanoparticles can be used to isolate and detect ECM-derived biomarkers, such as fragments of degraded ECM proteins or ECM-associated exosomes, which are often indicative of disease. These nanotechnology-enhanced liquid biopsy platforms offer a convenient and efficient way to monitor ECM changes in real time, providing a powerful tool for early diagnosis, disease staging, and treatment monitoring. This approach holds significant potential for personalized medicine, where disease-specific ECM alterations can guide therapeutic decisions. In addition to detection, nanotechnology enables the design of smart diagnostic platforms that can respond dynamically to ECM changes [29]. For instance, nano sensors embedded in hydrogels or scaffolds can be programmed to release diagnostic signals in response to specific ECM modifications, such as changes in pH or enzyme activity. These smart materials can be used in vivo to continuously monitor the ECM environment, providing real-time feedback on disease progression or therapeutic efficacy. This dynamic monitoring capability is particularly valuable in chronic diseases, where ongoing assessment of ECM status is crucial for adjusting treatment strategies.

The integration of nanotechnology with advanced computational tools, such as artificial intelligence (AI) and machine learning, further enhances the diagnostic capabilities of ECM analysis. AI algorithms can be used to analyze complex data generated from nano sensors or imaging techniques, identifying patterns and correlations that may not be apparent through traditional analysis methods. This data-driven approach improves the accuracy and predictive power of ECM diagnostics, enabling the identification of novel biomarkers and the development of more targeted and effective diagnostic tests. Nanotechnology is redefining ECM diagnostics by providing innovative tools that offer higher sensitivity, specificity, and functionality than conventional methods. The ability to detect and analyze ECM alterations at the nanoscale opens new avenues for early disease detection, personalized treatment, and continuous monitoring of disease progression.

6.7.2 Computational Models and AI in ECM Data Analysis

Computational models and artificial intelligence (AI) are transforming the field of extracellular matrix (ECM) data analysis by offering powerful tools to decode the complex dynamics of ECM biology. The ECM, a vital network of proteins and other biomolecules, plays a critical role in tissue function, cell behavior, and disease progression. However, the complexity and heterogeneity of ECM components pose significant challenges in understanding its role in health and disease. Computational models, including finite element analysis, systems biology models, and molecular dynamics simulations, allow researchers to simulate ECM structure and behavior under various physiological and pathological conditions. These models provide insights into how mechanical properties, molecular interactions, and structural changes in the ECM influence cellular responses, enabling more precise predictions of disease mechanisms. AI and machine learning (ML) have further enhanced ECM data analysis by automating the processing of large and complex datasets generated from imaging, proteomics, and genomics studies. AI algorithms can rapidly identify patterns and correlations in ECM data that might be missed by traditional analysis methods. For example, ML models can analyze high-resolution imaging data to detect subtle changes in ECM architecture associated with early-stage diseases such as cancer or fibrosis. By training on large datasets, these models improve their accuracy over time, enabling more precise identification of ECM alterations that are indicative of disease progression or therapeutic response.

One of the most impactful applications of AI in ECM data analysis is the integration of multi-omics data, which combines information from genomics, proteomics, and metabolomics to provide a comprehensive view of the ECM's role in disease

[30]. AI-driven platforms can analyze these diverse data types simultaneously, uncovering complex interactions between ECM components and cellular processes. This integrative approach helps identify novel ECM biomarkers and potential therapeutic targets, enhancing the development of personalized medicine strategies. For instance, AI algorithms can predict how specific ECM modifications affect cellular signaling pathways, guiding the design of targeted therapies that address the underlying ECM dysfunctions in diseases. Computational models and AI are instrumental in the design and optimization of ECM-based biomaterials for tissue engineering and regenerative medicine. By simulating how different ECM compositions and mechanical properties influence cell behavior, these tools assist in creating biomimetic scaffolds that closely replicate the natural ECM environment. AI can also predict the long-term behavior of engineered ECMs in vivo, aiding in the development of more effective and durable biomaterials for clinical applications.

6.7.3 Integrating Multi-omics Data for ECM Analysis

Integrating multi-omics data has become a powerful approach for analyzing the extracellular matrix (ECM), providing a comprehensive understanding of its complex and dynamic nature. The ECM is a highly intricate network of proteins, glycoproteins, and other biomolecules that play critical roles in tissue structure and cellular communication. Traditional methods of ECM analysis often focus on a single aspect, such as proteomics or genomics, which limits the ability to capture the full scope of ECM biology. Multi-omics integration combines data from various omics fields—genomics, transcriptomics, proteomics, metabolomics, and glycomics—to create a holistic view of the ECM. This approach enables researchers to uncover the intricate relationships between genetic variations, protein expression, metabolic changes, and ECM modifications, providing deeper insights into how the ECM contributes to health and disease.

Advantages of multi-omics integration in ECM analysis includes the ability to identify novel ECM biomarkers that are involved in disease progression. For instance, integrating proteomics and transcriptomics data allows researchers to link gene expression changes with alterations in ECM protein composition, revealing key drivers of ECM remodeling in diseases such as cancer, fibrosis, and cardiovascular disorders. By correlating these molecular changes, multi-omics analysis helps pinpoint specific ECM components that may serve as early diagnostic markers or therapeutic targets [31]. This approach is particularly valuable in complex diseases where ECM alterations are not solely driven by genetic factors but are also influenced by epigenetic, metabolic, and post-translational modifications. Multi-omics integration also enhances the understanding of how ECM composition and structure influence cellular behavior and tissue function. By combining data from different omics layers, researchers can map the interactions between ECM components and cell signaling pathways, shedding light on how ECM alterations affect cellular processes such as adhesion, migration, and proliferation. For example, metabolomics data can reveal how changes in ECM stiffness or composition alter cellular metabolism, leading to insights into disease mechanisms like cancer metastasis or tissue fibrosis. This comprehensive view enables a systems-level understanding of the ECM's role in regulating cell behavior, which is critical for developing targeted therapies that modulate ECM interactions.

In the field of personalized medicine, integrating multi-omics data for ECM analysis offers significant potential for tailoring treatments based on an individual's unique ECM profile. By analyzing patient-specific omics data, clinicians can identify ECM alterations that are specific to a particular disease state or patient subgroup. This personalized approach allows for the selection of targeted therapies that address the specific ECM dysfunctions present in a patient's tissue. For instance, in cancer treatment, multi-omics data can help identify ECM proteins that are driving tumor progression and predict which patients are likely to respond to ECM-targeted drugs, such as inhibitors of matrix metalloproteinases or integrin blockers. The integration of multi-omics data also plays a crucial role in advancing ECM-based biomaterial design for regenerative medicine and tissue engineering [32]. By understanding the molecular composition and functional properties of native ECMs through multi-omics analysis, researchers can better design synthetic or natural biomaterials that mimic the native ECM environment. This biomimetic approach enhances the performance of engineered tissues, improving their ability to support cell growth, differentiation, and integration with host tissues. Multi-omics data can guide the selection of ECM components that optimize the mechanical properties and bioactivity of scaffolds, leading to more effective treatments for tissue repair and regeneration.

Challenges in multi-omics integration, such as data complexity and variability, are being addressed through advanced computational tools and machine learning algorithms. These technologies enable the efficient analysis and integration of large-scale omics datasets, extracting meaningful patterns and correlations that provide new insights into ECM biology. Machine learning models can predict how specific ECM changes impact disease outcomes or treatment responses, facilitating the development of predictive biomarkers and personalized therapeutic strategies. The integration of multi-omics data represents a transformative approach in ECM analysis, offering a comprehensive view of the molecular and functional landscape of the ECM. By bridging the gaps between genomics, proteomics, metabolomics, and other omics fields, this approach provides unprecedented insights into the roles of the ECM in health and disease. The ability to capture the full spectrum of ECM dynamics opens new avenues for diagnostic innovation, personalized medicine, and the development of novel ECM-targeted therapies, ultimately advancing our understanding of tissue biology and improving patient care.

6.8 Clinical Applications and Future Directions

ECM-based diagnostics are rapidly gaining traction in clinical settings due to their potential to provide novel insights into the microenvironment of tissues, offering a unique approach to disease detection and monitoring. The extracellular matrix (ECM) is crucial in maintaining tissue structure, influencing cell behavior, and regulating various physiological processes. Abnormal ECM remodeling is a hallmark of several diseases, including cancer, fibrosis, cardiovascular diseases, and inflammatory disorders. By analyzing ECM components, such as proteins, glycoproteins, and enzymes, ECM-based diagnostics can detect early pathological changes that traditional diagnostic methods might overlook. This capability is particularly valuable in oncology, where the ECM plays a significant role in tumor progression, metastasis, and resistance to therapy. ECM-derived biomarkers, such as altered collagen isoforms or increased levels of matrix metalloproteinases (MMPs), can serve as early indicators of cancer, improving the accuracy of diagnosis and prognosis.

In cardiovascular diseases, ECM-based diagnostics have shown promise in detecting conditions such as atherosclerosis, myocardial fibrosis, and heart failure. Changes in ECM composition, like increased collagen deposition or altered elastin content, can indicate pathological remodeling of the heart and blood vessels. For instance, the detection of specific ECM proteins or degradation products in blood samples can provide insights into the extent of vascular damage or cardiac remodeling, aiding in the early diagnosis and monitoring of disease progression. These ECM biomarkers can also help assess the risk of adverse cardiovascular events, allowing for more personalized and timely interventions [33]. Additionally, ECM analysis can guide the management of patients with heart failure by evaluating the effectiveness of therapies targeting ECM remodeling, such as anti-fibrotic drugs. In the field of regenerative medicine, ECM-based diagnostics are crucial for evaluating tissue-engineered constructs and biomaterials designed for clinical applications. By analyzing the ECM composition of these constructs, clinicians can assess their suitability for implantation and their ability to support cell growth and integration with host tissues. This is particularly important in applications like wound healing, where ECM-based diagnostics can monitor the quality of newly formed ECM, ensuring that it closely mimics the native tissue environment. ECM analysis can help detect signs of adverse reactions, such as excessive fibrosis or immune rejection, guiding modifications to improve the performance and safety of regenerative therapies. The future of ECM-based diagnostics lies in integrating advanced technologies, such as nanotechnology, artificial intelligence (AI), and microfluidics, to enhance the sensitivity, specificity, and accessibility of these diagnostic tools. Nanotechnology enables the development of highly sensitive sensors and imaging agents that can detect minute changes in ECM composition, providing early and accurate disease diagnosis. AI and machine learning algorithms can analyze complex ECM datasets, uncovering patterns and biomarkers that might be missed by conventional methods. These computational approaches can also predict disease outcomes based on ECM alterations, facilitating the development of personalized treatment plans. Microfluidic devices, or lab-on-a-chip systems, allow for the miniaturization and automation of ECM analysis, enabling rapid, cost-effective testing at the point of care.

Looking ahead, ECM-based diagnostics are poised to play a central role in the era of precision medicine, where treatments are tailored to the specific molecular and structural characteristics of a patient's disease. By providing detailed information about the ECM's role in disease progression, these diagnostics can guide the selection of targeted therapies that modulate ECM interactions, such as drugs that inhibit ECM-degrading enzymes or alter ECM stiffness. In oncology, for example, ECM-based diagnostics could help identify patients who are likely to benefit from ECM-targeted therapies, such as inhibitors of tumor-associated fibroblasts or agents that normalize the tumor ECM to improve drug delivery.

As research continues to uncover new ECM biomarkers and refine diagnostic technologies, ECM-based diagnostics will likely expand into a broader range of clinical applications, including monitoring response to therapy and predicting disease recurrence. The integration of multi-omics data with ECM analysis will further enhance our understanding of ECM dynamics, allowing for the discovery of novel biomarkers and therapeutic targets.

6.8.1 Translational Potential of ECM-Based Diagnostics

The translational potential of ECM-based diagnostics is significant, offering promising pathways from research discoveries to clinical applications that can profoundly impact patient care. By focusing on the extracellular matrix (ECM), researchers can identify novel biomarkers and mechanistic insights into various diseases, including cancer, fibrosis, and cardiovascular disorders. The ability to detect subtle ECM changes and alterations in its composition or function enables earlier diagnosis, more precise disease staging, and more targeted therapies. As ECM-based diagnostics evolve from the laboratory to clinical settings, they hold the potential to enhance disease management by providing tools for early detection and monitoring of disease progression, ultimately improving patient outcomes. In addition to early diagnosis, ECM-based diagnostics offer substantial translational potential in the development of personalized medicine approaches. By analyzing individual patient ECM profiles, these diagnostics can guide the selection of tailored therapies based on specific ECM alterations. This personalized approach is particularly valuable in oncology, where different tumors exhibit distinct ECM characteristics that influence treatment response. For example, identifying ECM biomarkers associated with drug resistance or sensitivity can help clinicians choose the most effective treatment regimens, reducing trial-and-error and improving therapeutic efficacy. Similarly, in regenerative medicine, ECM-based diagnostics can evaluate the quality of tissue-engineered constructs and biomaterials, ensuring that they meet the required standards for successful implantation and integration.

The future of ECM-based diagnostics also includes the integration of advanced technologies such as nanotechnology, artificial intelligence (AI), and microfluidics, which enhance their clinical applicability and accessibility. Nanotechnology enables the creation of highly sensitive sensors and imaging agents, while AI and machine learning algorithms provide sophisticated data analysis capabilities to identify biomarkers and predict disease outcomes [34]. Microfluidic devices facilitate rapid, cost-effective testing, bringing ECM-based diagnostics closer to point-of-care settings. As these technologies advance, they will support the broader adoption of ECM-based diagnostics in routine clinical practice, bridging the gap between cutting-edge research and practical applications, and ultimately leading to improved diagnostics and patient care across a range of medical fields.

6.9 Summary

The chapter on ECM-based diagnostics innovations highlights the transformative impact of advanced technologies on the analysis of extracellular matrix (ECM) components. It underscores how novel diagnostic approaches, leveraging cutting-edge tools such as high-resolution imaging, nanotechnology, and multi-omics integration, are enhancing our ability to detect and monitor ECM-related changes associated with various diseases. These innovations allow for unprecedented precision in identifying biomarkers, understanding disease mechanisms, and guiding personalized treatment strategies. By focusing on the intricate role of the ECM in maintaining tissue integrity and regulating cellular processes, these advancements offer new avenues for early diagnosis and targeted therapy, particularly in complex diseases like cancer, cardiovascular disorders, and fibrosis. The chapter also explores the integration of computational models and artificial intelligence (AI) with ECM data analysis, emphasizing their role in advancing diagnostic capabilities. AI and machine learning algorithms facilitate the processing of large-scale, multi-omics datasets, uncovering patterns and relationships that traditional methods might overlook. This integration enables the development of predictive models for disease progression and treatment response, supporting the creation of more effective and personalized therapeutic approaches. Computational tools enhance our ability to analyze ECM alterations comprehensively, offering insights that drive innovations in both diagnostics and therapeutic development.

Looking to the future, the chapter anticipates the continued evolution of ECM-based diagnostics through further technological advancements and broader clinical applications. Innovations such as microfluidic devices and nanotechnology are expected to make ECM diagnostics more accessible and cost-effective, paving the way for point-of-care testing and real-time monitoring. As these technologies become more refined, they will play a crucial role in personalized medicine, enabling precise, individualized treatment plans and improving patient outcomes. The chapter concludes by emphasizing the potential of ECM-based diagnostics to bridge the gap between research and clinical practice, offering valuable tools for enhancing disease detection, monitoring, and management across a range of medical disciplines.

Points to Remember

- **ECM-based diagnostics have revealed that specific ECM proteins, such as tenascin-C and fibronectin, can serve as early indicators of disease progression in cancer, even before structural changes are visible using traditional imaging techniques.**

- **Nanotechnology has enabled the creation of novel ECM sensors using gold nanoparticles, which can be functionalized to bind specifically to ECM biomarkers. These sensors offer extremely high sensitivity and can detect biomarker levels at picogram quantities.**

- **By integrating transcriptomics, proteomics, and metabolomics with ECM analysis, researchers have discovered that changes in ECM components can reflect shifts in cellular metabolism and gene expression patterns, providing a more holistic understanding of disease mechanisms and enabling more precise diagnostics.**

- **AI algorithms have been developed to analyze ECM images and data, identifying patterns and changes that are imperceptible to the human eye. For example, AI can detect subtle ECM structural changes associated with early-stage diseases, enhancing early detection and intervention strategies.**

- **Microfluidic devices have been engineered to mimic the mechanical and biochemical properties of the ECM, allowing researchers to study cell-ECM interactions in a controlled environment.**

- **ECM-based diagnostics are paving the way for personalized medicine by identifying individual-specific ECM profiles. This information helps tailor treatment strategies based on the unique ECM alterations present in a patient's disease, improving the efficacy of targeted therapies and reducing the risk of adverse reactions.**

References

1. Pompili S, Latella G, Gaudio E, Sferra R, Vetuschi A (2021) The charming world of the extracellular matrix: a dynamic and protective network of the intestinal wall. Front Med 8:610189. https://doi.org/10.3389/fmed.2021.610189
2. Yue B (2014) Biology of the extracellular matrix: an overview. J Glaucoma 23(8 Suppl 1):S20–S23. https://doi.org/10.1097/IJG.0000000000000108
3. Papanicolaou M, Cox TR (2021) Extracellular matrix (ECM). In: Offermanns S. Rosenthal W (eds) Encyclopedia of molecular pharmacology. Springer, Cham. https://doi.org/10.1007/978-3-030-57401-7_5691
4. Kyriakopoulou K, Piperigkou Z, Tzaferi K, Karamanos NK (2023) Trends in extracellular matrix biology. Mol Biol Rep 50(1):853–863. https://doi.org/10.1007/s11033-022-07931-y
5. Soles A, Selimovic A, Sbrocco K, Ghannoum F, Hamel K, Moncada EL, Gilliat S, Cvetanovic M (2023) Extracellular matrix regulation in physiology and in brain disease. Int J Mol Sci 24(8):7049. https://doi.org/10.3390/ijms24087049

6. Bloksgaard M, Lindsey M, Martinez-Lemus LA (2018) Extracellular matrix in cardiovascular pathophysiology. Am J Physiol Heart Circ Physiol 315(6):H1687–H1690. https://doi.org/10.1152/ajpheart.00631.2018
7. Tewari BP, Conant K (2024) Editorial: Emerging roles of extracellular matrix in the physiology and pathophysiology of the central nervous system. Front Cell Neurosci 18:1400652. https://doi.org/10.3389/fncel.2024.1400652
8. Petersen EV, Chudakova DA, Skorova EY, Anikin V, Reshetov IV, Mynbaev OA (2020) The extracellular matrix-derived biomarkers for diagnosis, prognosis, and personalized therapy of malignant tumors. Front Oncol 10:575569. https://doi.org/10.3389/fonc.2020.575569
9. Yin Y, Yang X, Cheng Z, Wang H, Lei J, Wang D, Wang P, Li B, Mi J, Yuan Q (2024) Identification of extracellular matrix-related biomarkers in colon adenocarcinoma by bioinformatics and experimental validation. Front Immunol 15:1371584. https://doi.org/10.3389/fimmu.2024.1371584
10. Barbosa LC, Machado GC, Heringer M, Ferrer VP (2024) Identification of established and novel extracellular matrix components in glioblastoma as targets for angiogenesis and prognosis. Neurogenetics 25(3):249–262. https://doi.org/10.1007/s10048-024-00763-x
11. Bonnans C, Chou J, Werb Z (2014) Remodelling the extracellular matrix in development and disease. Nat Rev Mol Cell Biol 15(12):786–801. https://doi.org/10.1038/nrm3904
12. Huang J, Zhang L, Wan D et al (2021) Extracellular matrix and its therapeutic potential for cancer treatment. Sig Transduct Target Ther 6:153. https://doi.org/10.1038/s41392-021-00544-0
13. Sarohi V, Chakraborty S, Basak T (2022) Exploring the cardiac ECM during fibrosis: a new era with next-gen proteomics. Front Mol Biosci 9:1030226. https://doi.org/10.3389/fmolb.2022.1030226
14. Kim SJ, Mesquita FCP, Hochman-Mendez C (2023) New biomarkers for cardiovascular disease. Texas Heart Inst J 50(5):e238178. https://doi.org/10.14503/THIJ-23-8178
15. Grover VP, Tognarelli JM, Crossey MM, Cox IJ, Taylor-Robinson SD, McPhail MJ (2015) Magnetic resonance imaging: principles and techniques: lessons for clinicians. J Clin Exp Hepatol 5(3):246–255. https://doi.org/10.1016/j.jceh.2015.08.001
16. Rong J, Liu Y (2024) Advances in medical imaging techniques. BMC Methods 1:10. https://doi.org/10.1186/s44330-024-00010-7
17. Abhisheka B, Biswas SK, Purkayastha B et al (2024) Recent trend in medical imaging modalities and their applications in disease diagnosis: a review. Multimed Tools Appl 83:43035–43070. https://doi.org/10.1007/s11042-023-17326-1
18. Ong J, Zarnegar A, Corradetti G, Singh SR, Chhablani J (2022) Advances in optical coherence tomography imaging technology and techniques for choroidal and retinal disorders. J Clin Med 11(17):5139. https://doi.org/10.3390/jcm11175139
19. Glover, B., Teare, J., & Patel, N. (2020). The status of advanced imaging techniques for optical biopsy of colonic polyps. Clin Transl Gastroenterol 11(3):e00130. https://doi.org/10.14309/ctg.0000000000000130
20. Young MD, Field JJ, Sheetz KE, Bartels RA, Squier J (2015) A pragmatic guide to multiphoton microscope design. Adv Opt Photon 7(2):276–378. https://doi.org/10.1364/AOP.7.000276
21. Salarian M, Ibhagui OY, Yang JJ (2020) Molecular imaging of extracellular matrix proteins with targeted probes using magnetic resonance imaging. Wiley Interdiscip Rev Nanomed Nanobiotechnol 12(4):e1622. https://doi.org/10.1002/wnan.1622
22. Crosby D, Bhatia S, Brindle KM, Coussens LM, Dive C, Emberton M, Esener S, Fitzgerald RC, Gambhir SS, Kuhn P, Rebbeck TR, Balasubramanian S (2022) Early detection of cancer. Science (New York, N.Y.) 375(6586), eaay9040. https://doi.org/10.1126/science.aay9040
23. Li Y, Fan H, Ding J, Xu J, Liu C, Wang H (2022) Microfluidic devices: The application in TME modeling and the potential in immunotherapy optimization. Front Genet 13:969723. https://doi.org/10.3389/fgene.2022.969723
24. Ingber DE (2022) Human organs-on-chips for disease modelling, drug development and personalized medicine. Nat Rev Genet 23(8):467–491. https://doi.org/10.1038/s41576-022-00466-9

25. Savoji H, Mohammadi MH, Rafatian N, Toroghi MK, Wang EY, Zhao Y, Korolj A, Ahadian S, Radisic M (2019) Cardiovascular disease models: a game changing paradigm in drug discovery and screening. Biomaterials 198:3–26. https://doi.org/10.1016/j.biomaterials.2018.09.036
26. Buttkewitz MA, Heuer C, Bahnemann J (2023) Sensor integration into microfluidic systems: trends and challenges. Curr Opin Biotechnol 83:102978. https://doi.org/10.1016/j.copbio.2023.102978
27. Karamanos NK, Theocharis AD, Piperigkou Z, Manou D, Passi A, Skandalis SS, Vynios DH, Orian-Rousseau V, Ricard-Blum S, Schmelzer CEH, Duca L, Durbeej M, Afratis NA, Troeberg L, Franchi M, Masola V, Onisto M (2021) A guide to the composition and functions of the extracellular matrix. FEBS J 288(24):6850–6912. https://doi.org/10.1111/febs.15776
28. Nasir A, Khan A, Li J, Naeem M, Khalil AAK, Khan K, Qasim M (2021) Nanotechnology, a tool for diagnostics and treatment of cancer. Curr Top Med Chem 21(15):1360–1376. https://doi.org/10.2174/1568026621666210701144124
29. Savaliya R, Shah D, Singh R, Kumar A, Shanker R, Dhawan A, Singh S (2015) Nanotechnology in disease diagnostic techniques. Curr Drug Metab 16(8):645–661
30. Geng T, Zheng M, Wang Y, Reseland JE, Samara A (2023) An artificial intelligence prediction model based on extracellular matrix proteins for the prognostic prediction and immunotherapeutic evaluation of ovarian serous adenocarcinoma. Front Mol Biosci 10:1200354. https://doi.org/10.3389/fmolb.2023.1200354
31. Bersanelli M, Mosca E, Remondini D, Giampieri E, Sala C, Castellani G, Milanesi L (2016) Methods for the integration of multi-omics data: mathematical aspects. BMC Bioinform 17(Suppl 2), 15. https://doi.org/10.1186/s12859-015-0857-9
32. Wörheide MA, Krumsiek J, Kastenmüller G, Arnold M (2021) Multi-omics integration in biomedical research—a metabolomics-centric review. Anal Chim Acta 1141:144–162. https://doi.org/10.1016/j.aca.2020.10.038
33. Stremmel C, Breitschwerdt R (2023) Digital transformation in the diagnostics and therapy of cardiovascular diseases: comprehensive literature review. JMIR cardio 7:e44983. https://doi.org/10.2196/44983
34. Möller A, Lobb RJ (2020) The evolving translational potential of small extracellular vesicles in cancer. Nat Rev Cancer 20(12):697–709. https://doi.org/10.1038/s41568-020-00299-w

Chapter 7
Therapeutic Applications of ECM Mimicry

7.1 Introduction

Extracellular matrix (ECM) mimicry refers to the design and fabrication of biomaterials that replicate the structure, composition, and function of the natural ECM found within tissues. The ECM is a complex network of proteins, glycoproteins, and polysaccharides that provides structural and biochemical support to surrounding cells. Mimicking this environment is crucial in tissue engineering and regenerative medicine, as it can facilitate cell adhesion, proliferation, differentiation, and migration, leading to enhanced tissue repair and regeneration [1]. ECM mimicry is increasingly recognized as a pivotal approach in developing advanced therapeutic strategies for various diseases and injuries. The importance of ECM mimicry lies in its ability to provide a supportive scaffold that closely resembles the native cellular environment, thus promoting more natural cell behavior and function. In tissue engineering, replicating ECM is essential for creating scaffolds that can integrate seamlessly with host tissues. These biomimetic materials can also be engineered to release growth factors or other bioactive molecules, further enhancing their therapeutic potential [2]. The ability to mimic ECM properties has opened new avenues for developing therapies that can accelerate wound healing, regenerate damaged tissues, and improve the efficacy of drug delivery systems.

Historically, the concept of ECM mimicry has evolved alongside advancements in biomaterials and tissue engineering. The early focus in regenerative medicine was on developing materials that could physically replace damaged tissues. However, it soon became evident that simply replacing tissue was insufficient; the materials needed to interact dynamically with the surrounding biological environment. This realization led to the development of synthetic materials designed to mimic the natural ECM, enabling better integration with host tissues and more effective tissue regeneration. Over the decades, research has expanded to include a variety of approaches to replicate ECM, ranging from natural polymers like collagen and fibrin to synthetic polymers like poly (lactic-co-glycolic acid) (PLGA) [3]. Natural ECM is a highly

D. Suhag, *Extracellular Matrix biomimicking for Theranostics*, Biomedical Materials for Multi-functional Applications, https://doi.org/10.1007/978-981-95-7900-6_7

dynamic and complex structure, composed of a diverse array of proteins such as collagen, elastin, fibronectin, and laminin, as well as proteoglycans and glycosaminoglycans like hyaluronic acid. These components not only provide mechanical support but also play crucial roles in cell signaling and the regulation of cellular activities. The composition and organization of the ECM vary between different tissues, reflecting their unique functional requirements. For example, the ECM in bone tissue is rich in collagen and hydroxyapatite, providing both strength and flexibility, whereas the ECM in the liver is softer and more elastic to accommodate the organ's unique functions.

Synthetic ECMs, on the other hand, are typically engineered to replicate specific aspects of the natural ECM. These materials can be designed to match the mechanical properties, porosity, and biochemical cues of the tissue they aim to replace or support. Synthetic ECMs offer several advantages, including the ability to tailor their properties to specific applications and the potential for large-scale production. However, they often lack the full complexity of natural ECMs, which can limit their effectiveness in certain applications. To overcome this, researchers are exploring hybrid approaches that combine natural and synthetic components to create more sophisticated and functional ECM mimics [4]. The development of synthetic ECMs has been driven by the need to address the limitations of natural ECMs, such as their potential for immunogenicity and variability between batches. Synthetic materials can be engineered with precise control over their properties, allowing for the creation of scaffolds that are biocompatible, biodegradable, and capable of promoting specific cellular responses. These materials are often used in combination with natural ECM components or growth factors to enhance their bioactivity and functionality. Advances in nanotechnology and materials science have further expanded the possibilities for creating ECM mimics with highly tunable properties.

An overview of ECM components highlights the diversity and complexity of this essential biological structure. Collagen, the most abundant protein in the ECM, provides tensile strength and structural integrity, while elastin allows tissues to stretch and recoil. Glycoproteins like fibronectin and laminin are involved in cell adhesion and migration, mediating interactions between cells and the ECM. Proteoglycans and glycosaminoglycans, such as hyaluronic acid, contribute to the ECM's viscoelastic properties and play a role in cell signaling and hydration. Understanding the roles of these components is crucial for designing effective ECM mimics that can support tissue regeneration and repair. ECM mimicry is a rapidly evolving field that holds significant promise for advancing therapeutic applications in regenerative medicine. By replicating the structure and function of the natural ECM, researchers can develop biomaterials that not only support tissue regeneration but also enhance the body's natural healing processes. The ongoing exploration of both natural and synthetic ECMs, and their components, continues to drive innovation in the design of advanced scaffolds and therapeutic strategies, with the ultimate goal of improving patient outcomes in a wide range of medical conditions.

7.2 Principles of ECM Mimicry in Therapeutic Applications

The principles of extracellular matrix (ECM) mimicry in therapeutic applications revolve around the careful replication of the natural ECM's structural, mechanical, and biochemical properties to create biomaterials that can effectively support tissue regeneration and healing. At its core, ECM mimicry aims to create an environment that closely resembles the native tissue, allowing cells to interact with the scaffold in a manner similar to how they would within the body. This approach is critical for promoting cellular activities such as adhesion, proliferation, and differentiation, which are essential for tissue repair and regeneration. By integrating specific biochemical cues and mechanical properties, ECM mimics can guide cellular behavior, leading to improved outcomes in various therapeutic applications, including wound healing, cancer treatment, and organ regeneration. In therapeutic contexts, the success of ECM mimicry hinges on understanding the intricate interactions between cells and their microenvironment. The ECM provides not only physical support but also signals that regulate cellular processes, such as growth and migration. Mimicking these interactions requires a deep understanding of the ECM's role in tissue homeostasis and the factors that influence its function [5]. For instance, the stiffness of the ECM can affect stem cell differentiation, while the presence of specific proteins like collagen or laminin can enhance cell adhesion and tissue integration. By replicating these features in synthetic or hybrid materials, researchers can develop scaffolds that not only support tissue structure but also actively participate in the healing process, making ECM mimicry a powerful tool in regenerative medicine and therapeutic interventions.

7.2.1 Understanding ECM Structure and Function

The extracellular matrix (ECM) is a highly complex and dynamic structure that plays a crucial role in maintaining tissue integrity and function. Composed primarily of proteins, glycoproteins, and polysaccharides, the ECM provides both structural support and biochemical signals that regulate a wide range of cellular behaviors. Its structure is not uniform across different tissues but is instead tailored to meet the specific mechanical and functional requirements of each tissue type. For example, the ECM in bone tissue is mineralized to provide rigidity and strength, while in the skin, it is more elastic to allow for flexibility and resilience. Understanding the ECM's composition and organization is key to appreciating its role in health and disease, as well as its potential for therapeutic applications.

At the molecular level, the ECM is composed of various macromolecules, with collagen being the most abundant protein. Collagen fibers provide tensile strength and structural support, forming a scaffold that maintains tissue architecture. Elastin, another key ECM component, imparts elasticity to tissues, allowing them to stretch and return to their original shape. Glycoproteins like fibronectin and laminin are involved in cell adhesion, migration, and differentiation, serving as bridges between the ECM and cell surface receptors. These interactions are essential for tissue development, repair, and homeostasis, as they influence cell behavior through a process known as mechano transduction, where cells convert mechanical signals from the ECM into biochemical responses. The ECM is not just a passive scaffold but is actively involved in regulating cellular processes through the binding and presentation of growth factors and cytokines [6]. These bioactive molecules are sequestered within the ECM and can be released in response to environmental cues, such as tissue injury or changes in mechanical stress. This capability allows the ECM to modulate processes like wound healing, inflammation, and tissue remodeling. For instance, during wound healing, the ECM undergoes dynamic remodeling, with enzymes like matrix metalloproteinases (MMPs) breaking down the ECM to allow for the migration of repair cells, followed by the deposition of new matrix components to restore tissue structure.

The mechanical properties of the ECM, such as its stiffness and elasticity, play a significant role in influencing cell behavior. Cells can sense these mechanical cues through integrins, which are transmembrane receptors that connect the ECM to the cytoskeleton. The stiffness of the ECM can direct stem cell fate, with softer matrices promoting adipogenic (fat cell) differentiation and stiffer matrices favoring osteogenic (bone cell) differentiation. This mechano sensitivity is critical in tissue engineering and regenerative medicine, where creating scaffolds with appropriate mechanical properties can guide tissue regeneration in a controlled manner.

The ECM is a multifaceted structure that performs vital functions beyond providing physical support to tissues. Its ability to influence cellular behavior through biochemical signals and mechanical cues underscores its importance in maintaining tissue homeostasis and facilitating repair. Understanding the ECM's structure and function is essential for developing biomimetic materials that can replicate these features, offering significant potential for therapeutic applications in regenerative medicine and beyond.

Exercise 1

Fill Ups
(Natural, Proliferation, Structure, Migration, Composition, Proliferation, Fibronectin, Adhesion)

1. The principle of ECM mimicry involves designing materials that replicate the ______ and ______ of natural extracellular matrices.

2. ECM mimetics are often used in tissue engineering to promote cell ______ and ______.

3. The use of ECM mimicry in therapeutics can enhance drug delivery by providing a more ______ environment for targeted therapies.

4. One of the key components often incorporated into ECM mimics is ______, which plays a role in cell adhesion and growth.

5. ECM mimicry can improve wound healing by facilitating ______ and ______ of skin cells at the injury site.

7.2.2 Biochemical Cues in ECM

Biochemical cues within the extracellular matrix (ECM) are crucial in regulating cellular behavior and maintaining tissue homeostasis. These cues are primarily provided by the various proteins, glycoproteins, and proteoglycans that make up the ECM. These molecules interact with cell surface receptors, such as integrins, to influence processes like cell adhesion, migration, proliferation, and differentiation. For instance, specific domains within fibronectin and laminin bind to integrins on the cell surface, initiating signaling cascades that guide cell movement and positioning within the tissue. These biochemical interactions are essential for processes like wound healing, where cells need to migrate to the injury site and proliferate to replace damaged tissue. Growth factors and cytokines sequestered within the ECM also provide vital biochemical cues that regulate cellular responses to environmental changes. These signaling molecules are often bound to ECM components and can be released in response to mechanical stress or tissue injury, enabling the ECM to act as a reservoir of bioactive molecules. For example, transforming growth factor-beta (TGF-β), a key regulator of cell proliferation and differentiation, is stored within the ECM and becomes activated during tissue remodeling or repair [7]. This controlled release of growth factors is critical for coordinating complex biological processes like angiogenesis, inflammation, and tissue regeneration, ensuring that cells receive the right signals at the right time.

The biochemical composition of the ECM can vary between different types of tissue and stages of development, allowing for the precise regulation of cellular behavior in different contexts. In embryonic tissues, the ECM is rich in hyaluronic acid and other glycosaminoglycans, which promote cell proliferation and migration, facilitating rapid tissue growth and morphogenesis. In contrast, the ECM in mature tissues may contain higher levels of cross-linked collagen and elastin, providing structural stability while maintaining the ability to transmit biochemical signals. This dynamic nature of the ECM's biochemical landscape allows it to adapt to the changing needs of tissues, making it a powerful regulator of cellular function and a key target for therapeutic interventions in regenerative medicine.

7.2.3 Mechanical Properties of ECM

The mechanical properties of the extracellular matrix (ECM) play a pivotal role in influencing cell behavior and are a fundamental aspect of ECM mimicry in therapeutic applications (Fig. 7.1). The ECM's mechanical characteristics, such as stiffness, elasticity, and tensile strength, provide structural support to tissues and serve as crucial signals that cells interpret to guide their functions. Cells are highly sensitive to the mechanical environment, and they respond to these cues through a process called mechano transduction. This process involves the conversion of mechanical signals into biochemical responses, influencing key cellular activities like proliferation, differentiation, and migration. In tissue engineering, replicating these mechanical properties in synthetic or hybrid scaffolds is essential for creating environments that promote the desired cellular responses, thereby enhancing tissue repair and regeneration. Different tissues in the body exhibit distinct mechanical properties, tailored to their specific functional requirements, and these differences must be carefully considered in the design of ECM mimics. For instance, bone tissue has a highly rigid ECM to withstand compressive forces, while the ECM in soft tissues like the brain is much more compliant, allowing for flexibility and the accommodation of delicate neural structures [8]. In therapeutic applications, scaffolds designed to mimic these mechanical properties can significantly influence the success of tissue regeneration. For example, in bone tissue engineering, a scaffold that closely mimics the stiffness of natural bone can encourage stem cells to differentiate into osteoblasts, the cells responsible for bone formation. Similarly, softer scaffolds are more suitable for applications in neural tissue engineering, where they can support the growth and function of neurons without imposing undue mechanical stress.

The ability to modulate the mechanical properties of ECM mimics provides a powerful tool for tailoring therapeutic approaches to specific clinical needs. Advances in materials science have enabled the development of biomaterials with tunable mechanical properties, allowing researchers to create scaffolds that can be adjusted to match the unique requirements of different tissues or stages of healing. This adaptability is particularly important in dynamic environments, such as wound healing, where the mechanical demands on the tissue change over time. By designing ECM

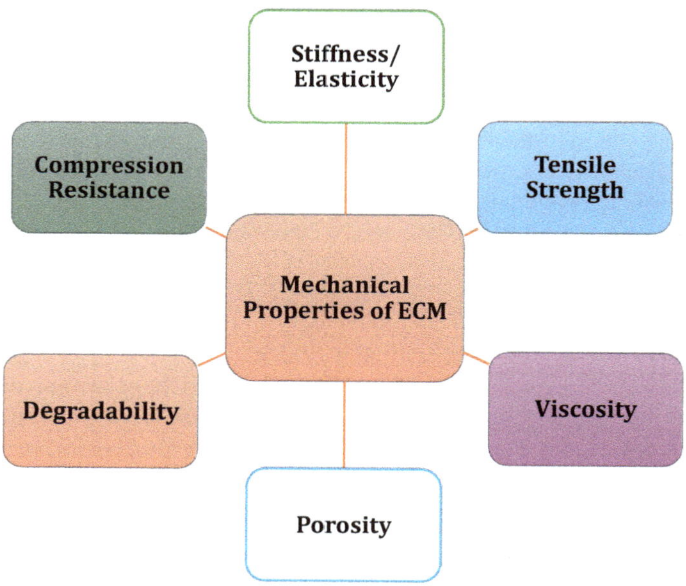

Fig. 7.1 Various mechanical properties of ECM

mimics that can evolve in response to these changes, it is possible to support more effective and sustained tissue regeneration, making mechanical properties a central principle in the design of ECM-based therapeutic applications.

7.2.4 Cellular Interactions with ECM

Cellular interactions with the extracellular matrix (ECM) are fundamental to tissue development, maintenance, and repair, making them a critical focus in ECM mimicry for therapeutic applications. These interactions are mediated by various cell surface receptors, most notably integrins, which bind to specific ECM proteins such as collagen, fibronectin, and laminin. Through these connections, cells can sense the biochemical and mechanical properties of the ECM, which in turn influences their behavior. For example, integrin-mediated signaling pathways regulate essential processes like cell adhesion, migration, proliferation, and differentiation. In therapeutic contexts, designing ECM mimics that effectively replicate these cellular interactions is crucial for promoting desired cellular responses, ultimately enhancing tissue regeneration and healing outcomes. The dynamic nature of cellular interactions with the ECM is also a key consideration in ECM mimicry. Cells constantly remodel the ECM through processes such as secretion of matrix metalloproteinases (MMPs), which degrade ECM components, and the deposition of new matrix materials. This remodeling allows the ECM to adapt to changing tissue conditions, such as during wound healing or in response to mechanical stress. In creating ECM mimics, it is essential to incorporate materials that can support this dynamic interaction, allowing cells to remodel the scaffold as needed for optimal tissue regeneration [9]. This ability to interact and modify the ECM ensures that the cells remain functionally active and that the scaffold integrates effectively with the surrounding tissue.

The specificity of cellular interactions with the ECM can be leveraged to direct cell behavior in therapeutic applications. By incorporating specific binding motifs or sequences within ECM mimics, such as the RGD (arginine-glycine-aspartic acid) sequence found in many ECM proteins, it is possible to enhance cell adhesion and promote specific cellular functions. For instance, scaffolds designed for cartilage regeneration might include chondrogenic cues that encourage stem cells to differentiate into chondrocytes, the cells responsible for cartilage formation. This targeted approach not only improves the efficacy of the therapeutic scaffold but also reduces the risk of undesired cellular responses, making the design of cell-ECM interactions a central principle in the development of ECM-based therapies.

Exercise 2

Fill Ups
(Cosmetic, Cell, Microenvironment, Immune, Collagen)

1. ECM mimicry in wound healing aims to recreate the natural ______ environment to support tissue repair.

2. Biomaterials used for ECM mimicry often include components like ______ and hyaluronic acid to promote healing.

3. The presence of an ECM mimic can enhance ______ migration, which is crucial for closing wounds.

4. ECM mimics can facilitate the recruitment of ______ cells, which play a vital role in the healing process.

5. Successful ECM mimicry in wound healing can lead to improved ______ outcomes and reduced scarring.

7.2.5 ECM Mimicry in Tissue Engineering

ECM mimicry plays a crucial role in tissue engineering, where the goal is to develop scaffolds that can guide and support the regeneration of damaged or diseased tissues. The extracellular matrix (ECM) in natural tissues provides not only a physical framework but also biochemical and mechanical cues that are essential for maintaining cellular functions and tissue homeostasis. Mimicking these properties in synthetic or bioengineered scaffolds is vital for creating environments that encourage the proper cell behavior needed for tissue regeneration. By replicating ECM's complex architecture and functional properties, ECM mimics can significantly enhance the effectiveness of tissue engineering strategies. One of the key aspects of ECM mimicry is the replication of the ECM's intricate three-dimensional structure, which is critical for providing the appropriate spatial context for cells. This structure supports cell adhesion, migration, and organization, all of which are essential for the formation of functional tissues. For example, in bone tissue engineering, scaffolds that mimic the porous and mineralized structure of bone ECM can support osteoblasts, the cells responsible for bone formation, leading to more effective bone regeneration [10]. Similarly, in cartilage tissue engineering, replicating the ECM's unique composition and structure can promote the formation of robust, elastic cartilage. The ability to recreate these specific structural features is a major reason why ECM mimicry is so important in tissue engineering.

In addition to structural support, ECM mimics are also designed to replicate the biochemical signals provided by the natural ECM. These signals, which include growth factors, cytokines, and specific protein domains, are crucial for regulating cell behavior and tissue development. For example, the incorporation of specific peptides or protein fragments that mimic the natural ECM can enhance cell attachment, proliferation, and differentiation. This biochemical mimicry is particularly important in guiding stem cell differentiation, where the presence of specific cues can direct stem cells to develop into the desired cell type, such as osteoblasts for bone or chondrocytes for cartilage. By carefully designing ECM mimics that incorporate these biochemical signals, researchers can create scaffolds that more effectively support tissue regeneration. Another critical aspect of ECM mimicry in tissue engineering is the replication of ECM's mechanical properties. The stiffness, elasticity, and viscoelasticity of the ECM vary between different tissues and play a significant role in influencing cell behavior [11]. For example, the stiffness of the ECM can direct stem cell differentiation; softer matrices tend to promote the development of fat cells, while stiffer matrices encourage the formation of bone cells. By designing scaffolds with mechanical properties that closely match those of the target tissue, researchers can create environments that are more conducive to successful tissue regeneration. This mechanical mimicry is particularly important in tissues that experience significant mechanical stress, such as bone, cartilage, and cardiovascular tissues.

The importance of ECM mimicry in tissue engineering extends to its potential for improving clinical outcomes in regenerative medicine. By creating scaffolds that closely mimic the natural ECM, it is possible to enhance the integration of engineered tissues with the host tissue, reduce the risk of immune rejection, and promote faster and more effective healing. This approach is particularly valuable in treating complex tissue defects or injuries where traditional therapies may be inadequate.

7.3 Design and Fabrication of ECM-Mimicking Scaffolds

The design and fabrication of extracellular matrix (ECM) mimicking scaffolds are central to the field of tissue engineering and regenerative medicine. These scaffolds aim to replicate the intricate architecture, biochemical composition, and mechanical properties of the natural ECM, providing an environment that supports and guides cellular processes essential for tissue regeneration. The design process involves a deep understanding of the specific tissue's ECM, including its structural and functional characteristics. This knowledge allows researchers to tailor scaffold properties such as porosity, stiffness, and bioactivity to closely match those of the native ECM, ensuring that the scaffold can effectively support cell adhesion, proliferation, differentiation, and tissue integration.

Fabrication techniques for ECM-mimicking scaffolds have evolved significantly, driven by advances in materials science and engineering [12]. Methods such as electrospinning, 3D bioprinting, and self-assembly enable the precise control of scaffold architecture at both the micro and nanoscale levels, allowing for the creation of complex structures that closely resemble the natural ECM. The incorporation of bioactive molecules, such as growth factors and adhesive peptides, into these scaffolds enhances their ability to interact with cells and guide tissue formation. The combination of innovative design and advanced fabrication techniques is key to developing effective ECM-mimicking scaffolds that can be used in various therapeutic applications, from wound healing to organ regeneration.

7.3.1 Material Selection for ECM Mimicry

Material selection is a crucial aspect of ECM mimicry, as the chosen materials must closely replicate the physical, chemical, and biological properties of the natural extracellular matrix (ECM) to support tissue regeneration effectively. The ideal materials for ECM mimicking scaffolds should possess biocompatibility, biodegradability, and mechanical properties that match those of the target tissue. These materials should facilitate cell adhesion, proliferation, and differentiation while promoting the integration of the scaffold with the surrounding tissue. The selection process often involves a careful balance between natural and synthetic materials, each offering distinct advantages that can be tailored to specific tissue engineering applications. Natural materials, such as collagen, gelatin, fibrin, and hyaluronic acid, are frequently used in ECM mimicry due to their inherent bioactivity and similarity to the components of the native ECM. These materials are highly biocompatible and can support cell attachment and growth without eliciting significant immune responses. Collagen, for instance, is widely used in bone and cartilage tissue engineering due to its structural role in natural ECM and its ability to promote osteogenic differentiation [13]. Similarly, hyaluronic acid, a key component of the ECM in skin and cartilage, is often used in scaffolds to enhance cell migration and proliferation. However, natural materials can present challenges, such as variability between batches and limited mechanical strength, which may require further modification or combination with synthetic materials.

Synthetic materials, such as poly (lactic-co-glycolic acid) (PLGA), polyethylene glycol (PEG), and polycaprolactone (PCL), offer several advantages in ECM mimicry, including tunable mechanical properties, controlled degradation rates, and ease of fabrication. These materials can be engineered to match the stiffness and elasticity of various tissues, making them suitable for applications ranging from soft tissues like skin and nerves to harder tissues like bone. The chemical structure of synthetic materials can be modified to incorporate functional groups or bioactive molecules, enhancing their interaction with cells and promoting specific cellular responses. Despite their versatility, synthetic materials may lack the inherent bioactivity of natural ECM components, necessitating their combination with natural materials or the addition of biochemical cues to enhance their performance. The combination of natural and synthetic materials, known as hybrid materials, is a promising approach in ECM mimicry, offering the best of both worlds. Hybrid scaffolds can be designed to provide the bioactivity and cellular interactions of natural materials while benefiting from the mechanical strength and customizability of synthetic polymers. For example, combining collagen with PLGA can result in a scaffold that supports cell adhesion and proliferation while providing the necessary mechanical properties for bone tissue engineering. Hybrid materials also allow for the incorporation of growth factors, peptides, and other bioactive molecules, further enhancing the scaffold's ability to guide tissue regeneration and repair.

The selection of materials for ECM mimicry is a critical factor that influences the success of tissue engineering and regenerative medicine applications. By carefully choosing and combining natural, synthetic, and hybrid materials, researchers can create scaffolds that closely mimic the structure, function, and bioactivity of the native ECM. These scaffolds can provide the necessary support for cellular processes, promote tissue integration, and ultimately lead to more effective and successful therapeutic outcomes.

7.3.2 Techniques for Scaffold Fabrication

Scaffold fabrication is a cornerstone of tissue engineering, with various techniques available to create structures that mimic the extracellular matrix (ECM) [14]. Each method has unique advantages, making it suitable for different applications. Techniques such as electrospinning, 3D printing, hydrogels, and self-assembly are among the most prominent, offering distinct ways to engineer scaffolds that support tissue regeneration and repair by replicating the natural ECM's physical, chemical, and biological properties.

- Electrospinning is a widely used technique for creating fibrous scaffolds that closely resemble the nanoscale architecture of natural ECM. In this process, a polymer solution is ejected through a needle under a high-voltage electric field, creating fine fibers that are collected on a grounded surface. The resulting scaffolds have a high surface area-to-volume ratio, which is ideal for cell attachment and proliferation. Electrospinning allows for precise control over fibre diameter, porosity, and alignment, making it suitable for engineering a wide range of tissues, from skin to neural tissue [15]. Bioactive molecules and growth factors can be incorporated into the electrospun fibers, further enhancing their functionality for tissue-specific applications.
- 3D printing, also known as additive manufacturing, has revolutionized scaffold fabrication by enabling the precise creation of complex, custom-designed structures. This technique involves layer-by-layer deposition of materials, guided by a digital model, to create scaffolds with highly controlled architecture and geometry. 3D printing allows for the fabrication of scaffolds with specific pore sizes, shapes, and interconnectivity, which are critical for promoting nutrient diffusion, waste removal, and cell migration within the scaffold. 3D printing can combine multiple materials within a single scaffold, creating gradients of mechanical and biochemical properties that better mimic the natural ECM of specific tissues. This versatility makes 3D printing an excellent choice for creating patient-specific scaffolds tailored to individual clinical needs.
- Hydrogels are another important scaffold fabrication technique, particularly valued for their high-water content and biocompatibility, which closely mimic the natural ECM's hydrated environment. Hydrogels are formed by cross-linking

hydrophilic polymers, resulting in a three-dimensional network that can encapsulate cells and bioactive molecules. Their tunable mechanical properties and ability to deliver sustained biochemical cues make hydrogels ideal for soft tissue engineering applications, such as in cartilage, brain, and wound healing. Hydrogels can be engineered to respond to environmental stimuli, such as pH or temperature, enabling controlled release of therapeutic agents and dynamic interactions with cells.
- Self-assembly is a biomimetic approach to scaffold fabrication, leveraging the inherent ability of molecules to spontaneously organize into structured forms. This technique relies on the natural affinity of certain peptides, proteins, or other biomolecules to form stable, organized structures without the need for external forces. Self-assembled scaffolds can mimic the nanoscale features of the ECM, providing an environment that closely resembles the native tissue's biochemical and structural cues. These scaffolds are particularly useful in applications where precise control over molecular interactions and scaffold architecture is required, such as in neural tissue engineering or the creation of biomimetic membranes.

The techniques for scaffold fabrication, including electrospinning, 3D printing, hydrogels, and self-assembly, offer a diverse array of tools for engineering scaffolds that mimic the ECM. Each technique brings its strengths to the table, allowing researchers to design and fabricate scaffolds tailored to the specific needs of different tissues.

7.3.3 Surface Modification Strategies

Surface modification strategies are critical in enhancing the functionality of scaffolds used in tissue engineering. By altering the surface properties of scaffolds, researchers can improve cell attachment, proliferation, and differentiation, ultimately leading to better integration with the surrounding tissue. Surface modifications can be physical, chemical, or biological, each offering unique benefits depending on the application. These strategies are designed to replicate the natural extracellular matrix (ECM) environment, providing the necessary cues for cells to perform their functions effectively.

Physical surface modification involves altering the scaffold's topography to create micro- and nanoscale features that mimic the ECM's structural complexity. Techniques such as plasma treatment, laser ablation, and nanopatterning are commonly used to introduce grooves, ridges, or pores onto the scaffold surface. These topographical features can influence cell behavior by guiding cell alignment, migration, and organization. For example, aligned nanofibers created through electrospinning can encourage the orientation of cells in nerve or muscle tissue engineering, promoting the formation of tissue structures that more closely resemble their natural counterparts. Chemical surface modification focuses on altering the scaffold's surface chemistry to enhance cell-material interactions. This can be achieved through techniques like plasma polymerization or chemical grafting which introduce functional groups or biomolecules onto the scaffold surface. By adding specific chemical groups, such as hydroxyl, carboxyl, or amine groups, researchers can improve the scaffold's hydrophilicity, thereby promoting better cell adhesion. Chemical modifications can be used to immobilize bioactive molecules like growth factors, peptides, or adhesive proteins on the scaffold surface, creating a more favorable environment for cell attachment and signaling.

Biological surface modification involves the incorporation of biological molecules directly onto the scaffold surface to mimic the ECM's natural composition. This approach often includes the use of peptides, proteins, or polysaccharides that are found in the ECM, such as fibronectin, laminin, or collagen. These biomolecules can be physically adsorbed or covalently bound to the scaffold, providing specific binding sites for cell surface receptors like integrins [16]. By presenting these biological cues, the scaffold can better support cell attachment, proliferation, and differentiation, leading to improved tissue integration and regeneration. Biological modifications are particularly useful in applications where specific cellular responses are required, such as in the differentiation of stem cells or the promotion of angiogenesis in vascular tissue engineering. Surface modification strategies are essential for enhancing the performance of tissue engineering scaffolds. Whether through physical, chemical, or biological means, these modifications allow for the fine-tuning of scaffold properties to better mimic the natural ECM and support tissue regeneration. By carefully selecting and applying the appropriate surface modification techniques, researchers can create scaffolds that provide the necessary structural, chemical, and biological cues for successful tissue repair and integration. As surface modification technologies continue to advance, they will play an increasingly important role in the development of more effective and functional tissue engineering scaffolds.

7.3.4 Integration of Growth Factors and Bioactive Molecules

The integration of growth factors and bioactive molecules into ECM-mimicking scaffolds is a critical aspect of their design and fabrication, aimed at enhancing their functionality for tissue engineering applications. Growth factors are proteins that play a pivotal role in regulating cellular processes such as proliferation, differentiation, and migration. By incorporating these molecules into scaffolds, researchers can create a more conducive environment for tissue regeneration, mimicking the dynamic signaling that occurs within the natural extracellular matrix (ECM) [17]. This strategy allows the scaffold to actively participate in the healing process, rather than merely providing structural support, thereby significantly improving the outcome of tissue engineering efforts. One of the key challenges in integrating growth factors into scaffolds is ensuring their controlled release over time. Growth factors are typically short-lived and can degrade quickly once introduced into the body. To address this, various techniques have been developed to encapsulate growth factors within the scaffold matrix or bind them to the scaffold surface. For instance, growth factors can be loaded into biodegradable microspheres or nanoparticles that are embedded within the scaffold, allowing for their gradual release as the scaffold degrades. Alternatively, growth factors can be tethered to the scaffold surface using chemical linkers, providing a sustained and localized delivery directly to the surrounding cells. These methods ensure that bioactive molecules are available at the right time and concentration to influence cell behavior effectively.

In addition to growth factors, other bioactive molecules such as peptides, proteins, and polysaccharides can be incorporated into ECM-mimicking scaffolds to enhance their biological activity. Peptides like the RGD (arginine-glycine-aspartic acid) sequence, which is found in many ECM proteins, can be used to promote cell adhesion by providing specific binding sites for cell surface receptors. Similarly, the inclusion of polysaccharides like heparin can enhance the scaffold's ability to bind and stabilize growth factors, further improving their bioavailability. By integrating a combination of these bioactive molecules, scaffolds can be designed to closely mimic the biochemical environment of the natural ECM, thereby promoting more effective tissue regeneration. The integration of growth factors and bioactive molecules into ECM-mimicking scaffolds also offers the potential for creating multifunctional scaffolds that address multiple aspects of tissue repair simultaneously. For example, a scaffold designed for bone regeneration might incorporate both osteoinductive growth factors, which encourage the differentiation of stem cells into bone-forming cells, and angiogenic factors, which promote the formation of new blood vessels to supply the growing tissue. This multi-faceted approach not only enhances the scaffold's ability to support complex tissue regeneration but also helps to overcome some of the limitations of traditional tissue engineering strategies.

7.3.5 Characterization of ECM-Mimicking Scaffolds

Characterization of ECM-mimicking scaffolds is a crucial step in ensuring that these engineered structures meet the necessary criteria for effective tissue regeneration. The characterization process involves a comprehensive assessment of the scaffold's physical, chemical, mechanical, and biological properties. This evaluation is essential for understanding how well the scaffold replicates the natural extracellular matrix (ECM) and how it will interact with cells and tissues in vivo. Through meticulous characterization, researchers can optimize scaffold design, ensuring it provides the appropriate environment for cell attachment, proliferation, differentiation, and tissue integration. One of the primary aspects of scaffold characterization is the analysis of its physical structure, including porosity, pore size, and overall architecture. Techniques such as scanning electron microscopy (SEM) and micro-computed tomography (micro-CT) are commonly used to visualize and measure the scaffold's microstructure. The porosity and pore size are particularly important because they influence nutrient diffusion, waste removal, and cell infiltration within the scaffold. An ideal scaffold should have an interconnected porous structure that allows cells to migrate throughout the scaffold while maintaining sufficient mechanical strength [18]. The surface roughness and topography of the scaffold are characterized to assess their impact on cell adhesion and morphology.

The chemical properties of ECM-mimicking scaffolds are also critical and are often characterized using techniques such as Fourier-transform infrared spectroscopy (FTIR), X-ray photoelectron spectroscopy (XPS), and mass spectrometry. These methods help identify the chemical composition and functional groups present on the scaffold surface, which can affect its interaction with cells and bioactive molecules. For instance, the presence of specific chemical groups can enhance the scaffold's hydrophilicity, improving cell attachment and proliferation. Chemical characterization can confirm the successful incorporation of bioactive molecules, such as growth factors or peptides, which are crucial for guiding cellular behavior and promoting tissue regeneration.

Mechanical characterization is another key aspect, as the scaffold must possess mechanical properties that match those of the target tissue. Techniques like tensile testing, compression testing, and dynamic mechanical analysis (DMA) are used to assess the scaffold's strength, elasticity, and stiffness. These properties are particularly important for scaffolds intended for load-bearing tissues, such as bone or cartilage, where the mechanical environment plays a significant role in guiding cell differentiation and tissue formation. The mechanical stability of the scaffold over time is also evaluated to ensure it can withstand the physiological forces it will encounter in the body without degrading prematurely or losing its structural integrity. Finally, the biological characterization of ECM-mimicking scaffolds involves evaluating their biocompatibility, biodegradability, and ability to support cell function. In vitro cell culture studies are typically conducted to assess cell adhesion, proliferation, and differentiation on the scaffold. These studies often use various cell types, such as stem cells or primary cells, relevant to the tissue being engineered. The scaffold's degradation rate is monitored to ensure it degrades at a pace that matches tissue formation, providing temporary support until the new tissue is fully developed. In vivo studies, often conducted in animal models, further assess the scaffold's integration with surrounding tissues, immune response, and overall efficacy in promoting tissue regeneration.

The characterization of ECM-mimicking scaffolds is a multifaceted process that encompasses physical, chemical, mechanical, and biological evaluations. Each aspect of characterization provides essential information about the scaffold's properties and performance, guiding the optimization of scaffold design for specific tissue engineering applications.

7.4 ECM Mimicry in Wound Healing

ECM mimicry plays a pivotal role in advancing wound healing strategies, particularly in challenging scenarios like chronic wounds and severe tissue injuries. The extracellular matrix (ECM) is a complex network of proteins and polysaccharides that

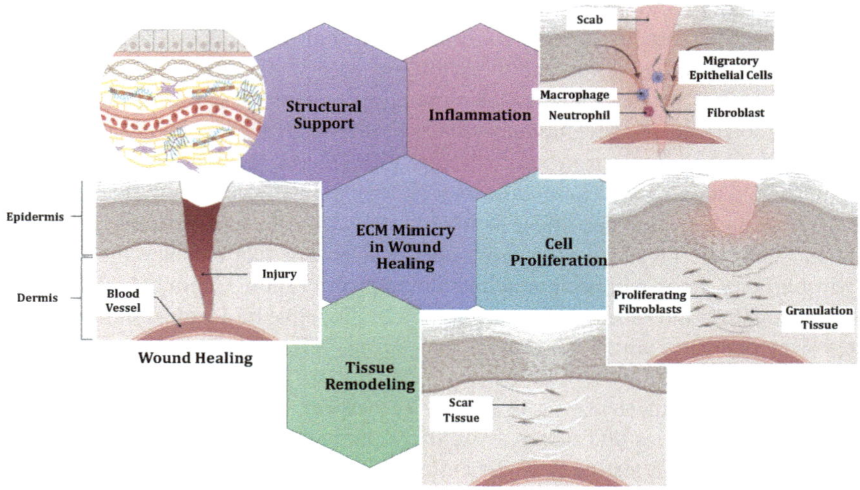

Fig. 7.2 Role of ECM mimicry in wound healing

provide structural support and biochemical cues essential for tissue repair (Fig. 7.2). In the natural wound healing process, the ECM not only acts as a scaffold for cellular infiltration but also regulates critical processes such as inflammation, cell proliferation, and tissue remodeling. However, in cases of impaired healing, such as diabetic ulcers or large burns, the natural ECM may be insufficient to support effective regeneration [19]. This is where ECM mimicry comes into play, offering a biomimetic approach to designing scaffolds that replicate the structure and function of the natural ECM, thereby enhancing the body's ability to heal.

The application of ECM-mimicking scaffolds in wound healing is grounded in their ability to provide a supportive environment that promotes cellular functions necessary for tissue repair. These scaffolds are designed to mimic the native ECM's composition, architecture, and mechanical properties, ensuring that they can integrate seamlessly with the wound site. By incorporating bioactive molecules, growth factors, and other signaling cues, ECM-mimicking scaffolds can accelerate the healing process, reduce scar formation, and improve the overall quality of the regenerated tissue. This biomimetic approach represents a significant advancement over traditional wound dressings, offering a more dynamic and effective solution for enhancing the body's natural healing capabilities.

7.4.1 *ECM-Mimicking Scaffolds in Acute Wound Healing and Skin Regeneration*

The extracellular matrix (ECM) is integral to skin regeneration, serving as a dynamic and complex scaffold that orchestrates the various stages of wound healing. In skin tissue, the ECM comprises a network of proteins such as collagen, elastin, fibronectin, and glycosaminoglycans, which provide structural support and regulate cellular behavior. During wound healing, the ECM plays a critical role in modulating inflammation, facilitating cell migration, and promoting the proliferation of skin cells, including keratinocytes and fibroblasts. These processes are essential for re-epithelialization and the formation of new connective tissue, ensuring the restoration of the skin's barrier function and overall integrity. In the context of acute wound healing, the ECM undergoes significant remodeling. Following an injury, the ECM at the wound site is degraded and reorganized to accommodate the influx of immune cells and the formation of granulation tissue. This dynamic restructuring is mediated by enzymes such as matrix metalloproteinases (MMPs), which break down damaged ECM components, allowing for the deposition of new matrix proteins. The newly formed ECM serves as a provisional matrix that supports angiogenesis, the growth of new blood vessels, and the migration of cells necessary for tissue repair. As healing progresses, this provisional matrix is further remodeled into a more organized structure that closely resembles the native skin ECM, contributing to the restoration of normal tissue architecture.

ECM-mimicking scaffolds have emerged as a promising tool in enhancing acute wound healing by replicating the natural ECM's functions. These scaffolds are engineered to mimic the structural and biochemical properties of the ECM, providing a supportive environment that promotes the rapid regeneration of skin tissue [20]. One of the key benefits of ECM-mimicking scaffolds is their ability to support cell adhesion and proliferation. By closely resembling the natural ECM's topography and composition, these scaffolds facilitate the attachment of skin cells, which is crucial for the formation of new tissue. The controlled release of bioactive molecules and growth factors from ECM-mimicking scaffolds can further accelerate the wound healing process, promoting faster and more effective tissue regeneration. The role of ECM-mimicking scaffolds in acute wound healing extends beyond simply providing structural support. These scaffolds can also modulate the wound healing environment, influencing cellular responses such as migration, differentiation, and apoptosis. For example, scaffolds that incorporate collagen or hyaluronic acid can enhance the migration of keratinocytes to the wound site, speeding up re-epithelialization. Similarly, the inclusion of elastin in these scaffolds can improve the elasticity and mechanical properties of the newly formed tissue, reducing the likelihood of scar formation. By fine-tuning the composition and properties of ECM-mimicking scaffolds, researchers can create tailored solutions for different types of acute wounds, optimizing the healing process and improving patient outcomes.

Another significant advantage of ECM-mimicking scaffolds in acute wound healing is their potential to reduce inflammation and prevent infection. Inflammation is a natural part of the wound healing process, but excessive or prolonged inflammation can lead to chronic wounds and impaired healing. ECM-mimicking scaffolds can be designed to deliver anti-inflammatory agents or to modulate the local immune response, helping to maintain a balanced inflammatory environment that supports healing. These scaffolds can be loaded with antimicrobial agents to prevent infection, a common complication in acute wounds that can delay healing and lead to more severe tissue damage.

The integration of ECM-mimicking scaffolds into clinical practice represents a significant advancement in the treatment of acute wounds. Traditional wound dressings, while effective in protecting the wound site, do not actively participate in the healing process. In contrast, ECM-mimicking scaffolds offer a more interactive approach, providing both physical support and biochemical signals that guide tissue regeneration. This biomimetic strategy has the potential to improve the speed and quality of wound healing, particularly in cases where the natural healing process is compromised or where rapid tissue regeneration is essential, such as in burns or surgical wounds.

The ECM plays a pivotal role in skin regeneration by providing the structural framework and biochemical signals necessary for wound healing. ECM-mimicking scaffolds build on this natural blueprint, offering an advanced approach to acute wound healing that combines structural support with bioactive guidance. Through the careful design and application of these scaffolds, it is possible to enhance the body's natural healing capabilities, reduce complications such as inflammation and infection, and ultimately achieve better outcomes in skin regeneration.

7.4.2 Chronic Wound Healing and ECM Mimicry

Chronic wounds, such as diabetic ulcers, pressure sores, and venous leg ulcers, present significant challenges in clinical management due to their prolonged and often incomplete healing process. Unlike acute wounds, which follow a predictable sequence of healing stages, chronic wounds are characterized by a persistent inflammatory state, impaired re-epithelialization, and dysfunctional ECM remodeling. The failure to progress through the normal stages of healing is often associated with a degraded or dysfunctional ECM, which fails to provide the necessary structural and biochemical cues for tissue repair. As a result, chronic wounds remain a significant burden on healthcare systems and adversely affect patients' quality of life. ECM mimicry offers a promising strategy for addressing the complexities of chronic wound healing by providing scaffolds that replicate the natural ECM's supportive environment. In chronic wounds, the native ECM is often compromised due to excessive enzymatic degradation by matrix metalloproteinases (MMPs) and a lack of balanced ECM synthesis. ECM-mimicking scaffolds can help restore this balance by providing a temporary matrix that supports cell adhesion, migration, and proliferation, crucial for effective wound closure. These scaffolds can be designed to resist premature degradation while allowing for controlled turnover, which is essential for proper tissue remodeling in chronic wounds.

One of the key advantages of ECM-mimicking scaffolds in chronic wound healing is their ability to modulate the wound microenvironment, particularly by reducing inflammation and promoting a more regenerative healing process. Chronic wounds are often trapped in a state of persistent inflammation, which impedes the healing process [21]. ECM-mimicking scaffolds can be engineered to deliver anti-inflammatory agents or to sequester pro-inflammatory cytokines, thereby shifting the wound environment towards a more conducive state for healing. Additionally, by incorporating bioactive molecules such as growth factors and cytokines, these scaffolds can actively promote cellular activities like angiogenesis, which is often impaired in chronic wounds, facilitating better blood supply and tissue regeneration.

Another critical aspect of ECM mimicry in chronic wound healing is the potential to address the issue of impaired cell signaling and migration. In chronic wounds, the cells responsible for healing, such as fibroblasts and keratinocytes, often exhibit reduced responsiveness to the signals that typically guide their behavior. ECM-mimicking scaffolds can be tailored to include specific binding sites and signaling molecules that re-establish these crucial cell-ECM interactions. For instance, scaffolds that present the RGD (arginine-glycine-aspartic acid) peptide sequence can enhance cell adhesion and migration, encouraging the re-epithelialization process that is often stalled in chronic wounds. This tailored approach helps to "jump-start" the healing process, enabling the wound to progress through the stages of healing that have been otherwise blocked. The use of ECM-mimicking scaffolds in chronic wound healing not only aims to restore the structural integrity of the tissue but also to prevent the recurrence of wounds. Chronic wounds have a high rate of recurrence due to the underlying pathological conditions that often remain unaddressed. By promoting a more organized and functional ECM, these scaffolds contribute to the formation of a stronger and more resilient tissue structure. This not only facilitates initial wound closure but also strengthens the repaired tissue, reducing the likelihood of wound reopening. The incorporation of sustained-release bioactive agents within the scaffold can provide ongoing support to the newly formed tissue, helping to maintain its integrity over the long term.

ECM mimicry represents a powerful approach to overcoming the challenges associated with chronic wound healing. By providing scaffolds that replicate the natural ECM's functions, researchers and clinicians can better manage the persistent issues that characterize chronic wounds, such as excessive inflammation, impaired cell signaling, and inadequate tissue remodeling. The use of ECM-mimicking scaffolds offers the potential not only to accelerate the healing of chronic wounds but also to improve the quality and durability of the regenerated tissue, ultimately leading to better patient outcomes.

7.5 ECM Mimicry in Cancer Therapy

ECM mimicry in cancer therapy represents an innovative approach that leverages the principles of tissue engineering to influence tumor behavior and improve treatment outcomes. The extracellular matrix (ECM) is a critical component of the tumor microenvironment, playing a significant role in cancer progression, metastasis, and resistance to therapy. In the context of cancer, the ECM undergoes substantial remodeling, leading to a stiffened matrix that supports tumor growth and shields cancer cells from immune surveillance and therapeutic agents. By mimicking the ECM, researchers aim to create engineered scaffolds or materials that can either replicate the tumor ECM for in vitro studies or disrupt the pathological ECM in vivo, thereby enhancing the efficacy of cancer treatments.

The application of ECM mimicry in cancer therapy is particularly promising in the development of more accurate tumor models for drug testing and personalized medicine (Fig. 7.3). By creating biomimetic scaffolds that closely resemble the tumor ECM, researchers can cultivate cancer cells in a more physiologically relevant environment, allowing for better prediction of how these cells will respond to different therapies [22]. Furthermore, in vivo, ECM-mimicking strategies can be used to modify the tumor microenvironment, making it less supportive of tumor growth and more accessible to drugs. This approach not only offers the potential to improve the delivery and effectiveness of current cancer therapies but also opens new avenues for developing treatments that target the tumor ECM itself, potentially inhibiting cancer progression and metastasis.

7.5.1 Tumor Microenvironment and ECM

The tumor microenvironment (TME) is a complex and dynamic ecosystem composed of various cellular and non-cellular components, including cancer cells, stromal cells, immune cells, blood vessels, and the extracellular matrix (ECM). The ECM within the TME is not merely a structural scaffold but plays a pivotal role in modulating tumor behavior. It influences processes such as cell proliferation, migration, invasion, and metastasis, making it a critical factor in cancer progression. The ECM within the TME undergoes extensive remodeling due to the interaction between cancer cells and stromal cells, leading to changes in its composition, stiffness, and architecture. These changes often create a more supportive environment for tumor growth and spread, complicating treatment efforts.

One of the key characteristics of the tumor ECM is its increased stiffness compared to normal tissue. This stiffening is primarily due to the excessive deposition and cross-linking of ECM proteins like collagen, driven by the overactivity of enzymes such as lysyl oxidase (LOX). The altered mechanical properties of the ECM can enhance cancer cell survival and proliferation by activating mechano transduction pathways, which involve the conversion of mechanical signals into biochemical cues.

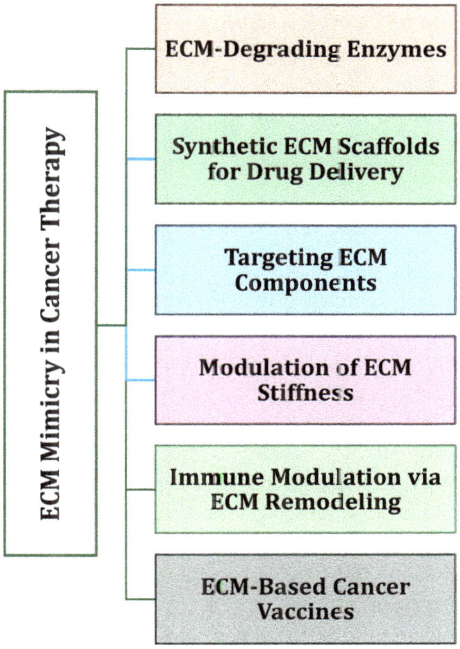

Fig. 7.3 Applications of ECM mimicry in cancer therapy

These pathways often promote oncogenic signaling, leading to more aggressive tumor behavior. The dense and rigid ECM can physically impede the penetration of therapeutic agents, contributing to drug resistance and reducing the effectiveness of chemotherapy and immunotherapy [23]. In addition to its physical properties, the biochemical composition of the tumor ECM is also significantly altered. ECM in the TME is rich in growth factors, cytokines, and other signaling molecules that cancer cells exploit to promote their survival and invasion. For instance, ECM components such as fibronectin and laminin can bind to cell surface receptors like integrins, triggering pathways that support cell adhesion, migration, and invasion. These interactions between cancer cells and the ECM create a feedback loop that not only sustains tumor growth but also facilitates metastasis by enabling cancer cells to break away from the primary tumor, invade surrounding tissues, and eventually colonize distant organs. Thus, ECM in the TME is a major driver of the malignant phenotype and a significant barrier to successful cancer treatment.

The tumor ECM plays a crucial role in modulating the immune response within the TME. ECM can influence the infiltration and function of immune cells, often creating an immunosuppressive environment that allows the tumor to evade immune detection. For example, certain ECM components can recruit regulatory T cells (Tregs) and myeloid-derived suppressor cells (MDSCs), both of which suppress the activity of cytotoxic T cells and natural killer (NK) cells that are critical for antitumor immunity. Additionally, the dense ECM can serve as a physical barrier that prevents immune cells from effectively reaching and attacking cancer cells. This immunosuppressive role of ECM is a major challenge in cancer immunotherapy, as it can significantly reduce the efficacy of treatments designed to stimulate the immune system against tumors. Understanding and targeting ECM within the TME, therefore, holds great promise for developing more effective cancer therapies.

7.5.2 ECM Mimicry for Targeted Drug Delivery

ECM mimicry offers a promising approach for targeted drug delivery, leveraging the structural and biochemical characteristics of the extracellular matrix (ECM) to enhance the precision and effectiveness of therapeutic interventions. Natural ECM provides a highly specific environment where cells interact with a complex network of proteins, glycoproteins, and polysaccharides. By mimicking these properties, drug delivery systems can be designed to target specific tissues or pathological sites more effectively. This approach allows for the localized delivery of drugs, reducing systemic side effects and improving the therapeutic index of various treatments, particularly in diseases like cancer where targeted intervention is crucial. One of the key strategies in ECM mimicry for targeted drug delivery involves designing drug carriers that interact with ECM components or ECM-mimicking scaffolds at the site of interest. These carriers can be engineered to recognize and bind to specific ECM proteins, such as collagen, fibronectin, or laminin, which are overexpressed in certain pathological conditions like tumors or fibrotic tissues. By incorporating ECM-binding ligands or peptides into drug delivery systems, researchers can create carriers that home in on diseased tissues with high specificity. This targeted approach ensures that the therapeutic agents are concentrated at the site of action, maximizing their efficacy while minimizing exposure to healthy tissues. Another significant aspect of ECM mimicry in drug delivery is the use of ECM-mimicking hydrogels or scaffolds that can be implanted or injected directly into the target tissue [24]. These materials are designed to replicate the mechanical properties and biochemical cues of the native ECM, providing a supportive environment for sustained and controlled drug release. The porous nature of these hydrogels allows for the encapsulation of drugs, growth factors, or other therapeutic agents, which can be released in response to specific stimuli, such as changes in pH, temperature, or the presence of certain enzymes. This stimulus-responsive release mechanism ensures that the drug is delivered in a controlled manner, corresponding to the needs of the tissue, thereby enhancing the therapeutic outcome.

ECM mimicry in targeted drug delivery holds great potential in the treatment of cancers, where the tumor microenvironment (TME) plays a critical role in drug resistance. By designing drug delivery systems that mimic ECM, it is possible to penetrate the dense and often impenetrable tumor ECM, overcoming barriers that typically hinder drug access to cancer cells. These ECM-mimicking systems can also be tailored to disrupt the pathological ECM within tumors, making the microenvironment less conducive to cancer cell survival and more permeable to therapeutic agents. As a result, ECM mimicry not only enhances drug delivery to the tumor site but also increases the overall effectiveness of cancer therapies by altering the tumor's supportive environment.

7.5.3 Impact on Tumor Progression and Metastasis

The extracellular matrix (ECM) plays a profound role in influencing tumor progression and metastasis, acting as both a physical scaffold and a biochemical regulator within the tumor microenvironment. Tumor cells interact with ECM to facilitate their growth, survival, and dissemination. The composition and architecture of the ECM are often significantly altered in cancer, contributing to an environment that supports tumor progression. For instance, the overproduction and cross-linking of ECM components like collagen lead to a stiffer matrix, which can promote tumor cell proliferation by activating mechano transduction pathways. These pathways, when triggered, can enhance oncogenic signaling, thereby driving the malignant transformation of cells and supporting their unchecked growth.

As tumors grow, the remodeling of ECM becomes more pronounced, with enzymes like matrix metalloproteinases (MMPs) playing a key role in breaking down the existing matrix to make way for new, tumor-supportive ECM. This remodeling process is crucial for tumor progression as it facilitates angiogenesis—the formation of new blood vessels—which is essential for providing the growing tumor with nutrients and oxygen. Moreover, the breakdown of ECM components releases bioactive fragments known as matricryptins, which can further influence tumor behavior by promoting angiogenesis, cell survival, and invasion. This dynamic interplay between tumor cells and ECM not only supports the growth of the primary tumor but also prepares the tumor microenvironment for metastasis. Metastasis, the spread of cancer cells from the primary tumor to distant organs, is heavily influenced by the interactions between tumor cells and the ECM. The altered ECM in tumors often provides the necessary cues for cancer cells to acquire invasive properties, allowing them to breach the basement membrane and enter the bloodstream or lymphatic system [25]. ECM components such as fibronectin and laminin play critical roles in this process by interacting with integrins and other cell surface receptors, promoting the detachment, migration, and invasion of tumor cells. Once in circulation, these cells must then navigate new microenvironments in distant tissues. The ability of metastatic cells to colonize these new sites is often dependent on their interaction with ECM, which can either support or inhibit their growth depending on the composition and state of ECM in the target tissue.

The impact of ECM on tumor progression and metastasis highlights the potential of targeting ECM in cancer therapy. By disrupting the tumor ECM or modifying its properties, it may be possible to inhibit tumor growth and prevent metastasis. For example, therapies aimed at inhibiting ECM remodeling enzymes like MMPs or reducing ECM stiffness could impair the tumor's ability to grow and spread. Additionally, strategies that alter the ECM to make it less supportive of tumor cell invasion and more permeable to immune cells could enhance the effectiveness of existing treatments, such as chemotherapy and immunotherapy.

7.5.4 ECM-Based Therapies in Clinical Trials

ECM-based therapies are gaining significant attention in clinical trials as innovative approaches to treating various diseases, particularly in oncology and regenerative medicine. These therapies leverage the biological functions of ECM to either restore normal tissue architecture or disrupt pathological ECM that supports disease progression. In cancer treatment, for instance, ECM-based therapies aim to modify the tumor microenvironment to make it less conducive to tumor growth and more accessible to therapeutic agents. One promising avenue is the use of ECM-targeting drugs that inhibit enzymes like matrix metalloproteinases (MMPs), which are responsible for ECM remodeling in tumors. Clinical trials are exploring the potential of these inhibitors to slow tumor progression, reduce metastasis, and enhance the efficacy of conventional therapies such as chemotherapy and radiation. In addition to oncology, ECM-based therapies are also being explored in the context of tissue engineering and regenerative medicine. Clinical trials are investigating the use of ECM-mimicking scaffolds and hydrogels to promote the healing of chronic wounds, repair damaged organs, and regenerate tissues. These scaffolds are designed to replicate the native ECM's structure and biochemical properties, providing a supportive environment for cell adhesion, migration, and differentiation [26]. For example, ECM-derived products such as decellularized matrices are being tested for their ability to facilitate the regeneration of heart tissue after myocardial infarction or to improve the healing of skin grafts in burn patients. The outcomes of these trials could pave the way for new, more effective treatments that harness the body's natural regenerative processes.

Another exciting area of ECM-based therapies in clinical trials involves their application in orthopedics and musculoskeletal disorders. ECM-derived biomaterials are being used to develop implants and grafts that can support bone regeneration, cartilage repair, and tendon healing. These materials not only provide the necessary structural support but also deliver bioactive molecules that promote tissue regeneration and reduce inflammation. Ongoing clinical trials are assessing the safety and efficacy of these ECM-based implants in patients with conditions such as osteoarthritis, tendon injuries, and bone defects. If successful, these therapies could offer substantial improvements over current treatments, providing long-lasting solutions for musculoskeletal injuries and degenerative diseases.

7.6 ECM Mimicry in Bone and Cartilage Regeneration

ECM mimicry in bone and cartilage regeneration represents a cutting-edge approach to addressing the challenges associated with repairing and regenerating these critical tissues. The extracellular matrix (ECM) in bone and cartilage provides essential structural support and biochemical signals that regulate cell behavior, including proliferation, differentiation, and matrix synthesis. In cases of injury or degenerative diseases, such as osteoarthritis or osteoporosis, the natural ECM is often damaged or insufficient, leading to impaired healing and function. By mimicking the natural ECM, researchers aim to create scaffolds and materials that replicate the key features of the native bone and cartilage ECM, promoting effective tissue regeneration and restoring normal function. The design of ECM-mimicking materials for bone and cartilage regeneration focuses on replicating the unique composition and mechanical properties of these tissues [27]. Bone ECM, for instance, is highly mineralized and provides a rigid scaffold that supports load-bearing functions, while cartilage ECM is rich in proteoglycans and collagen, offering the flexibility and resilience necessary for joint movement. By developing biomaterials that closely resemble these properties, ECM mimicry can facilitate the proper integration of new tissue with the surrounding environment, encourage the formation of healthy bone or cartilage, and support the long-term stability of the regenerated tissue. This approach holds significant promise for improving the outcomes of treatments for bone fractures, cartilage injuries, and degenerative joint diseases.

7.6.1 Scaffolds for Bone Regeneration

Scaffolds for bone regeneration are at the forefront of regenerative medicine, providing a structural framework that supports the growth and differentiation of cells necessary for bone repair. These scaffolds are engineered to mimic the natural extracellular matrix (ECM) of bone, which is crucial for guiding the regeneration process. An ideal bone scaffold should be biocompatible, promoting cell adhesion, proliferation, and differentiation without eliciting an immune response. Additionally, it must be biodegradable, breaking down at a controlled rate to allow the gradual replacement of the scaffold with new bone tissue. The mechanical properties of the scaffold are equally important; it must possess sufficient strength to support the load-bearing functions of bone, especially in weight-bearing areas, while maintaining the porous architecture that facilitates nutrient and oxygen diffusion. The materials used to fabricate bone scaffolds range from natural polymers, such as collagen and chitosan, to synthetic polymers like polylactic acid (PLA) and polyglycolic acid (PGA). Natural polymers offer the advantage of being inherently biocompatible and often contain bioactive molecules that promote cell activity. However, they may lack the mechanical strength required for bone regeneration in certain applications. In contrast, synthetic polymers can be engineered to have tailored mechanical properties, but they may require surface modifications to enhance their bioactivity and support cellular interactions [28]. Hybrid scaffolds, combining both natural and synthetic materials, are also being developed to leverage the benefits of each type, offering a balance between mechanical strength and biological functionality.

In recent years, advancements in scaffold fabrication techniques, such as 3D printing and electrospinning, have allowed for the creation of highly customized and intricate scaffold designs that closely mimic the architecture of native bone. 3D printing, in particular, enables the precise control of scaffold geometry, pore size, and interconnectivity, which are critical factors in promoting vascularization and bone in-growth. These advanced scaffolds can be designed to match the patient's specific anatomical requirements, facilitating better integration with existing bone and reducing the risk of implant rejection or failure. 3D-printed scaffolds can incorporate gradients of materials and bioactive molecules, creating a more natural transition between the scaffold and the native bone tissue.

Another exciting development in the field of bone regeneration scaffolds is the incorporation of bioactive molecules, such as growth factors and peptides, to enhance osteogenesis—the process of new bone formation. Scaffolds can be functionalized with these bioactive agents to create a conducive environment for stem cells or osteoblasts, the cells responsible for bone formation. For example, the inclusion of bone morphogenetic proteins (BMPs) within the scaffold has been shown to significantly accelerate bone healing and improve the quality of the regenerated bone. Additionally, scaffolds can be engineered to release these bioactive molecules in a controlled manner, ensuring that the therapeutic effects are sustained over the course of healing. This combination of structural support and biochemical stimulation makes ECM-mimicking scaffolds a powerful tool in the treatment of bone defects, fractures, and diseases like osteoporosis.

7.6.2 ECM-Mimicking Strategies for Cartilage Repair

- **Biomimetic Scaffolds**
  - Development of scaffolds that replicate the native cartilage ECM, primarily using collagen, hyaluronic acid, and chondroitin sulfate to support chondrocyte proliferation and matrix production.
  - Use of synthetic polymers like polycaprolactone (PCL) and polylactic-co-glycolic acid (PLGA) for structural support, combined with natural ECM components to enhance biocompatibility.
- **Hydrogels**
  - Design of hydrogels that mimic the hydrated environment of cartilage ECM, supporting cell viability and nutrient diffusion.

- Incorporation of ECM-mimicking molecules such as glycosaminoglycans (GAGs) to promote cartilage-specific matrix production and retention of mechanical properties.

- **Growth Factor Incorporation**
  - Functionalization of scaffolds and hydrogels with growth factors like transforming growth factor-beta (TGF-β) and insulin-like growth factor (IGF) to stimulate chondrocyte activity and cartilage regeneration.
  - Controlled release systems within the scaffolds to provide sustained delivery of growth factors, enhancing cartilage repair.
- **3D Bioprinting**
  - Use of 3D bioprinting technology to create customized, patient-specific scaffolds that mimic the zonal architecture of cartilage.
  - Integration of bioactive materials and cells during the printing process to replicate the natural ECM's structural and functional properties.
- **Nanofiber Scaffolds**
  - Fabrication of nanofiber-based scaffolds using techniques like electrospinning to create a fibrous structure that closely resembles the collagen network in native cartilage.
  - Enhancement of scaffold properties with the addition of nanomaterials or nanoparticles that mimic the nanoscale features of cartilage ECM.
- **Cell-seeded Constructs**
  - Engineering of scaffolds pre-seeded with chondrocytes or stem cells that can differentiate into cartilage-forming cells in response to ECM-mimicking cues.
  - Use of dynamic culture systems like bioreactors to provide mechanical stimuli that further mimic the in vivo environment and promote cartilage formation.
- **Decellularized Cartilage ECM**
  - Utilization of decellularized cartilage as a natural scaffold, preserving the native ECM components and structure.
  - Recellularization of these scaffolds with patient-derived cells to enhance integration and functional recovery in cartilage repair applications.

7.7 ECM Mimicry in Cardiovascular Applications

The extracellular matrix (ECM) plays a crucial role in maintaining the structural integrity and functional dynamics of cardiovascular tissues. It provides biochemical and mechanical signals that regulate cell behavior, which is essential for the development, repair, and normal function of the heart and blood vessels (Fig. 7.4). In cardiovascular applications, ECM mimicry is increasingly being recognized as a

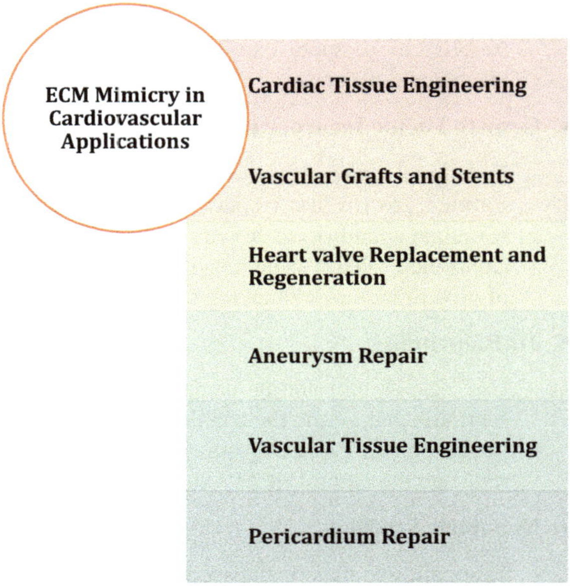

Fig. 7.4 Applications of ECM mimicry in cardiovascular applications

vital strategy for tissue engineering and regenerative medicine. By replicating the complex composition and architecture of native ECM, researchers aim to create biomaterials that support cell adhesion, migration, proliferation, and differentiation, thereby promoting the repair or replacement of damaged cardiovascular tissues. ECM mimicking approaches in cardiovascular applications often involve the design of scaffolds that can emulate the native ECM's physical properties, such as elasticity and stiffness, as well as its biochemical composition, including the presence of key proteins like collagen and elastin [29]. These scaffolds not only provide a supportive environment for cell growth but also influence cellular activities crucial for tissue regeneration. For instance, incorporating bioactive molecules that are naturally found in ECM can enhance endothelial cell function and smooth muscle cell alignment, which are critical for the formation of functional blood vessels. As a result, ECM mimicry holds significant potential in advancing treatments for conditions such as myocardial infarction, atherosclerosis, and congenital heart defects.

7.7.1 Role of ECM in Cardiovascular Health

The extracellular matrix (ECM) is a complex network of proteins and polysaccharides that provides structural and biochemical support to cells within tissues. In the cardiovascular system, ECM plays a critical role in maintaining the integrity and function of the heart and blood vessels. It provides mechanical strength, elasticity, and a scaffold that supports cellular organization, critical for normal heart function.

Key ECM components like collagen, elastin, fibronectin, and laminin are crucial for regulating the biomechanical properties of the heart, ensuring that it can withstand the continuous stress of blood flow and pressure changes.

ECM dynamics are essential for cardiovascular development, maintenance, and repair. During normal heart function, ECM undergoes constant remodeling to adapt to physiological changes such as growth or exercise. This remodeling process is tightly regulated to maintain a balance between ECM synthesis and degradation, allowing the heart to function efficiently. Dysregulation of ECM remodeling, however, can lead to pathological conditions such as fibrosis, where excessive ECM deposition stiffens the heart, impairing its function and leading to conditions like heart failure. Thus, the balance of ECM remodeling is crucial for cardiovascular health.

In addition to its structural role, ECM also regulates cellular behavior through biochemical signaling. ECM interacts with cells via integrins and other receptors, influencing cell adhesion, migration, proliferation, and differentiation. This regulation is vital in processes such as angiogenesis, where the formation of new blood vessels is necessary to supply oxygen and nutrients to the heart muscle, especially after injury [30]. ECM provides not only the physical scaffold but also the signals needed for vascular repair and regeneration, highlighting its importance in maintaining cardiovascular health. ECM's role in cardiovascular disease is increasingly recognized, particularly in atherosclerosis, myocardial infarction, and heart failure. In these conditions, ECM components undergo abnormal modifications that disrupt normal heart and vessel function. For instance, altered ECM composition in atherosclerosis contributes to plaque formation, vessel stiffening, and increased risk of rupture. In myocardial infarction, ECM degradation occurs as part of the inflammatory response, followed by excessive scar formation, which impairs the contractility of the heart. Understanding these processes has led to the exploration of ECM-targeted therapies, aiming to restore normal ECM function and improve cardiovascular outcomes.

ECM is not just a passive scaffold but an active regulator of cardiovascular health. Its dynamic and complex nature allows it to maintain heart structure, facilitate tissue repair, and regulate cellular activities essential for heart function. Abnormal ECM remodeling and signaling are at the core of many cardiovascular diseases, making it a crucial focus for research and therapeutic intervention. Maintaining ECM homeostasis is, therefore, fundamental to preserving cardiovascular health and preventing disease progression.

7.7.2 *Scaffold Design for Vascular Tissue Engineering*

Scaffold design is a cornerstone of vascular tissue engineering, aiming to create functional blood vessels that can integrate with the body's natural tissues. The primary goal of these scaffolds is to mimic the native extracellular matrix (ECM) of blood vessels, providing a structural framework that supports cell attachment, proliferation, and differentiation. Ideal vascular scaffolds must possess appropriate mechanical properties, such as elasticity and strength, to withstand physiological blood flow and pressure. Materials used for scaffold fabrication range from synthetic polymers like polycaprolactone (PCL) and polylactic acid (PLA) to natural biomaterials such as collagen and fibrin, each offering unique advantages in replicating the native vascular environment. Biocompatibility and bioactivity are crucial considerations in scaffold design, as they directly influence cell behavior and tissue regeneration. Scaffolds should be designed to promote endothelial cell adhesion and smooth muscle cell alignment, which are essential for recreating the layered structure of blood vessels. Bioactive coatings or incorporation of growth factors can enhance cell-scaffold interactions, improving endothelialisation and promoting the formation of functional vascular tissue. Scaffolds are often designed with specific surface chemistries and topography that mimic the biochemical cues found in the native ECM, guiding cellular responses critical for successful vascular tissue engineering.

The architecture of vascular scaffolds is another vital aspect, particularly the pore size, porosity, and interconnectivity, which affect nutrient diffusion, waste removal, and cell migration. Scaffolds with highly interconnected porous structures facilitate uniform cell distribution and vascularization within the scaffold, enhancing tissue integration and function. Advanced fabrication techniques such as electrospinning, 3D bioprinting, and freeze-drying are commonly employed to create scaffolds with precise control over their microstructure, enabling the development of complex geometries that closely resemble natural blood vessels [31].

Scaffold degradation is an essential factor that must be tailored to match the rate of new tissue formation. Biodegradable scaffolds gradually break down as the newly formed tissue takes over, eliminating the need for surgical removal. The degradation rate must be carefully controlled to prevent premature loss of scaffold support or, conversely, delayed degradation that could hinder tissue growth. Materials like polyglycolic acid (PGA) and poly (lactic-co-glycolic acid) (PLGA) are frequently used due to their tunable degradation rates and proven biocompatibility, making them suitable for vascular tissue engineering applications.

The design of scaffolds for vascular tissue engineering involves a complex interplay of material properties, structural design, and bioactivity to recreate the native vascular environment. Successful scaffolds not only provide the necessary mechanical support but also actively participate in the healing process by guiding cell behavior and tissue formation. Innovations in scaffold design, including the use of bioactive materials and advanced manufacturing techniques, continue to advance the field, bringing us closer to developing functional, implantable vascular grafts that can repair or replace damaged blood vessels in clinical settings.

7.7.3 ECM-Mimicking Approaches in Heart Tissue Repair

Extracellular matrix (ECM)-mimicking approaches have emerged as a promising strategy in heart tissue repair, aiming to replicate the natural environment of cardiac tissues to support regeneration after injury, such as myocardial infarction. ECM in the heart provides structural support, biochemical signals, and mechanical properties that are crucial for maintaining heart function. Mimicking these characteristics in engineered scaffolds or biomaterials helps create a favorable environment for cardiac cells, promoting their survival, alignment, and function. These approaches involve using biomaterials that closely resemble the native cardiac ECM in terms of composition, architecture, and mechanical properties. One key approach in ECM mimicry involves using natural biomaterials such as collagen, gelatin, fibrin, and decellularized ECM derived from heart tissues. These materials offer inherent bioactivity and compatibility, making them ideal for cardiac repair. Decellularized heart ECM, for instance, retains the complex composition of native cardiac ECM, including essential proteins, growth factors, and glycosaminoglycans, which support cell attachment and function [32]. When used as a scaffold, it provides a natural template that encourages the repopulation of cardiac cells and supports tissue remodeling, enhancing the heart's ability to recover after damage.

Synthetic biomaterials, often modified to mimic ECM properties, are also widely used in heart tissue repair. Polymers like polycaprolactone (PCL), polylactic acid (PLA), and poly (lactic-co-glycolic acid) (PLGA) can be engineered to replicate the mechanical properties of heart tissue while allowing for controlled degradation. To enhance bioactivity, these synthetic scaffolds are frequently combined with ECM components or functionalized with peptides and growth factors that promote cell adhesion and proliferation. Hybrid scaffolds that combine synthetic and natural materials offer the benefits of both, balancing structural integrity with bioactivity to support cardiac tissue repair.

Advanced fabrication techniques such as electrospinning and 3D bioprinting have revolutionized ECM-mimicking scaffold design for cardiac repair. Electrospun fibers can be tailored to mimic the nanoscale architecture of the cardiac ECM, providing cues that guide cell alignment and promote the formation of organized cardiac tissue. 3D bioprinting enables precise control over scaffold architecture, allowing for the creation of complex, tissue-like structures that replicate the spatial organization of heart tissue. These techniques provide the flexibility to design scaffolds that closely mimic the physical and biochemical environment of the native heart. ECM-mimicking approaches in heart tissue repair represent a vital area of research, offering innovative ways to enhance cardiac regeneration following injury. By replicating the native ECM's structural and functional properties, these approaches provide a supportive environment for cardiac cells, promoting tissue integration and function. The combination of natural and synthetic materials, along with advanced fabrication methods, continues to push the boundaries of what is possible in heart tissue engineering, bringing us closer to effective treatments for heart disease.

7.8 ECM Mimicry in Nervous System and Neural Tissue Regeneration

- **Overview of ECM Mimicry in Nervous System Regeneration**

The extracellular matrix (ECM) is critical in maintaining the structural and functional integrity of neural tissues. It provides a complex, dynamic environment that regulates cellular behavior, including differentiation, migration, and survival, which are crucial for nervous system function and regeneration. ECM mimicry in regenerative medicine seeks to replicate the natural ECM's biochemical and physical cues to promote healing and tissue repair, particularly in nervous system injuries [33]. By mimicking the ECM, researchers aim to create scaffolds that closely resemble the native neural environment, fostering neural cell adhesion, growth, and communication.

- **ECM Components in Neural Tissue**

ECM of neural tissue is composed of a unique combination of proteins, glycoproteins, proteoglycans, and glycosaminoglycans. Key components include collagen, laminin, fibronectin, hyaluronic acid, and chondroitin sulfate proteoglycans, which provide both structural support and signaling cues for neural cells. ECM in neural tissue is highly specialized, differing from that in other tissues due to the presence of neuro-specific proteins like tenascin-C and brevican. These components play a pivotal role in regulating synaptic plasticity, axonal guidance, and neural cell proliferation, which are essential for maintaining neural tissue homeostasis and facilitating regeneration following injury.

- **Role of ECM in Neural Tissue Regeneration**

ECM influences neural tissue regeneration by modulating the cellular microenvironment through biochemical and biomechanical signals. During nervous system injuries, such as spinal cord injuries, the ECM undergoes significant remodeling, which can either support or inhibit regeneration [34]. ECM mimicry aims to overcome the inhibitory aspects of the native ECM, such as the presence of inhibitory molecules like chondroitin sulfate proteoglycans, by designing scaffolds that promote regenerative processes. By incorporating growth factors and other bioactive molecules, ECM-mimicking scaffolds can enhance cell survival, differentiation, and axonal regrowth, thus promoting neural repair.

- **ECM Mimicry Strategies in Spinal Cord Injury (SCI) Repair**

Spinal cord injuries are challenging to treat due to the limited regenerative capacity of the central nervous system and the formation of a glial scar, which acts as a physical and chemical barrier to regeneration. ECM mimicry strategies in SCI focus on developing scaffolds that can bridge the injury site, support neural cell attachment, and guide axonal regrowth. These scaffolds are often designed to release growth factors like nerve growth factor (NGF) and brain-derived neurotrophic factor (BDNF), which further enhance the regenerative potential of the damaged neural tissue. Moreover, scaffolds can be functionalized with ECM proteins like laminin to mimic the natural environment, encouraging neural cell migration and synapse formation [35].

- **Advances in Biomaterials for ECM Mimicry in Neural Applications**

Recent advances in biomaterials have enabled the development of sophisticated ECM-mimicking scaffolds for neural regeneration. Hydrogels, electrospun fibers, and 3D-printed scaffolds made from natural and synthetic polymers are widely used due to their ability to mimic the ECM's physical and biochemical properties. Natural polymers like collagen, fibrin, and hyaluronic acid are often used due to their excellent biocompatibility and ability to support cell growth. Synthetic polymers, such as polylactic acid (PLA) and polyethylene glycol (PEG), are also popular due to their tunable properties and ability to be functionalized with bioactive molecules, enhancing their regenerative capabilities in neural tissues.

- **Applications in Spinal Cord Injury Treatment**

ECM-mimicking scaffolds have shown promising results in preclinical models of spinal cord injury. These scaffolds can reduce inflammation, prevent glial scar formation, and support neural regeneration. By closely mimicking the native ECM, these biomaterials provide a favorable microenvironment for neural stem cells, Schwann cells, and oligodendrocytes, which are critical for axonal repair and remyelination. Additionally, these scaffolds can be engineered to deliver therapeutic agents, such as anti-inflammatory drugs or neurotrophic factors, directly to the injury site, further enhancing the regenerative process and functional recovery.

- **Future Directions and Challenges**

While ECM mimicry holds great potential for nervous system regeneration, several challenges remain. Achieving precise control over scaffold degradation, ensuring long-term biocompatibility, and integrating these materials with host tissues without triggering adverse immune responses are critical factors that need to be addressed. Future research will focus on optimizing scaffold design, incorporating advanced materials, and developing personalized approaches that account for individual patient variability. As our understanding of the neural ECM and its role in regeneration improves, ECM mimicry could pave the way for novel therapies that offer hope for patients with spinal cord injuries and other nervous system disorders.

7.9 ECM Mimicry in Organ Regeneration

Extracellular matrix (ECM) mimicry is an emerging strategy in tissue engineering and regenerative medicine that aims to replicate the complex biological environment of tissues to support organ regeneration. ECM is a dynamic and intricate network of proteins, glycoproteins, and polysaccharides that provides structural support, biochemical cues, and regulates cell behavior, such as proliferation, differentiation, and migration. Mimicking ECM involves designing scaffolds that closely replicate their structural and functional properties, thereby enhancing the regenerative capacity of damaged organs. This approach has shown significant promise in regenerating complex organs such as the lung, kidney, and liver, where native ECM components are critical for maintaining tissue-specific functions.

In lung regeneration, ECM scaffolds play a pivotal role in reconstructing the delicate architecture necessary for gas exchange and respiratory function. Lung tissue engineering faces challenges due to the organ's complex structure, including branching airways, alveoli, and a vascular network. Decellularized lung scaffolds, derived from animal or cadaveric sources, retain the native ECM composition, including collagen, elastin, and laminin, which are crucial for maintaining lung-specific cellular functions. By preserving the lung's structural integrity, these scaffolds provide an optimal environment for cell attachment, growth, and differentiation of epithelial and endothelial cells. The introduction of ECM-mimicking hydrogels, loaded with growth factors and bioactive molecules, further enhances tissue regeneration by promoting angiogenesis and cellular integration, thus improving the overall functionality of regenerated lung tissue.

Kidney regeneration using ECM mimicry has gained attention due to the kidney's complex filtration system, which is vital for maintaining fluid and electrolyte balance. ECM of the kidney supports nephron function, including glomerular filtration, tubular reabsorption, and secretion. Decellularized kidney scaffolds preserve the intricate ECM composition, providing a supportive framework for cellular repopulation. These scaffolds promote the regeneration of nephrons, blood vessels, and interstitial cells, crucial for kidney function. Advanced ECM-mimicking biomaterials, such as hydrogels and nanofibrous scaffolds, have been engineered to replicate the kidney ECM's biochemical and mechanical properties, promoting stem cell differentiation into renal lineages [36]. These scaffolds can be customized with bioactive cues, such as heparin-binding growth factors, to enhance vascularization and restore kidney function, highlighting the potential of ECM mimicry in renal tissue engineering.

Liver regeneration using ECM scaffolds addresses the liver's complex lobular architecture and extensive vascular network, essential for its metabolic, detoxification, and synthetic functions. The liver ECM is rich in glycoproteins, collagen, and proteoglycans, which provide structural support and biochemical signals that regulate hepatocyte behavior. Decellularized liver scaffolds retain these ECM components, offering a native-like environment that promotes hepatocyte survival, proliferation, and function. ECM-mimicking scaffolds, such as collagen-based hydrogels, have been utilized to support the co-culture of hepatocytes and endothelial cells, mimicking the liver's microenvironment. These biomimetic scaffolds facilitate the formation of functional liver tissue, supporting essential processes such as albumin synthesis, ammonia detoxification, and bile production, making ECM mimicry a promising approach for liver regeneration. ECM mimicry not only provides structural support but also modulates cellular responses through biochemical and biomechanical cues. The dynamic interactions between cells and ECM influence tissue remodeling, immune responses, and vascularization, all critical for effective organ regeneration. By tailoring ECM scaffolds with specific proteins, growth factors, and mechanical properties, researchers can create biomimetic environments that closely resemble native tissues. This customization enhances cellular integration and functional recovery, which is crucial for the successful regeneration of complex organs like the lung, kidney, and liver.

Recent advancements in ECM mimicry involve the use of 3D bioprinting and bio fabrication techniques to create organ-specific ECM scaffolds with precise spatial organization of ECM components. These technologies enable the fabrication of scaffolds with controlled porosity, stiffness, and bioactive molecule gradients, closely replicating the native tissue microenvironment. In lung regeneration, for example, 3D bio printed scaffolds can recreate the alveolar structure, while in kidney regeneration, they can mimic the glomerular and tubular architecture. These bio-fabricated ECM scaffolds, when combined with patient-derived cells, offer a personalized approach to organ regeneration, reducing the risk of immune rejection and improving functional outcomes. The integration of ECM mimicry into regenerative medicine holds immense potential for addressing the limitations of traditional organ transplantation, such as donor shortages and immunological complications. By harnessing the regenerative capabilities of ECM scaffolds, researchers can develop bioengineered tissues that mimic the native structure and function of organs, providing a sustainable solution for organ failure. Ongoing research is focused on optimizing ECM scaffolds to enhance their biocompatibility, mechanical properties, and integration with host tissues, paving the way for future clinical applications in lung, kidney, and liver regeneration.

7.10 Smart ECM Mimicking Scaffolds

Smart ECM-mimicking scaffolds represent an advanced approach in tissue engineering, designed to not only replicate the structural and biochemical properties of the native extracellular matrix (ECM) but also to respond dynamically to environmental stimuli. These scaffolds are engineered to provide an interactive platform that can modulate cellular responses in real time, enhancing tissue regeneration and healing. By incorporating responsive materials, such as stimuli-sensitive polymers, bioactive molecules, and nano sensors, smart ECM scaffolds can adapt to changes in pH, temperature, mechanical stress, or the presence of specific enzymes, creating a highly customized microenvironment that promotes cell survival, differentiation, and function. One of the key features of smart ECM-mimicking scaffolds is their ability to deliver bioactive signals in a controlled manner, tailored to the specific needs of regenerating tissues. These scaffolds can be embedded with growth factors, cytokines, or other bioactive agents that are released in response to specific triggers, such as inflammation or changes in the local biochemical environment. For instance, pH-responsive hydrogels can release anti-inflammatory agents in acidic conditions, commonly associated with injured or infected tissues. This controlled release mechanism ensures that therapeutic agents are delivered precisely when and where they are needed, reducing the risk of side effects and enhancing the overall efficacy of the treatment [12]. Another innovative aspect of smart ECM scaffolds is their capacity for real-time monitoring and feedback. By integrating nano sensors or conductive materials into the scaffold matrix, these biomaterials can provide valuable information about the local tissue environment, such as changes in oxygen levels, pH, or mechanical forces. This data can be used to adjust the scaffold's properties, such as stiffness or drug release rate, optimizing the regenerative process. For example, scaffolds designed with piezoelectric materials can generate electrical stimuli in response to mechanical loading, which can enhance cellular activities like proliferation and differentiation, particularly in musculoskeletal tissues. This level of interactivity makes smart ECM-mimicking scaffolds a powerful tool for personalized and adaptive regenerative therapies. The future of smart ECM-mimicking scaffolds lies in their potential to be fully integrated into bio fabrication technologies, such as 3D bioprinting, allowing for the precise design and assembly of complex tissue structures. These scaffolds can be customized to match the anatomical and functional requirements of specific organs, enhancing the potential for successful tissue regeneration. By combining smart materials, advanced fabrication techniques, and bioactive cues, researchers aim to develop next-generation scaffolds that not only mimic ECM but actively participate in the healing process, offering new possibilities for treating complex injuries and organ failures.

7.10.1 Customizable ECM Mimicry Through AI and Machine Learning

Customizable ECM mimicry using AI and machine learning is revolutionizing the design of biomimetic scaffolds in tissue engineering. These technologies enable the precise modelling and simulation of ECM's complex properties, including its composition, structure, and mechanical characteristics, to create personalized scaffolds tailored to specific patient needs. Machine learning algorithms can analyze vast datasets of ECM components, cellular responses, and tissue dynamics, allowing researchers to predict how different scaffold designs will interact with cells. This data-driven approach accelerates the optimization of scaffold properties, such as stiffness, porosity, and biochemical cues, enhancing their ability to support tissue-specific regeneration. AI-powered tools are particularly valuable in customizing ECM-mimicking scaffolds for complex organs like the heart, liver, and lungs, where the microenvironment varies significantly across different tissue regions. By integrating patient-specific data, such as imaging scans and genomic information, AI algorithms can design scaffolds that closely replicate the unique ECM architecture of the target tissue. For example, AI can help create scaffolds with gradient structures that mimic the varying stiffness found in the heart or the hierarchical organization of the liver's lobules. This level of customization improves cell-scaffold interactions, promotes the formation of functional tissue, and reduces the likelihood of adverse immune responses, paving the way for more effective regenerative therapies. Moreover, AI and machine learning facilitate the development of adaptive ECM scaffolds that can respond to changes in the tissue environment. Predictive models can guide the incorporation of smart materials within the scaffolds, such as stimuli-responsive polymers, which adjust their properties in response to biochemical signals or mechanical forces[37]. For instance, AI algorithms can be used to predict the optimal conditions for releasing growth factors or other bioactive molecules in response to specific tissue cues, enhancing the scaffold's regenerative capabilities. This intelligent design approach not only improves the performance of ECM-mimicking scaffolds but also offers the potential for real-time adjustments during the healing process, aligning with the dynamic nature of tissue repair and regeneration.

Exercise 3

Multiple Choice Questions

1. What is the primary purpose of using ECM mimicry in therapeutic applications?
a) To enhance mechanical strength
b) To replicate biological functions
c) To reduce costs
d) To increase toxicity

2. Which of the following is a common material used in ECM mimicry for tissue engineering?
a) Polyethylene
b) Collagen
c) Silicon
d) Glass

3. ECM mimics can be utilized in regenerative medicine to:
a) Prevent inflammation
b) Support cell adhesion and proliferation
c) Block nutrient absorption
d) Induce apoptosis

4. Which therapeutic area benefits significantly from ECM mimicry for wound healing?
a) Cardiovascular
b) Dermatology
c) Neurology
d) Endocrinology

5. In ECM mimicry, the inclusion of growth factors is important because they:
a) Enhance material strength
b) Promote cell migration and differentiation
c) Reduce inflammation
d) Prevent infection

6. The design of ECM mimetics often aims to mimic the ______ of natural tissues.
a) Color
b) Texture
c) Viscosity
d) Temperature

7. ECM mimetics can be applied in the development of:
a) Drug delivery systems
b) Diagnostic tools
c) Vaccines
d) All of the above

8. One of the challenges in using ECM mimicry for therapeutic applications is:
a) High cost of materials
b) Ensuring biocompatibility
c) Excessive complexity
d) Lack of research

9. ECM mimicry can enhance the effectiveness of ______ therapies by improving targeted delivery.
a) Chemotherapy
b) Radiotherapy
c) Immunotherapy
d) All of the above

10. Which of the following cell types is particularly influenced by ECM mimicry in therapeutic applications?
a) Red blood cells
b) Neurons
c) Stem cells
d) Platelets

7.11 Summary

The chapter "Therapeutic Applications of ECM Mimicry" delves into the fundamental principles and innovative strategies for replicating the extracellular matrix (ECM) in various therapeutic contexts. It begins by defining ECM mimicry, highlighting its importance in regenerative medicine and tissue engineering, and contrasting natural and synthetic ECMs. The chapter explores the structural, biochemical, and mechanical properties of ECM that are crucial for guiding cellular behavior, focusing on how these principles are harnessed in the design of ECM-mimicking scaffolds. Advanced fabrication techniques, such as electrospinning, 3D bioprinting, and hydrogel formation, are discussed, along with strategies for surface modification and the integration of bioactive molecules to enhance scaffold functionality. ECM mimicry is examined across a range of therapeutic applications, including wound healing, cancer therapy, and regeneration of bone, cartilage, and cardiovascular tissues. The chapter highlights the role of ECM-based scaffolds in skin regeneration, chronic wound healing, and tumor progression, emphasizing the potential of ECM mimicry in improving targeted drug delivery and disrupting cancer metastasis. Furthermore, it covers the use of ECM-mimicking scaffolds in bone and cartilage repair, cardiovascular applications, and neural regeneration, showcasing clinical case studies and outcomes. The chapter concludes by exploring future directions in ECM mimicry, including smart scaffolds, AI-driven customizations, and ongoing challenges in clinical translation, positioning ECM mimicry as a transformative approach in the field of regenerative medicine and beyond.

Points to Remember

- **ECM mimicry allows for the design of patient-specific scaffolds using advanced technologies like 3D bioprinting. These scaffolds can be tailored to match the unique anatomy and pathology of an individual, leading to personalized treatment strategies in regenerative medicine and cancer therapy.**

- **ECM-mimicking scaffolds can be engineered with "smart" features, such as responsive drug delivery systems that release therapeutic agents in response to specific stimuli, like pH changes or enzymatic activity within a wound or tumor microenvironment. This capability makes these scaffolds not just passive supports but active players in healing and disease treatment.**

- **One of the most innovative ECM mimicry strategies involves using decellularized organs—real tissues from donors that have had their cells removed, leaving behind a natural ECM scaffold. These decellularized matrices retain the complex architecture and biochemical cues of the original tissue, making them ideal for organ regeneration.**

- **Beyond supporting normal tissue function, ECM mimicry is being explored as a weapon against cancer. By altering the tumor ECM to make it less supportive of cancer cell growth and more accessible to immune cells, researchers are developing ECM-based therapies that could turn the tumor microenvironment against the cancer itself, enhancing the effectiveness of existing treatments.**

- **ECM-mimicking materials can be multifunctional, combining structural support with bioactive properties. These materials can simultaneously guide tissue regeneration, deliver drugs, and provide mechanical strength, positioning them as highly versatile tools in medical treatments that go beyond traditional approaches.**

References

1. Yue B (2014) Biology of the extracellular matrix: an overview. J Glaucoma 23(8 Suppl 1):S20–S23. https://doi.org/10.1097/IJG.0000000000000108
2. Nicolas J, Magli S, Rabbachin L, Sampaolesi S, Nicotra F, Russo L (2020) 3D Extracellular Matrix mimics: fundamental concepts and role of materials chemistry to influence stem cell fate. Biomacromol 21(6):1968–1994. https://doi.org/10.1021/acs.biomac.0c00045
3. Frantz C, Stewart KM, Weaver VM (2010) The extracellular matrix at a glance. J Cell Sci 123(24):4195–4200. https://doi.org/10.1242/jcs.023820
4. Ligorio C, Mata A (2023) Synthetic extracellular matrices with function-encoding peptides. Nat Rev Bioeng 1:518–536. https://doi.org/10.1038/s44222-023-00055-3
5. Pellowe AS, Gonzalez AL (2016) Extracellular matrix biomimicry for the creation of investigational and therapeutic devices. Wiley Interdiscip Rev Nanomed Nanobiotechnol 8(1):5–22. https://doi.org/10.1002/wnan.1349
6. Karamanos NK, Theocharis AD, Piperigkou Z, Manou D, Passi A, Skandalis SS, Vynios DH, Orian-Rousseau V, Ricard-Blum S, Schmelzer CEH, Duca L, Durbeej M, Afratis NA, Troeberg L, Franchi M, Masola V, Onisto M (2021) A guide to the composition and functions of the extracellular matrix. FEBS J 288(24):6850–6912. https://doi.org/10.1111/febs.15776
7. Nakayama KH, Hou L, Huang NF (2014) Role of extracellular matrix signaling cues in modulating cell fate commitment for cardiovascular tissue engineering. Adv Healthcare Mater 3(5):628–641. https://doi.org/10.1002/adhm.201300620

8. Shebanova O, Hammer DA (2012) Biochemical and mechanical extracellular matrix properties dictate mammary epithelial cell motility and assembly. Biotechnol J 7(3):397–408. https://doi.org/10.1002/biot.201100188
9. Sainio A, Järveläinen H (2020) Extracellular matrix-cell interactions: focus on therapeutic applications. Cell Signal 66:109487. https://doi.org/10.1016/j.cellsig.2019.109487
10. Raspa A, Gelain F (2021) Mimicking extracellular matrix via engineered nanostructured biomaterials for neural repair. Curr Neuropharmacol 19(12):2110–2124. https://doi.org/10.2174/1570159X18666201111111102
11. Hama R, Reinhardt JW, Ulziibayar A, Watanabe T, Kelly J, Shinoka T (2023) Recent tissue engineering approaches to mimicking the extracellular matrix structure for skin regeneration. Biomimetics (Basel, Switzerland) 8(1):130. https://doi.org/10.3390/biomimetics8010130
12. Salerno A, Netti PA (2023) Review on bioinspired design of ECM-mimicking scaffolds by computer-aided assembly of cell-free and cell laden micro-modules. J Funct Biomater 14(2):101. https://doi.org/10.3390/jfb14020101
13. Cai L, Heilshorn SC (2014) Designing ECM-mimetic materials using protein engineering. Acta Biomater 10(4):1751–1760. https://doi.org/10.1016/j.actbio.2013.12.028
14. Collins MN, Ren G, Young K, Pina S, Reis RL, Oliveira JM (2021) Scaffold fabrication technologies and Structure/Function properties in bone tissue engineering. Adv Funct Mater 31(21). https://doi.org/10.1002/adfm.202010609
15. Jun I, Han HS, Edwards JR, Jeon H (2018) Electrospun fibrous scaffolds for tissue engineering: viewpoints on architecture and fabrication. Int J Mol Sci 19(3):745. https://doi.org/10.3390/ijms19030745
16. Teimouri R, Abnous K, Taghdisi SM, Ramezani M, Alibolandi M (2023) Surface modifications of scaffolds for bone regeneration. J Mater Res Technol 24:7938–7973. https://doi.org/10.1016/j.jmrt.2023.05.076
17. Liu J, Song Q, Yin W, Li C, An N, Le Y, Wang Q, Feng Y, Hu Y, Wang Y (2024) Bioactive scaffolds for tissue engineering: a review of decellularized extracellular matrix applications and innovations. Exploration. https://doi.org/10.1002/exp.20230078
18. Koka P, Chandramohan Y, Perumal E, Kavarthapu A, Dhanasekaran A, Chandran A, Gunasekaran K (2023) Fabrication of ECM mimicking bioactive scaffold: a regenerative approach for MSC mediated applications. Stem Cells Int 2023:6282987. https://doi.org/10.1155/2023/6282987
19. Diller RB, Tabor AJ (2022) The role of the extracellular matrix (ECM) in wound healing: a review. Biomimetics (Basel, Switzerland) 7(3):87. https://doi.org/10.3390/biomimetics7030087
20. Da LC, Huang YZ, Xie HQ, Zheng BH, Huang YC, Du SR (2021) Membranous extracellular matrix-based scaffolds for skin wound healing. Pharmaceutics 13(11):1796. https://doi.org/10.3390/pharmaceutics13111796
21. Agren MS, Werthén M (2007) The extracellular matrix in wound healing: a closer look at therapeutics for chronic wounds. Int J Low Extrem Wounds 6(2):82–97. https://doi.org/10.1177/1534734607301394
22. Huang J, Zhang L, Wan D, Zhou L, Zheng S, Lin S, Qiao Y (2021) Extracellular matrix and its therapeutic potential for cancer treatment. Signal Transduct Target Ther 6(1):153. https://doi.org/10.1038/s41392-021-00544-0
23. Nazemi M, Rainero E (2020) Cross-talk between the tumor microenvironment, extracellular matrix, and cell metabolism in cancer. Front Oncol 10:239. https://doi.org/10.3389/fonc.2020.00239
24. Hwang J, Sullivan MO, Kiick KL (2020) Targeted drug delivery via the use of ECM-mimetic materials. Front Bioeng Biotechnol 8:69. https://doi.org/10.3389/fbioe.2020.00069
25. Eble JA, Niland S (2019) The extracellular matrix in tumor progression and metastasis. Clin Exp Metas 36(3):171–198. https://doi.org/10.1007/s10585-019-09966-1
26. Parmaksiz M, Dogan A, Odabas S, Elçin AE, Elçin YM (2016) Clinical applications of decellularized extracellular matrices for tissue engineering and regenerative medicine. Biomed Mater (Bristol, England) 11(2):022003. https://doi.org/10.1088/1748-6041/11/2/022003

27. Benders KE, van Weeren PR, Badylak SF, Saris DB, Dhert WJ, Malda J (2013) Extracellular matrix scaffolds for cartilage and bone regeneration. Trends Biotechnol 31(3):169–176. https://doi.org/10.1016/j.tibtech.2012.12.004
28. Todd EA, Mirsky NA, Silva BLG, Shinde AR, Arakelians ARL, Nayak VV, Marcantonio RAC, Gupta N, Witek L, Coelho PG (2024) Functional scaffolds for bone tissue regeneration: a comprehensive review of materials, methods, and future directions. J Funct Biomater 15(10):280. https://doi.org/10.3390/jfb15100280
29. Bloksgaard M, Lindsey M, Martinez-Lemus LA (2018) Extracellular matrix in cardiovascular pathophysiology. Am J Physiol Heart Circ Physiol 315(6):H1687–H1690. https://doi.org/10.1152/ajpheart.00631.2018
30. Silva AC, Pereira C, Fonseca ACRG, Pinto-do-Ó P, Nascimento DS (2021) Bearing my heart: the role of extracellular matrix on cardiac development, homeostasis, and injury response. Front Cell Dev Biol 8:621644. https://doi.org/10.3389/fcell.2020.621644
31. Thottappillil N, Nair PD (2015) Scaffolds in vascular regeneration: current status. Vasc Health Risk Manag 11:79–91. https://doi.org/10.2147/VHRM.S50536
32. Kato B, Wisser G, Agrawal DK, Wood T, Thankam FG (2021) 3D bioprinting of cardiac tissue: current challenges and perspectives. Journal of materials science. Mater Med 32(5):54. https://doi.org/10.1007/s10856-021-06520-y
33. Estrada V, Tekinay A, Müller HW (2014) Neural ECM mimetics. Prog Brain Res 214:391–413. https://doi.org/10.1016/B978-0-444-63486-3.00016-5
34. Chelyshev YA, Kabdesh IM, Mukhamedshina YO (2022) Extracellular matrix in neural plasticity and regeneration. Cell Mol Neurobiol 42:647–664. https://doi.org/10.1007/s10571-020-00986-0
35. Haggerty AE, Marlow MM, Oudega M (2016) Extracellular matrix components as therapeutics for spinal cord injury. Neurosci Lett 652:50–55. https://doi.org/10.1016/j.neulet.2016.09.053
36. Gilpin A, Yang Y (2017) Decellularization strategies for regenerative medicine: from processing techniques to applications. Biomed Res Int 2017:9831534. https://doi.org/10.1155/2017/9831534
37. Kim J, McKee JA, Fontenot JJ, Jung JP (2020) Engineering tissue fabrication with machine intelligence: generating a blueprint for regeneration. Front Bioeng Biotechnol 7:443. https://doi.org/10.3389/fbioe.2019.00443

Chapter 8
Advanced Technologies in ECM Mimicry

8.1 Introduction

The extracellular matrix (ECM) plays a crucial role in maintaining tissue architecture and cellular function across various biological systems. It provides structural support, mediates biochemical signals, and influences cell behavior such as proliferation, migration, and differentiation. The ECM is composed of a complex network of proteins, glycoproteins, proteoglycans, and polysaccharides, which vary in composition depending on the tissue type. In tissue engineering, the ability to replicate the ECM is paramount, as it allows the creation of environments that closely mimic the native tissue's mechanical and biochemical cues. ECM mimicry, therefore, aims to replicate these properties to enhance tissue repair, regeneration, and drug delivery applications. The importance of ECM in tissue engineering cannot be overstated [1]. As the primary environment in which cells reside, the ECM influences nearly every aspect of cell behavior. It acts as a physical scaffold and provides the biochemical signals that dictate cell fate and function. Successful tissue engineering endeavors rely on mimicking the native ECM to create scaffolds that encourage tissue growth and integration. By simulating the dynamic nature of the ECM, researchers can develop biomaterials that not only support cells but also guide their development into functional tissue. ECM mimicry in tissue engineering offers the potential to address various medical challenges, from wound healing to organ regeneration.

Mimicking the native ECM, however, presents several challenges. The ECM is not a static entity; its composition, stiffness, and structure change depending on the physiological context and the stage of development or healing. Replicating the intricate and dynamic nature of the ECM is a complex task, requiring materials that can adjust their properties in response to environmental signals [2]. Furthermore, each tissue type has its own unique ECM composition, making it difficult to develop a one-size-fits-all scaffold. Biomimetic scaffolds must, therefore, be tissue-specific, capable of replicating not only the physical but also the biochemical environment of the target tissue.

D. Suhag, *Extracellular Matrix biomimicking for Theranostics*, Biomedical Materials for Multi-functional Applications, https://doi.org/10.1007/978-981-95-7900-6_8

Another challenge in ECM mimicry is the integration of biological signals that control cell behavior. The ECM contains numerous growth factors, cytokines, and other bioactive molecules that regulate processes such as cell adhesion, migration, and differentiation. Designing scaffolds that can deliver these signals in a controlled and sustained manner is essential for successful tissue engineering. The mechanical properties of ECM scaffolds, such as stiffness and elasticity, need to be carefully tuned to match the properties of the target tissue. Imbalances in these properties can lead to inappropriate cell behavior or scaffold failure. To overcome these challenges, researchers have developed various advanced technologies in ECM mimicry. One approach involves using natural biomaterials, such as collagen, fibrin, and alginate, which closely resemble the components of native ECM [3]. These materials can be processed into hydrogels or fibrous scaffolds that mimic the structural and biochemical properties of the ECM. Synthetic polymers, such as poly (lactic-co-glycolic acid) (PLGA) and polyethylene glycol (PEG), are often combined with bioactive molecules to create tunable, ECM-like environments that can be customized for specific applications. Another technology gaining prominence is 3D bioprinting, which allows the precise fabrication of ECM-mimicking scaffolds with high spatial control. By incorporating cells and ECM-like materials into bioinks, 3D bioprinting can create complex tissue structures that resemble the native ECM architecture [4]. This technology also enables the creation of multi-layered structures that can simulate the heterogeneity of tissues, providing more accurate models for tissue repair and drug testing. Decellularization techniques, which involve removing cellular components from donor tissues while preserving the ECM, offer another route for generating scaffolds that retain the native tissue's mechanical and biochemical properties.

Nanotechnology has also emerged as a valuable tool in ECM mimicry. Nanoscale fibers, particles, and scaffolds can replicate the fine structure of the ECM, creating a microenvironment that supports cellular processes. Nanofibers, for example, can be electrospun to form scaffolds that closely resemble the fibrous nature of ECM, while nanoparticles can be used to deliver growth factors and other bioactive molecules. The integration of nanotechnology with ECM mimicry has opened new avenues for the development of highly functional and responsive scaffolds that promote tissue regeneration.

8.2 Biomaterials for ECM Mimicry

Biomaterials play a pivotal role in ECM mimicry, serving as the foundation for creating scaffolds that replicate the structural, mechanical, and biochemical properties of the native extracellular matrix [5]. In tissue engineering, these materials are designed to provide a supportive environment that fosters cell growth, differentiation, and tissue regeneration. Biomaterials can be derived from natural sources, such as collagen and fibrin, or synthesized from polymers like polyethylene glycol (PEG) and poly (lactic-co-glycolic acid) (PLGA), each offering distinct advantages depending on the application. The ability to tailor these materials to mimic the ECM's complexity has advanced the development of scaffolds for a wide range of biomedical applications, including wound healing, organ repair, and drug delivery. By integrating bioactive molecules and engineering properties like stiffness and porosity, biomaterials are essential in creating functional ECM-like structures that enhance cellular function and tissue integration.

8.2.1 Natural Versus Synthetic Biomaterials for ECM Mimicry

In ECM mimicry, the choice between natural and synthetic biomaterials plays a significant role in scaffold design, as each offers distinct advantages and limitations. Natural biomaterials, derived from sources like collagen, fibrin, and hyaluronic acid, closely resemble the native ECM, providing a biological environment conducive to cell attachment, migration, and differentiation. These materials typically offer superior biocompatibility and bioactivity, which makes them highly effective in promoting tissue regeneration [6]. However, natural biomaterials often face challenges such as batch variability, limited mechanical strength, and rapid degradation, which can limit their functionality in long-term or load-bearing applications.

Synthetic biomaterials, such as poly (lactic acid) (PLA), poly (glycolic acid) (PGA), and polycaprolactone (PCL), provide greater control over mechanical properties, degradation rates, and reproducibility. Unlike natural biomaterials, synthetic options can be engineered to meet specific tissue requirements, offering tunable properties for various applications. They are also more scalable and customizable, making them suitable for large-scale production and use in diverse tissue types. However, synthetic biomaterials often lack the intrinsic bioactivity of natural materials, which can hinder cell interactions and integration unless they are functionalized with bioactive molecules. The trade-off between natural and synthetic materials continues to drive innovation in ECM mimicry, with hybrid approaches emerging to combine the best attributes of both types.

8.2.2 Polymers in ECM Scaffold Development

Polymers, both natural and synthetic, serve as foundational materials in ECM scaffold development due to their versatility in replicating ECM properties. Natural polymers, such as chitosan, alginate, and silk fibroin, offer bioactive platforms that support cellular activities critical for tissue regeneration. These polymers can form hydrogels, fibers, or films that mimic the soft, fibrous nature of ECM, providing a conducive microenvironment for cells. Their intrinsic biological activity promotes cellular adhesion, proliferation, and differentiation, making them particularly useful in wound healing and cartilage repair applications. Their mechanical properties often require enhancement, particularly in load-bearing tissues such as bone or muscle. Synthetic polymers, including PEG, PLGA, and polyurethane, are frequently used to engineer ECM scaffolds with precise mechanical and degradation properties. These polymers can be easily modified to create scaffolds that match the specific stiffness, porosity, and architecture of target tissues. Moreover, their long degradation times can support tissue regeneration over extended periods, making them ideal for applications where prolonged scaffold support is required [7]. To enhance bioactivity, synthetic polymers are often functionalized with peptides, growth factors, or other bioactive molecules that mimic ECM components. This integration of synthetic polymers with bioactive signals creates a balance between mechanical strength and biological function, making polymer-based scaffolds essential in advanced tissue engineering.

Exercise 1

Fill Ups
(Alginate, Hydrated, Biological, Polylactic acid (PLA), Mechanical, Chitosan, Growth factors)

1. Biomaterials used for ECM mimicry often include natural polymers like ______ and ______ to provide biocompatibility.

2. Synthetic biomaterials such as ______ are commonly used in ECM mimicry due to their tunable properties.

3. The incorporation of ______ in biomaterials can enhance cell adhesion and promote tissue regeneration.

4. Hydrogels are frequently utilized in ECM mimicry because they can maintain a ______ environment that supports cell survival.

5. One of the main challenges in developing biomaterials for ECM mimicry is achieving the right balance between ______ and ______ properties.

8.2.3 Bioactive Molecules for Functional Mimicry

Bioactive molecules are critical for functional ECM mimicry, as they replicate the biochemical cues that regulate cell behavior within the ECM. Growth factors, cytokines, and adhesion proteins such as fibronectin and laminin are commonly incorporated into scaffolds to mimic the native ECM's role in guiding cellular processes (Fig. 8.1). These bioactive molecules promote cell adhesion, migration, proliferation, and differentiation, enhancing the scaffold's ability to support tissue regeneration. For example, growth factors like vascular endothelial growth factor (VEGF) and bone morphogenetic proteins (BMPs) can be incorporated into ECM-mimicking scaffolds to stimulate angiogenesis and osteogenesis, respectively. Controlled release of these bioactive molecules ensures that cells receive appropriate signals throughout the healing process.

In addition to growth factors, short peptides derived from ECM proteins, such as RGD (arginine-glycine-aspartate), can be functionalized onto scaffolds to enhance cell attachment and interaction with the material. These peptides mimic natural

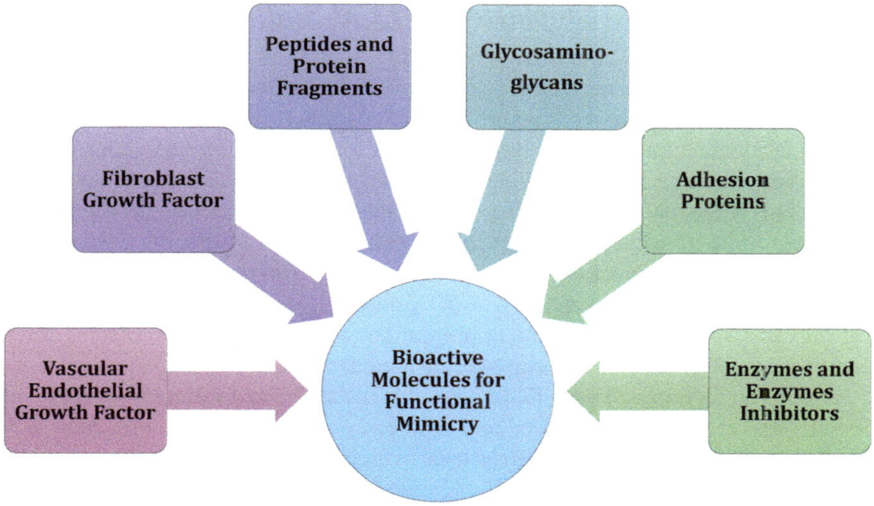

Fig. 8.1 Different bioactive molecules for functional mimicry

binding sites in the ECM, improving cellular responses and facilitating tissue integration. Bioactive molecules can be used to modulate the immune response, creating an environment that is more favorable for tissue repair by reducing inflammation and preventing foreign body reactions [8]. By carefully selecting and incorporating bioactive molecules into ECM-mimicking scaffolds, researchers can create advanced biomaterials that not only provide structural support but also actively guide tissue regeneration, accelerating the healing process and improving therapeutic outcomes.

8.3 3D Printing and Bioprinting Techniques

3D printing and bioprinting techniques have revolutionized the field of tissue engineering by enabling the precise fabrication of complex, tissue-like structures that mimic the architecture and function of the extracellular matrix (ECM). These advanced technologies allow for the layer-by-layer deposition of biomaterials, cells, and bioactive molecules, creating scaffolds that closely replicate the native environment of tissues. 3D printing primarily focuses on fabricating ECM-mimicking scaffolds, while bioprinting goes a step further by incorporating living cells directly into the printed structures [9]. These techniques offer unmatched control over scaffold geometry, porosity, and material composition, making them ideal for developing personalized, tissue-specific constructs for regenerative medicine, wound healing, and drug testing applications. The ability to create multi-material, cell-laden constructs open up new possibilities for more accurate tissue models and enhanced therapeutic strategies.

8.3.1 3D Printing Technologies for ECM Scaffolds

3D printing technologies have emerged as powerful tools for fabricating ECM scaffolds with precise control over structure and composition (Fig. 8.2). Unlike traditional scaffold fabrication methods, 3D printing allows for the layer-by-layer construction of scaffolds, enabling intricate designs that closely mimic the architecture of native tissues. The precision of 3D printing ensures that scaffold porosity, geometry, and mechanical properties can be customized to suit the needs of specific tissue types, which is critical for mimicking the complex microenvironment of the extracellular matrix (ECM). With advancements in 3D printing, researchers can now produce scaffolds that support cell attachment, migration, and differentiation, making them ideal for tissue regeneration and engineering applications. One of the key advantages of 3D printing technologies for ECM scaffolds is the ability to use a wide range of biomaterials, including both natural and synthetic polymers [10]. Biomaterials such as collagen, gelatin, and alginate can be used in their natural form, providing biocompatibility and biological cues for cells. On the other hand, synthetic polymers like poly (lactic acid) (PLA) and polycaprolactone (PCL) offer enhanced mechanical strength and can be tailored to specific degradation profiles. These materials can be printed in combinations to create hybrid scaffolds that offer both the biological functionality of natural ECM components and the durability of synthetic materials. This versatility in material selection expands the potential applications of 3D-printed ECM scaffolds, from bone regeneration to skin repair.

3D printing technologies allow for the incorporation of bioactive molecules, such as growth factors, into the scaffolds, promoting tissue-specific cellular responses. By embedding these molecules within the printed structure, scaffolds can release them in a controlled and sustained manner, mimicking the ECM's role in regulating cell behavior. This controlled release can enhance processes such as angiogenesis, cell differentiation, and wound healing. Multi-material printing capabilities enable the

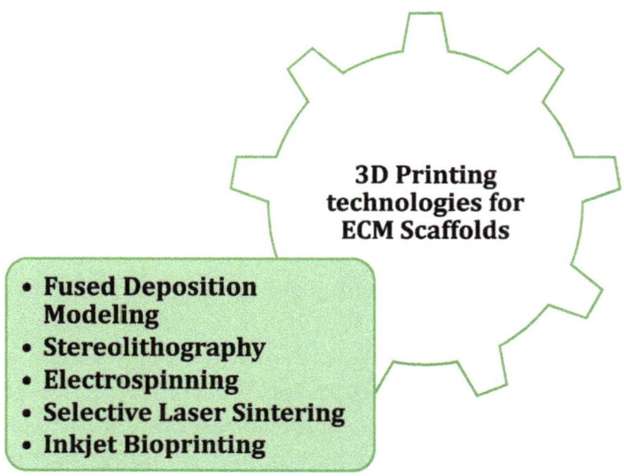

Fig. 8.2 3D printing technologies for ECM scaffolds

fabrication of scaffolds with gradient properties, simulating the varying mechanical and biochemical environments found in tissues like bone, cartilage, and muscle. These innovations make 3D printing a promising approach for developing next-generation ECM scaffolds tailored to complex tissue engineering challenges.

8.3.2 Bioprinting of Cells and ECM Components

Bioprinting has revolutionized tissue engineering by allowing the precise deposition of living cells and ECM components, creating tissue constructs that closely mimic the native microenvironment. Unlike traditional 3D printing, which focuses on fabricating scaffolds using synthetic or natural materials, bioprinting incorporates living cells directly into the printed structure, along with bioinks composed of ECM components like collagen, hyaluronic acid, or alginate. This method enables the creation of tissue scaffolds with not only the appropriate physical architecture but also the biological complexity needed for cellular growth, differentiation, and function. By integrating cells and ECM components, bioprinting provides a more holistic approach to replicating the native tissue environment, offering the potential for more effective tissue repair and regeneration [11]. One of the major advantages of bioprinting is the ability to precisely place multiple cell types within specific regions of the scaffold, which is crucial for mimicking the cellular organization of tissues. For example, in the creation of vascularized tissues, endothelial cells can be bio printed to form blood vessel networks alongside other cell types, such as fibroblasts or stem cells, to support tissue formation. This spatial control over cell placement ensures that the printed tissues better replicate the structure and function of native tissues, improving their potential for clinical applications. ECM components in the bioinks provide the necessary biochemical cues that guide cell behavior, including migration, adhesion, and differentiation, further enhancing the scaffold's ability to mimic the native ECM.

Bioprinting allows for the creation of gradient structures that mirror the natural variation in ECM composition and stiffness found in tissues. For example, certain tissues like cartilage or skin have layers with varying mechanical properties and ECM composition, which can be replicated through bioprinting by gradually changing the bioink formulation during the printing process. This capability to create functionally graded scaffolds enhances the scaffold's ability to integrate with native tissue, promoting more successful tissue repair. As a result, bioprinting offers an advanced platform for engineering complex tissues, enabling the development of more accurate tissue models for drug testing, regenerative medicine, and personalized treatments.

Exercise 2

Fill Ups
(3D, Water, Extracellular matrix (ECM), Water, Composition)

1. Hydrogels are composed primarily of ______ and are used to create a 3D environment that mimics the natural ECM.

2. One common property of hydrogels is their ability to retain large amounts of ______, which is essential for cell survival and function.

3. Bioinks are specialized formulations of hydrogels that are designed for use in ______ printing applications.

4. The mechanical properties of hydrogels can be tuned by altering their ______ and cross-linking density.

5. Incorporating ______ proteins into bioinks can enhance cellular interactions and promote tissue development in ECM mimicry.

8.4 Hydrogels and Bioinks for ECM Mimicry

Hydrogels and bioinks have emerged as vital components in tissue engineering and regenerative medicine, particularly in mimicking the extracellular matrix (ECM). Hydrogels, composed of water-swollen networks of polymers, exhibit a structure that closely resembles the natural ECM, providing a hydrated and biocompatible environment for cell growth and differentiation. Their ability to encapsulate bioactive molecules and living cells makes them ideal for applications where tissue regeneration and healing are required. Hydrogels mimic the mechanical and biochemical properties of the ECM, facilitating cell adhesion, migration, and proliferation, which are crucial for tissue repair and engineering [12].

Bioinks, on the other hand, are materials specifically formulated for 3D bioprinting, incorporating living cells, bioactive factors, and ECM-like components to fabricate complex tissue structures. Bioinks often include polymers like alginate, gelatin, and hyaluronic acid, which are naturally derived from the ECM or are chemically modified to mimic its properties. The primary aim of using bioinks in tissue engineering is to recreate the intricate architecture of tissues, including the microenvironment provided by the ECM. The adaptability of bioinks allows for the printing of diverse tissue types, ranging from soft tissues like skin and cartilage to more complex structures like blood vessels. Both hydrogels and bioinks can be functionalized with bioactive cues such as growth factors, peptides, or ECM-derived proteins to guide cellular behavior and promote tissue regeneration. These biofunctionalized materials enhance the natural processes of wound healing and tissue repair by mimicking ECM signals that regulate cellular activities. For example, hydrogels with incorporated growth factors can stimulate cell proliferation, while bioinks with ECM proteins can improve cell adhesion and differentiation, which is crucial for tissue-specific regeneration. This bioactivity ensures that the scaffold materials are not merely passive carriers but actively participate in tissue repair processes [13]. In addition, the tunable properties of hydrogels and bioinks enable the customization of scaffold stiffness, porosity, and degradation rates, which are essential for matching the specific requirements of different tissues. For instance, softer hydrogels are ideal for mimicking the ECM in neural tissues, while stiffer hydrogels may be better suited for bone or cartilage applications. The versatility in designing these materials allows researchers to create scaffolds that not only mimic the mechanical properties of the ECM but also degrade at rates that match the pace of new tissue formation, ensuring seamless tissue integration.

The use of hydrogels and bioinks for ECM mimicry is a promising approach in regenerative medicine, particularly for creating biomimetic environments that support tissue growth and healing. By integrating structural, mechanical, and biochemical properties of the ECM into scaffold design, these materials help overcome challenges in tissue engineering, such as cell integration, nutrient transport, and tissue functionality.

8.4.1 Functional Hydrogels as ECM Mimetics

Hydrogels are three-dimensional, water-swollen networks of polymers that can be engineered to replicate the physical and biochemical properties of the native ECM. Their ability to absorb large amounts of water while maintaining structural integrity makes them ideal for creating environments that support cell proliferation, migration, and differentiation, closely resembling the hydrated nature of the ECM. Moreover, their tunable mechanical properties allow researchers to adjust stiffness and elasticity to match specific tissue types, from soft tissues like skin and brain to more rigid tissues such as cartilage and bone. This adaptability makes hydrogels particularly suitable for a wide range of tissue engineering and regenerative medicine applications. In addition to their mechanical tunability, functional hydrogels can be designed to incorporate bioactive molecules, such as growth factors, peptides, or proteins, which are essential for guiding cell behavior. These bioactive molecules can be either physically entrapped within the hydrogel matrix or chemically conjugated to the polymer chains, allowing for controlled and sustained release over time [14]. This mimics the ECM's role in providing biochemical cues to cells, promoting processes like angiogenesis, wound healing, and tissue regeneration. For instance, hydrogels embedded with vascular endothelial growth factor (VEGF) can encourage the formation of blood vessels in tissue-engineered constructs, while other hydrogels may release bone morphogenetic proteins (BMPs) to support bone formation. By integrating bioactive molecules, hydrogels go beyond serving as passive scaffolds, becoming active participants in tissue healing and repair.

One of the major advantages of functional hydrogels as ECM mimetics is their ability to support cell encapsulation and 3D cell culture, which more accurately replicates the in vivo cellular environment compared to traditional 2D culture systems. Cells encapsulated within hydrogels experience a more natural, three-dimensional environment that promotes cell–cell and cell–matrix interactions, which are critical for proper tissue development. This 3D environment enhances cell function, promoting differentiation and the formation of tissue-specific structures. Hydrogels can be modified to allow for cell-mediated degradation, enabling cells to remodel the scaffold as they grow and proliferate, like how they interact with the native ECM [15]. This dynamic interaction between cells and the hydrogel matrix is essential for long-term tissue integration and function.

Hydrogels can also be engineered to respond to external stimuli, such as temperature, pH, or light, making them "smart" materials that adapt to changing conditions in the biological environment. For example, temperature-sensitive hydrogels can undergo sol–gel transitions, allowing for easy injection into the body as a liquid before solidifying into a gel at physiological temperatures. This property is particularly useful for minimally invasive delivery of cells and therapeutic agents. pH-sensitive hydrogels, on the other hand, can be designed to release bioactive molecules in response to changes in the local environment, such as the acidic conditions present in injured or cancerous tissues. These smart hydrogels offer an additional layer of functionality, allowing for more precise control over the timing and location of therapeutic interventions. Functional hydrogels provide an exceptional platform for mimicking ECM due to their tunable physical properties, ability to deliver bioactive molecules, and capacity to support 3D cell culture and encapsulation. Their versatility in design and functionality allows them to be tailored for specific tissue engineering applications, from soft tissue repair to bone regeneration. By mimicking both the structural and biochemical roles of the native ECM, functional hydrogels are instrumental in advancing the field of tissue engineering and regenerative medicine, offering promising solutions for complex tissue repair, drug delivery, and disease modeling.

8.4.2 Bioinks for Tissue-Specific ECM Mimicry

Bioinks have become essential tools for tissue-specific ECM mimicry in the rapidly evolving field of bioprinting. Bioinks are specialized materials composed of biomolecules, cells, and supportive scaffolds, designed to replicate the unique characteristics of ECM found in different tissues. The primary function of bioinks is to create a conducive environment for cells to thrive, differentiate, and develop into functional tissue (Fig. 8.3). This requires bioinks to be highly customizable to meet the distinct physical, chemical, and biological demands of various tissue types. By providing the appropriate biochemical cues, mechanical properties, and structural architecture, bioinks enable precise tissue-specific ECM mimicry, which is critical for effective tissue regeneration, repair, and function. One of the reasons bioinks are crucial for tissue-specific ECM mimicry is their ability to incorporate specific ECM components that are native to the target tissue. For instance, cartilage bioinks may be rich in collagen type II and chondroitin sulfate, which are key components of cartilage ECM, while bioinks for bone tissue may include hydroxyapatite or calcium phosphate to provide the necessary mineral structure. This tissue-specific composition ensures that the cells in the printed construct are exposed to the correct biochemical signals, encouraging them to adopt the desired phenotype [16]. Without such specificity, cells may not function optimally, reducing the efficacy of the printed tissue. This fine-tuning of bioinks is critical for ensuring that printed tissues closely replicate native counterparts, from structure to function. The mechanical properties of bioinks are equally important for tissue-specific ECM mimicry, as different tissues have vastly different requirements for stiffness, elasticity, and porosity. For example, bioinks designed for soft tissues like skin or liver need to be more elastic and flexible, while those intended for bone or cartilage require greater stiffness and mechanical stability. Bioinks can be tailored to match the mechanical properties of the target tissue by adjusting the concentration of biomaterials or by blending natural and synthetic polymers. Achieving the correct mechanical environment is vital for promoting cellular behaviors such as migration, adhesion, and differentiation, which are heavily influenced by the stiffness and structural integrity of ECM. Thus, the ability to design bioinks with tissue-specific mechanical characteristics is fundamental for successful tissue engineering.

In addition to their structural and biochemical roles, bioinks are essential for delivering bioactive molecules that further enhance tissue-specific ECM mimicry. Growth factors, cytokines, and other signaling molecules can be incorporated into bioinks

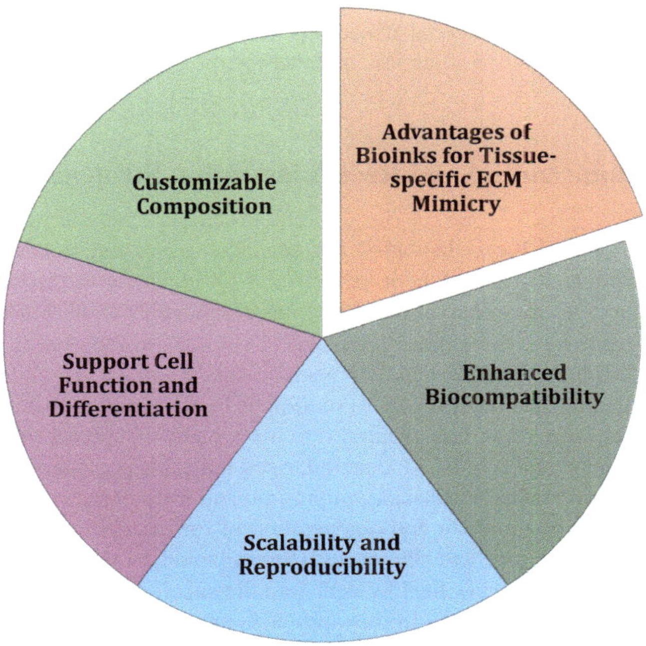

Fig. 8.3 Advantages of bioinks for tissue specific ECM mimicry

to provide additional cues that guide cellular development. For instance, bioinks for vascular tissues might include vascular endothelial growth factor (VEGF) to promote blood vessel formation, while bone tissue bioinks could contain bone morphogenetic proteins (BMPs) to stimulate osteogenesis. These bioactive molecules can be released in a controlled manner within the printed construct, mimicking the natural release of signals from ECM in vivo. This ability to control the microenvironment not only enhances tissue development but also improves the integration of the printed construct with native tissues in regenerative medicine applications.

Finally, bioinks play a pivotal role in creating complex, multi-tissue constructs by allowing for the co-printing of different cell types and ECM components in a spatially controlled manner. In tissues such as the liver or skin, different layers or regions have distinct ECM compositions and cellular populations. Bioinks can be precisely formulated and deposited to recreate these gradients, enabling the formation of layered tissues with proper cell–cell and cell-ECM interactions. This capability is particularly important for developing complex tissues and organoids that more accurately model the functionality of human organs [4]. By replicating the heterogeneity of native tissues, bioinks not only enhance the fidelity of the printed structures but also expand the potential applications of bioprinting in drug testing, disease modeling, and personalized medicine.

Bioinks are integral to achieving tissue-specific ECM mimicry, as they provide the necessary biochemical, mechanical, and structural cues that guide tissue formation. Through careful customization of their composition and properties, bioinks enable the creation of tissue-engineered constructs that closely replicate native tissues, promoting better integration, function, and therapeutic outcomes.

8.5 Decellularization and Recellularization Strategies

Decellularization and recellularization are advanced tissue engineering strategies aimed at creating biologically functional scaffolds for regenerative medicine. Decellularization involves the removal of all cellular components from donor tissues or organs, leaving behind the extracellular matrix (ECM), which retains the natural architecture, mechanical properties, and biochemical cues of the original tissue. This process ensures that the remaining ECM scaffold is free of antigens that could trigger an immune response when transplanted into a recipient. By preserving the native structure and composition of ECM, decellularized scaffolds provide an ideal framework for supporting cellular adhesion, migration, and differentiation, making them highly suitable for tissue repair and regeneration in various applications, including heart valves, blood vessels, and skin grafts. Recellularization complements decellularization by introducing new, healthy cells into the decellularized ECM scaffold to restore tissue functionality [17]. In this approach, stem cells or differentiated cells are seeded onto the scaffold, where they can repopulate and integrate with the matrix, eventually developing into functional tissue. Recellularization leverages the biocompatibility and native cues of the ECM to guide cellular behavior, such as proliferation and differentiation, thereby recreating a tissue that mimics its natural counterpart. This strategy holds great potential for the development of bioengineered organs and tissues for transplantation, as it combines the advantages of naturally derived ECM scaffolds with the regenerative capacity of patient-specific or pluripotent stem cells, reducing the risk of rejection and improving long-term outcomes.

8.5.1 Principles of Decellularization

The principles of decellularization revolve around the effective removal of cellular components from tissues or organs while preserving the structural, mechanical, and biochemical integrity of the extracellular matrix (ECM). This process is critical to ensuring that the remaining scaffold is biocompatible and non-immunogenic, reducing the risk of immune rejection when used for tissue engineering or transplantation. The first principle is the use of various physical, chemical, and enzymatic methods to break down and remove cellular materials. Physical methods, such as freeze–thaw cycles and agitation, help disrupt cell membranes, while chemical agents like detergents (e.g., SDS or Triton X-100) solubilize cell membranes and nucleic acids. Enzymatic treatments, such as the use of DNase and RNase, further degrade and eliminate genetic material. Each method must be carefully selected and optimized based on the tissue type to ensure complete decellularization without damaging the ECM. A key principle in decellularization is balancing the removal of cellular components with the preservation of ECM structure and function. ECM plays a crucial role in supporting cellular processes, and its architecture must remain intact for proper recellularization and tissue integration [18]. Therefore, decellularization protocols aim to minimize ECM disruption by carefully controlling the intensity and duration of treatments. For example, harsh chemical agents can effectively remove cells but may also damage important ECM proteins like collagen and elastin, which are essential for maintaining tissue strength and elasticity. Similarly, overexposure to enzymatic treatments can degrade growth factors and glycoproteins, reducing the bioactivity of the scaffold. The challenge lies in finding a balance where cellular components are sufficiently removed while maintaining the biological cues and mechanical properties of ECM.

Another principle is the importance of evaluating the effectiveness of decellularization. After treatment, it is essential to assess whether the tissue is fully decellularized and whether the ECM has been preserved. Histological staining techniques, such as Hematoxylin and Eosin (H&E) staining, can be used to visualize the absence of cellular nuclei, while DNA quantification assays measure the amount of residual genetic material. Immunohistochemistry can help identify the retention of ECM components like collagen, laminin, and fibronectin. Mechanical testing is also important to ensure that the scaffold retains its structural integrity, especially for load-bearing tissues like bone and cartilage. A thorough assessment of the decellularized scaffold ensures that it meets the necessary standards for use in tissue engineering and transplantation.

Lastly, the principle of tissue-specific decellularization recognizes that different tissues and organs have unique structural and cellular characteristics, requiring customized decellularization protocols. For example, soft tissues like skin or liver may require milder detergents or shorter exposure times, while dense tissues like heart valves or tendons might need more aggressive treatments to achieve complete decellularization. Vascularized tissues, such as the lungs or kidneys, often necessitate perfusion-based decellularization techniques to ensure that cellular material is removed from deep within the tissue while preserving the vascular network.

8.5.2 Recellularization with Native Cells and Stem Cells

Recellularization involves the process of repopulating a decellularized extracellular matrix (ECM) scaffold with either native cells or stem cells to restore the biological function of tissues or organs. The principle behind recellularization is to harness the inherent bioactivity of the ECM scaffold to guide cell behavior, such as adhesion, proliferation, and differentiation. Native cells are often used for recellularization when attempting to replicate a specific tissue's function, as these cells are already specialized and can integrate seamlessly with ECM. For example, hepatocytes might be used to recellularize liver scaffolds, while cardiomyocytes could be introduced to restore the function of decellularized heart tissues. The goal is to reintroduce the functional cell types that will perform the essential physiological roles of the tissue or organ being regenerated. Stem cells, particularly pluripotent stem cells and multipotent adult stem cells, are also widely used for recellularization due to their ability to differentiate into various cell types. Mesenchymal stem cells (MSCs), derived from sources like bone marrow or adipose tissue, have become popular for this purpose because they can differentiate into a range of cell types, including osteocytes, chondrocytes, and adipocytes, making them highly versatile for use in recellularizing different tissue types [19]. The combination of stem cells with a bioactive ECM scaffold can create an environment conducive to tissue regeneration, as ECM provides the structural and biochemical cues necessary for directing stem cell differentiation. This capacity to repopulate a scaffold with stem cells not only restores tissue function but also allows for the generation of patient-specific tissues that minimize the risk of immune rejection.

Exercise 3

Multiple Choice Questions

1. Which technology is commonly used to create 3D ECM scaffolds for tissue engineering?
a) Electrophoresis
b) 3D Bioprinting
c) Photolithography
d) Chromatography

2. Nanotechnology in ECM mimicry primarily focuses on:
a) Increasing the weight of materials
b) Modifying the surface properties at the nanoscale
c) Reducing material costs
d) Enhancing color

3. Which of the following advanced techniques is used to evaluate the mechanical properties of ECM mimetics?
a) Microscopy
b) Rheology
c) Spectroscopy
d) Electrophoresis

4. Decellularization is a method used to create ECM mimics by:
a) Increasing cell density
b) Removing cellular components from tissue
c) Adding synthetic materials
d) Enhancing blood flow

5. Incorporating bioactive molecules into ECM mimetics is done to:
a) Improve color
b) Promote cellular responses
c) Decrease mechanical strength
d) Increase weight

6. What role does electrospinning play in ECM mimicry?
a) It enhances color retention
b) It creates fibrous structures for scaffolds
c) It increases material rigidity
d) It reduces production time

7. Which imaging technique is used to visualize ECM structures at a cellular level?
a) MRI
b) Confocal microscopy
c) CT scan
d) X-ray

8. Biomimetic materials in ECM mimicry are designed to:
a) Replicate the aesthetic qualities of natural tissue
b) Simulate the mechanical and biochemical properties of the ECM
c) Reduce the cost of manufacturing
d) Increase the thermal stability

9. Which of the following is a potential application of advanced technologies in ECM mimicry?
a) Drug delivery systems
b) Cosmetic products
c) Energy storage
d) All of the above

10. The combination of ECM mimicry with ______ technologies can lead to enhanced tissue regeneration.
a) Information
b) Gene editing
c) Food processing
d) Virtual reality

The success of recellularization depends heavily on the method of cell delivery and the ability of the cells to adhere, survive, and integrate into the ECM scaffold. Various techniques are employed to enhance the efficiency of recellularization, including static seeding, dynamic perfusion, and bioreactor systems that provide controlled environments for cell growth. Bioreactors have shown promise in creating optimal conditions for stem cell differentiation and proliferation by providing mechanical stimuli, nutrient exchange, and oxygenation similar to those in the body. Additionally, the timing and distribution of cells within the scaffold are critical to ensuring even cell coverage and successful tissue formation [20]. When recellularization with native or stem cells is carefully optimized, the resulting tissue constructs have the potential to replicate the structure and function of the original tissue, offering transformative solutions for regenerative medicine and organ transplantation.

8.6 Microfluidics and Organ-on-a-Chip Models

Microfluidics and organ-on-a-chip models represent cutting-edge technologies that offer unprecedented insights into human biology, disease progression, and drug testing. Microfluidics involves the manipulation of small volumes of fluids through channels at the micrometer scale, allowing researchers to precisely control the cellular microenvironment. By mimicking physiological fluid flow, nutrient exchange, and mechanical forces, microfluidic platforms enable more accurate models of human tissues and organs compared to traditional cell culture methods. These systems are especially valuable for studying cell behavior, tissue interactions, and response to external stimuli in a highly controlled and reproducible manner. As a result, microfluidics has become a cornerstone of research in tissue engineering, regenerative medicine, and pharmaceutical development.

Organ-on-a-chip models are a specific application of microfluidic technology, designed to replicate the structure and function of human organs on a miniaturized platform. These models typically consist of living cells cultured within microfluidic channels that mimic the physical, chemical, and mechanical cues present in native tissues. By integrating multiple cell types and mimicking organ-specific architecture, organ-on-a-chip devices can simulate complex biological processes such as tissue barrier function, organ-specific metabolism, and inter-organ communication [21]. These systems offer a powerful alternative to traditional animal testing, providing more accurate, human-relevant data while reducing ethical concerns and experimental variability. Organ-on-a-chip technology is poised to revolutionize drug discovery, toxicity testing, and personalized medicine by offering a more predictive and scalable platform for modeling human health and disease.

8.6.1 Microfluidic Platforms for ECM Analysis

Microfluidic platforms have emerged as advanced tools for analyzing the extracellular matrix (ECM) due to their ability to replicate the cellular microenvironment with high precision (Fig. 8.4). ECM plays a crucial role in regulating cell behavior, including migration, proliferation, and differentiation, making its accurate analysis essential for understanding tissue development, disease progression, and wound healing. Traditional methods for ECM analysis, such as biochemical assays and microscopy, are often limited in their ability to replicate dynamic, in vivo-like conditions. Microfluidic systems, on the other hand, offer a controlled and tunable environment where various aspects of ECM, including its mechanical properties, molecular composition, and interactions with cells, can be systematically studied in real-time. This makes microfluidic platforms uniquely suited for dissecting the complexities of the ECM and its role in tissue biology [22]. One of the key advantages of microfluidic platforms for ECM analysis is their ability to simulate physiological fluid flows, which are critical for mimicking the in vivo environment. In tissues like blood vessels, skin, and lungs, ECM is constantly exposed to fluid shear stress, nutrient gradients, and waste removal processes, all of which influence its structure and function. Microfluidic devices allow researchers to replicate these dynamic conditions, enabling more accurate studies of how cells interact with the ECM under various flow regimes. This is especially important for understanding diseases such as cancer, where fluid flow can affect tumor progression and metastasis, or in wound healing, where fluid movement impacts the deposition of new ECM components. The ability to incorporate these factors into ECM analysis makes microfluidic platforms a significant advancement over static culture models.

Another reason microfluidic platforms are important for ECM analysis is their capacity for high-throughput experimentation. Due to their small scale, microfluidic devices can be used to create arrays of experiments in parallel, each with different ECM compositions, stiffness levels, or biochemical cues. This capability allows researchers to systematically test how different ECM properties influence cellular responses, such as migration or differentiation, in a highly efficient manner. High-throughput ECM analysis is particularly useful for drug screening and tissue engineering, where identifying the optimal combination of ECM components and mechanical cues is crucial for developing effective treatments or engineered tissues. The integration of sensors within microfluidic platforms enables real-time monitoring of cellular and ECM dynamics, providing valuable insights into how ECM evolves during processes like tissue repair or disease progression.

The ability to manipulate and measure ECM properties at a microscale also makes microfluidic platforms ideal for studying ECM remodeling. ECM remodeling is a dynamic process where the matrix is constantly being synthesized, degraded, and reorganized by cells in response to mechanical forces or biochemical signals. This process is crucial for tissue regeneration, tumor progression, and fibrosis, yet it is difficult to observe in traditional cell culture systems. Microfluidic devices can be designed to apply mechanical forces, such as stretching or compression, to mimic

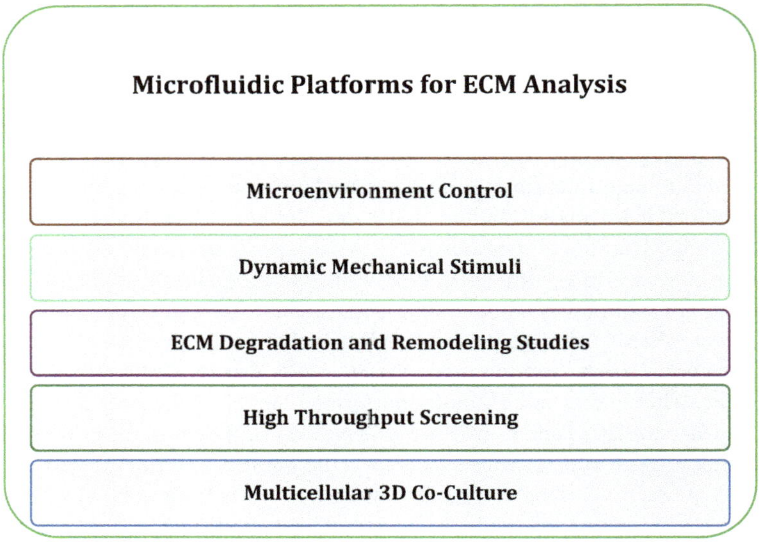

Fig. 8.4 Various microfluidic platforms for ECM analysis

the in vivo mechanical environment and observe how these forces influence ECM remodeling. Microfluidics allows for the precise control of the cellular microenvironment, enabling researchers to study how different ECM-degrading enzymes or growth factors affect ECM structure and function over time. Microfluidic platforms provide a powerful and advanced method for ECM analysis by offering precise control over the cellular microenvironment, incorporating physiological fluid dynamics, enabling high-throughput experimentation, and allowing real-time monitoring of ECM remodeling. These systems represent a significant leap forward in our ability to study ECM's role in health and disease, offering deeper insights into tissue biology, regenerative medicine, and cancer research. As microfluidic technologies continue to evolve, they will undoubtedly become even more integral to the development of innovative therapeutic strategies and tissue engineering approaches that depend on a deep understanding of ECM dynamics.

8.6.2 Organ-on-a-Chip for Disease Modeling

- **Miniaturized Human Organ Replication:** Organ-on-a-chip devices replicate human organ functions on a small scale using microfluidic technology, providing a more accurate physiological environment for disease modeling compared to traditional in vitro methods.
- **Human-Relevant Disease Models**: These systems mimic the architecture, cell types, and microenvironments of specific human organs, enabling the study of

diseases in conditions that closely resemble the human body, reducing reliance on animal models.

- **Dynamic Microenvironment**: Organ-on-a-chip platforms provide a dynamic environment, simulating blood flow, nutrient supply, and mechanical forces, which are crucial for understanding how diseases develop and progress under real-life conditions.
- **Complex Disease Interactions:** By incorporating multiple cell types and organ-specific ECM components, these chips allow researchers to study complex interactions between cells and tissues, providing insights into multifaceted diseases such as cancer, diabetes, and neurodegenerative disorders.
- **Precision and Personalization**: Disease models on a chip can be customized to reflect patient-specific genetic and biochemical profiles, enabling personalized medicine approaches by testing how individual patients may respond to specific treatments.
- **High-Resolution Monitoring**: These platforms often integrate sensors and imaging technologies that allow real-time monitoring of disease progression, drug efficacy, and tissue responses, offering valuable insights into the mechanisms of diseases.
- **Multiorgan Systems for Disease Progression**: By connecting multiple organ-on-a-chip systems, researchers can simulate disease progression across different organs, such as how cancer metastasizes or how liver and kidney function impact drug metabolism and toxicity.

8.6.3 ECM Role in Microfluidic Systems for Drug Testing

The extracellular matrix (ECM) plays a pivotal role in microfluidic systems for drug testing, as it provides the structural and biochemical support needed to replicate the native tissue environment. In microfluidic systems, ECM is often engineered to mimic the physical and chemical characteristics of tissues, enabling the accurate evaluation of drug efficacy and toxicity. ECM's ability to influence cell behavior—such as adhesion, proliferation, and differentiation—is critical for creating realistic tissue models within microfluidic devices. By integrating ECM components like collagen, elastin, and fibronectin into these systems, researchers can simulate the complex 3D architecture and mechanical properties of tissues, ensuring that the cellular responses to drugs reflect real-life conditions [23]. Advantages of using ECM in microfluidic systems for drug testing is its capacity to create tissue-specific environments that enhance the predictability of drug responses. Different tissues have distinct ECM compositions and mechanical properties, which can significantly influence how cells respond to drugs. For instance, ECM in liver tissue contains specific proteins and stiffness levels that regulate hepatocyte function, which is crucial for studying liver metabolism and drug-induced hepatotoxicity. By replicating these tissue-specific ECM characteristics, microfluidic systems allow for more accurate assessments of how drugs interact with different organs, offering insights that are difficult to achieve with traditional 2D cell cultures or animal models.

The dynamic nature of ECM also plays an essential role in simulating drug responses in microfluidic systems. In the human body, ECM is continuously remodeled by cells in response to mechanical stress, biochemical signals, and drug interventions. Microfluidic platforms allow researchers to study these interactions in real time by exposing cells embedded in ECM-mimicking scaffolds to controlled fluid flow and mechanical forces. This ability to simulate the dynamic interplay between cells and the ECM under the influence of drugs is vital for understanding how drug treatments may alter tissue structure and function, particularly in chronic diseases where ECM remodeling plays a significant role, such as cancer or fibrosis. The integration of ECM into microfluidic systems enhances the ability to study cell-ECM interactions that are critical for evaluating drug toxicity and side effects. Many drugs can have unintended consequences on ECM, such as altering its composition or stiffness, which in turn affects cell behavior. For example, certain chemotherapy drugs can lead to ECM degradation, resulting in impaired tissue function and increased metastasis in cancer patients. By incorporating ECM into drug testing platforms, researchers can observe these side effects and better predict drug-induced tissue damage or fibrosis. This insight helps in the development of drugs that are not only effective but also minimize adverse impacts on ECM and overall tissue health.

The role of ECM in microfluidic systems for drug testing is fundamental to achieving realistic and reliable models of human tissue. By providing the necessary structural, biochemical, and mechanical cues, ECM enables the replication of complex tissue environments, facilitating more accurate assessments of drug efficacy and toxicity. The ability to study dynamic ECM remodeling and cell-ECM interactions in microfluidic systems offers significant advantages over traditional methods, making these platforms essential tools for advancing drug discovery and personalized medicine.

8.7 Integration of AI and Machine Learning in ECM Scaffold Design

The integration of artificial intelligence (AI) and machine learning (ML) in the design of extracellular matrix (ECM) scaffolds represents a transformative approach that enhances the efficiency and precision of tissue engineering. Traditional scaffold design often relies on empirical methods and trial-and-error approaches, which can be time-consuming and resource-intensive. However, AI and ML algorithms can analyze large datasets derived from previous studies, computational simulations, and experimental results to identify patterns and predict optimal scaffold properties. By leveraging these advanced computational techniques, researchers can streamline the design process, significantly reducing development time and costs while increasing the likelihood of success in creating functional ECM scaffolds. Employing AI and

ML in ECM scaffold design is very helpful as their ability to optimize the composition and architecture of the scaffolds. These algorithms can assess various parameters such as pore size, material composition, mechanical properties, and degradation rates, all of which are critical for supporting cell behavior and tissue integration. For instance, ML models can be trained on datasets that include successful and unsuccessful scaffold designs, allowing them to learn the relationships between scaffold features and their biological performance [24]. This data-driven approach enables researchers to make informed decisions about scaffold design, leading to the creation of highly tailored ECM scaffolds that better mimic the native tissue environment.

AI and ML can facilitate the incorporation of biophysical and biochemical cues into scaffold design. By analyzing complex interactions between cells and ECM, these technologies can help identify the optimal conditions that promote cell adhesion, proliferation, and differentiation. For example, AI algorithms can simulate how various growth factors, signaling molecules, and mechanical forces influence cellular responses within ECM scaffolds. By integrating this knowledge into the design process, researchers can create scaffolds that not only provide structural support but also actively guide tissue regeneration and functional integration. This level of sophistication in scaffold design can significantly enhance the efficacy of tissue engineering applications. Lastly, the integration of AI and ML in ECM scaffold design paves the way for personalized medicine approaches in regenerative therapies. As individual patients may have unique physiological and biological characteristics, customizable scaffolds that are tailored to specific needs can lead to improved outcomes. AI algorithms can analyze patient-specific data, such as genetic information or imaging results, to design ECM scaffolds that match the unique requirements of each patient's tissue type or condition. This personalized approach holds great promise for advancing regenerative medicine, making it possible to create bespoke scaffolds that enhance healing and functional recovery in a way that standard designs cannot achieve.

8.7.1 Personalized ECM Mimicry for Precision Medicine

Personalized extracellular matrix (ECM) mimicry is an innovative approach in precision medicine that tailors scaffold design and composition to meet the specific biological and physiological needs of individual patients. This concept recognizes that each patient's unique genetic makeup, disease state, and tissue characteristics significantly influence their response to therapies. By creating ECM scaffolds that closely replicate the native environment of a patient's tissue, researchers can enhance cellular interactions, promote tissue regeneration, and improve the effectiveness of regenerative therapies. This personalized approach moves beyond one-size-fits-all solutions, aiming to optimize patient outcomes through customized treatments. The importance of personalized ECM mimicry lies in its potential to enhance the efficacy of tissue engineering and regenerative medicine. Different tissues possess distinct ECM compositions, mechanical properties, and biological cues that dictate how cells behave and respond to external stimuli. For example, ECM of cardiac tissue differs significantly from that of bone or cartilage, each requiring tailored scaffolds for successful regeneration. By utilizing patient-specific ECM data, including information on cellular behavior, protein expression, and mechanical properties, researchers can design scaffolds that not only support cell survival but also actively guide differentiation and integration of transplanted cells, leading to improved functional outcomes.

Personalized ECM mimicry can address the challenges associated with immune rejection and graft failure. When scaffolds are designed to closely match the biochemical and structural properties of a patient's native ECM, the risk of immune responses is minimized. This is particularly important in applications such as organ transplantation and stem cell therapy, where compatibility between the scaffold and host tissue is crucial for acceptance and functionality [25]. By employing techniques such as decellularization to derive ECM from the patient's own tissues or using bioengineering approaches to design scaffolds that incorporate patient-specific ECM components, the likelihood of successful integration and reduced inflammation is significantly enhanced.

Finally, the implementation of personalized ECM mimicry aligns with the broader goals of precision medicine by promoting more effective and targeted therapies. As healthcare shifts towards individualized treatment plans, the ability to customize ECM scaffolds based on a patient's specific condition and biological profile becomes increasingly vital. This approach not only improves the potential for successful tissue regeneration but also contributes to a deeper understanding of disease mechanisms and patient responses. By integrating personalized ECM designs into clinical practice, healthcare providers can enhance therapeutic efficacy, minimize adverse effects, and ultimately improve patient quality of life, marking a significant advancement in the field of regenerative medicine.

8.8 Summary

This chapter has explored the transformative role of extracellular matrix (ECM) mimicry in advancing tissue engineering and regenerative medicine. By highlighting the integration of advanced technologies such as microfluidics, artificial intelligence, and personalized approaches, the chapter illustrates how these innovations enhance the design and functionality of ECM scaffolds. The ability to replicate the complex architecture and biochemical properties of native ECM is critical for creating scaffolds that support cell behavior, promote tissue regeneration, and improve therapeutic outcomes.

Additionally, the chapter discusses the significance of personalized ECM mimicry in precision medicine, emphasizing its potential to tailor scaffold designs to individual patients' unique biological profiles. This customization not only enhances cellular interactions but also addresses challenges related to immune rejection and graft failure. By using patient-specific ECM data, researchers can create scaffolds that closely match the characteristics of the target tissue, ensuring more effective integration and functionality. This personalized approach represents a significant shift toward individualized therapies that improve the efficacy of regenerative treatments. Overall, the chapter underscores the importance of combining various technological advancements with a deeper understanding of ECM biology to drive innovation in tissue engineering. As these strategies continue to evolve, they hold the promise of revolutionizing regenerative medicine by providing more effective, patient-centered solutions. The ongoing integration of personalized ECM mimicry and advanced technologies will likely pave the way for novel therapeutic strategies that significantly improve patient outcomes in a variety of clinical applications.

Points to Remember

- **3D bioprinting allows for the layer-by-layer fabrication of ECM scaffolds with precise control over spatial organization and cell distribution. This technology can create complex tissue structures that closely resemble native tissues, enabling better integration and functionality.**

- **Advanced ECM mimetics include smart hydrogels that respond to environmental stimuli, such as pH, temperature, or light. These hydrogels can change their properties dynamically, facilitating controlled drug release or promoting cell migration and differentiation in response to specific signals.**

- **Microfluidic platforms equipped with ECM mimics enable real-time monitoring of cellular responses to drugs in a controlled microenvironment. This technology significantly enhances drug screening efficiency and accuracy, allowing researchers to assess drug efficacy and toxicity with high precision.**

- **AI algorithms can analyze vast datasets from previous studies to identify patterns and optimize ECM scaffold designs. This capability enables researchers to predict how different material properties will influence cell behavior, leading to more successful scaffold designs.**

- **Advanced technologies enable the creation of decellularized ECM from donor tissues, which retains the natural architecture and biochemical cues of the original tissue. These decellularized matrices can be repopulated with patient-specific cells, promoting better integration and reducing the risk of immune rejection.**

References

1. Karamanos NK, Theocharis AD, Piperigkou Z, Manou D, Passi A, Skandalis SS, Vynios DH, Orian-Rousseau V, Ricard-Blum S, Schmelzer CEH, Duca L, Durbeej M, Afratis NA, Troeberg L, Franchi M, Masola V, Onisto M (2021) A guide to the composition and functions of the extracellular matrix. FEBS J 288(24):6850–6912. https://doi.org/10.1111/febs.15776
2. Dzobo K, Dandara C (2023) The extracellular matrix: its composition, function, remodeling, and role in tumorigenesis. Biomimetics 8(2):146. https://doi.org/10.3390/biomimetics8020146
3. Khanna A, Oropeza BP, Jain I, Huang NF (2023) Extracellular matrix bioactive molecules and cell behavior modeling. In: Maia FRA, Oliveira JM, Reis RL (eds) Handbook of the extracellular matrix. Springer, Cham. https://doi.org/10.1007/978-3-030-92090-6_4-1

4. Wang H, Yu H, Zhou X, Zhang J, Zhou H, Hao H, Ding L, Li H, Gu Y, Ma J, Qiu J, Ma D (2022) An overview of extracellular matrix-based bioinks for 3D bioprinting. Front Bioeng Biotechnol 10:905438. https://doi.org/10.3389/fbioe.2022.905438
5. Pellowe AS, Gonzalez AL (2016) Extracellular matrix biomimicry for the creation of investigational and therapeutic devices. Wiley Interdiscip Rev Nanomed Nanobiotechnol 8(1):5–22. https://doi.org/10.1002/wnan.1349
6. Reddy MSB, Ponnamma D, Choudhary R, Sadasivuni KK (2021) A comparative review of natural and synthetic biopolymer composite scaffolds. Polymers 13(7):1105. https://doi.org/10.3390/polym13071105
7. Goyal R, Guvendiren M, Freeman O, Mao Y, Kohn J (2017) Optimization of polymer-ECM composite scaffolds for tissue engineering: effect of cells and culture conditions on polymeric nanofiber mats. J Funct Biomater 8(1):1. https://doi.org/10.3390/jfb8010001
8. Rais A, Husain A, Hasan GM, Hassan MI (2023) A review on regulation of cell cycle by extracellular matrix. Int J Biol Macromol 232:123426. https://doi.org/10.1016/j.ijbiomac.2023.123426
9. Dey M, Ozbolat IT (2020) 3D bioprinting of cells, tissues and organs. Sci Rep 10:14023. https://doi.org/10.1038/s41598-020-70086-y
10. Da Silva K, Kumar P, Choonara YE, du Toit LC, Pillay V (2020) Three-dimensional printing of extracellular matrix (ECM)-mimicking scaffolds: a critical review of the current ECM materials. J Biomed Mater Res Part A 108(12):2324–2350. https://doi.org/10.1002/jbm.a.36981
11. Reis DP, Domingues B, Fidalgo C, Reis RL, Gasperini L, Marques AP (2022) Bioinks enriched with ECM components obtained by supercritical extraction. Biomolecules 12(3):394. https://doi.org/10.3390/biom12030394
12. Behan K, Dufour A, Garcia O, Kelly D (2022) Methacrylated cartilage ECM-based hydrogels as injectables and bioinks for cartilage tissue engineering. Biomolecules 12(2):216. https://doi.org/10.3390/biom12020216
13. Nam SY, Park SH (2018) ECM based bioink for tissue mimetic 3D bioprinting. Adv Exp Med Biol 1064:335–353. https://doi.org/10.1007/978-981-13-0445-3_20
14. Spang MT, Christman KL (2018) Extracellular matrix hydrogel therapies: in vivo applications and development. Acta Biomater 68:1–14. https://doi.org/10.1016/j.actbio.2017.12.019
15. González-Díaz EC, Varghese S (2016) Hydrogels as extracellular matrix analogs. Gels (Basel, Switzerland) 2(3):20. https://doi.org/10.3390/gels2030020
16. Park W, Gao G, Cho DW (2021) Tissue-specific decellularized extracellular matrix bioinks for musculoskeletal tissue regeneration and modeling using 3D bioprinting technology. Int J Mol Sci 22(15):7837. https://doi.org/10.3390/ijms22157837
17. Gilpin A, Yang Y (2017) Decellularization strategies for regenerative medicine: from processing techniques to applications. Biomed Res Int 2017:9831534. https://doi.org/10.1155/2017/9831534
18. Fernández-Pérez J, Ahearne M (2019) The impact of decellularization methods on extracellular matrix derived hydrogels. Sci Rep 9:14933. https://doi.org/10.1038/s41598-019-49575-2
19. Ahmed E, Saleh T, Xu M (2021) Recellularization of native tissue derived acellular scaffolds with mesenchymal stem cells. Cells 10(7):1787. https://doi.org/10.3390/cells10071787
20. Barthold JE, Martin BM, Sridhar SL, Vernerey F, Schneider SE, Wacquez A, Ferguson V, Calve S, Neu CP (2021) Recellularization and integration of dense extracellular matrix by percolation of tissue microparticles. Adv Funct Mater 31(35):2103355. https://doi.org/10.1002/adfm.202103355
21. Cao UMN, Zhang Y, Chen J, Sayson D, Pillai S, Tran SD (2023) Microfluidic organ-on-a-chip: a guide to biomaterial choice and fabrication. Int J Mol Sci 24(4):3232. https://doi.org/10.3390/ijms24043232
22. Cho AN, Jin Y, An Y, Kim J, Choi YS, Lee JS, Kim J, Choi WY, Koo DJ, Yu W, Chang GE, Kim DY, Jo SH, Kim J, Kim SY, Kim YG, Kim JY, Choi N, Cheong E, Kim YJ, Cho SW et al (2021) Microfluidic device with brain extracellular matrix promotes structural and functional maturation of human brain organoids. Nat Commun 12(1):4730. https://doi.org/10.1038/s41467-021-24775-5

23. Mancera-Andrade EI, Parsaeimehr A, Arevalo-Gallegos A, Ascencio-Favela G, Parra Saldivar R (2018) Microfluidics technology for drug delivery: a review. Front Biosci (Elite Ed) 10(1):74–91. https://doi.org/10.2741/e809
24. Pemmada R, Gowtham N, Xia Y, Basu B, Thomas V (2023) ML and AI approaches for design of tissue scaffolds. In: Elsevier eBooks, pp 29–56. https://doi.org/10.1016/b978-0-443-18498-7.00008-9
25. Delpierre C, Lefèvre T (2023) Precision and personalized medicine: what their current definition says and silences about the model of health they promote. Implication for the development of personalized health. Front Sociol 8:1112159. https://doi.org/10.3389/fsoc.2023.1112159

Chapter 9
Clinical Case Studies and Trials

9.1 Introduction

The extracellular matrix (ECM) is a complex network of proteins and polysaccharides that provide structural and biochemical support to surrounding cells. In the human body, the ECM plays a critical role in tissue organization, wound healing, and maintaining cellular functions. Given its importance in physiological processes, researchers have developed numerous ECM-based therapies aimed at harnessing its regenerative potential. These therapies involve the use of natural or synthetic ECM components to repair damaged tissues, promote wound healing, or even replace diseased organs [1]. ECM-based therapies have gained attention in fields like regenerative medicine, wound care, and tissue engineering due to their ability to mimic the natural environment that supports cell growth and differentiation. In regenerative medicine, ECM-based therapies offer promising solutions for the repair and replacement of damaged tissues. By using ECM scaffolds derived from tissues or created synthetically, scientists can facilitate the growth of new, healthy cells. These scaffolds act as a biological template, guiding cells to restore tissue structure and function. For example, ECM scaffolds have been successfully employed in skin regeneration, bone repair, and organ reconstruction. The dynamic nature of ECM components, which interact with cells to promote migration, proliferation, and differentiation, makes them ideal candidates for driving tissue repair processes. As a result, ECM-based therapies are considered a cornerstone of modern regenerative medicine.

Beyond tissue repair, ECM-based therapies also hold potential for treating chronic wounds, such as diabetic ulcers, and improving the healing of surgical wounds [2]. Traditional wound treatments often fall short in cases where the body's natural healing process is impaired, such as in patients with diabetes or those undergoing radiation therapy. ECM-based wound dressings, developed from decellularized tissues or synthetic materials, provide the necessary biochemical cues to stimulate tissue regeneration and support angiogenesis. These therapies help accelerate the healing process, reduce the risk of infection, and improve overall patient outcomes. With

D. Suhag, *Extracellular Matrix biomimicking for Theranostics*, Biomedical Materials for Multi-functional Applications, https://doi.org/10.1007/978-981-95-7900-6_9

advancements in ECM-based biomaterials, wound healing therapies have entered a new era of effectiveness and precision.

One of the most exciting areas of ECM-based therapies is their application in cancer treatment. ECM plays a critical role in the tumor microenvironment, influencing cancer progression and metastasis. Researchers have developed biomimetic ECM models that replicate the tumor microenvironment to better understand cancer behavior and test therapeutic interventions. ECM-based drug delivery systems are being explored to target tumors more effectively, potentially improving the efficacy of chemotherapy and reducing systemic toxicity [3]. By harnessing the unique properties of the ECM, these therapies could revolutionize cancer treatment and provide more personalized, targeted approaches to combating the disease. While the potential of ECM-based therapies is vast, translating research into clinical applications requires robust validation through clinical case studies and trials. Clinical trials are crucial for demonstrating the safety and efficacy of ECM-based therapies in real-world settings. They provide the necessary data to ensure that these treatments perform as intended, with minimal adverse effects on patients. In addition, clinical case studies offer valuable insights into the challenges and successes encountered when using ECM-based therapies in diverse patient populations. By documenting individual patient outcomes, researchers can refine therapeutic approaches and address any limitations observed during the trial phase. The importance of clinical case studies and trials in ECM-based therapies cannot be overstated. They serve as a bridge between laboratory research and clinical practice, ensuring that novel therapies are rigorously tested before they are widely adopted. Clinical trials not only assess the effectiveness of ECM-based treatments but also reveal potential complications, such as immune responses to biomaterials or unforeseen interactions with existing medical conditions. Through this process, researchers can optimize treatment protocols, adjust dosages, and identify patient subgroups that may benefit the most from ECM-based interventions. Without the validation provided by clinical trials, promising therapies may never reach patients in need.

ECM-based therapies represent a transformative approach to treating a wide range of medical conditions, from tissue regeneration to cancer. However, their successful implementation depends heavily on the thorough evaluation provided by clinical trials and case studies. By translating ECM research into clinically viable therapies, these trials play an essential role in advancing medical science and improving patient care. Through continued research and clinical validation, ECM-based therapies have the potential to reshape the future of medicine, offering innovative solutions for some of the most challenging health problems of our time.

9.2 Success Stories in ECM-Based Therapies

ECM-based therapies have made remarkable strides in recent years, with several success stories demonstrating their potential to revolutionize modern medicine. From tissue engineering and wound healing to cancer treatment, these therapies have shown significant promise in restoring damaged tissues and enhancing patient outcomes [4]. Notable success has been achieved in areas such as skin grafting for burn patients, where ECM scaffolds have facilitated faster healing and reduced scarring. In cardiovascular regeneration, ECM-based therapies have been pivotal in promoting tissue repair following heart attacks. These real-world applications of ECM research underscore the transformative power of these therapies in both clinical and preclinical settings, paving the way for future innovations.

Exercise 1

Fill Ups
(Chemotherapy, Invasion, Supportive, Cytokines, Scaffolds)

1. ECM mimics can be used in cancer treatments to create a more ______ environment that resembles the tumor microenvironment.

2. By using ECM mimics, researchers aim to improve the efficacy of ______ therapies by enhancing drug delivery to cancer cells.

3. One approach in utilizing ECM mimics for cancer treatment is to design ______ that can inhibit tumor growth and metastasis.

4. ECM mimics can help in studying cancer cell behavior by providing a platform for ______ and migration assays.

5. The incorporation of specific ______ into ECM mimics can stimulate immune responses against cancer cells.

9.2.1 Breakthrough Cases in Wound Healing

Breakthroughs in ECM-based therapies for wound healing have opened new avenues for treating complex wounds that traditional methods often fail to address. One such breakthrough case involves the use of decellularized ECM scaffolds to treat chronic wounds, such as diabetic ulcers, which are notoriously difficult to heal. These scaffolds provide the necessary structural support and biochemical cues to promote tissue regeneration and accelerate the healing process. By mimicking the natural ECM, these scaffolds encourage cell migration, proliferation, and differentiation, which are crucial steps in wound closure. In many clinical cases, patients who had not responded to conventional therapies experienced significant improvement after ECM-based treatment, with faster wound closure and reduced complications [5]. A significant development in ECM-based wound healing therapies is the use of dermal templates for full-thickness burn wounds. Traditional skin grafts often result in scarring and limited mobility, but ECM-based dermal templates provide a more natural scaffold for skin regeneration. In one notable case, a patient with extensive third-degree burns underwent treatment with an ECM-based dermal template, which facilitated the regeneration of both the epidermis and dermis. This approach significantly reduced the need for autografts, minimized scarring, and improved the overall aesthetic and functional outcome. The ability of ECM-based therapies to promote more natural tissue regeneration has been a game-changer in the field of burn wound care.

Another groundbreaking case involves the use of ECM-based hydrogels in treating radiation-impaired surgical wounds. Radiation therapy often impairs wound healing due to damage to the surrounding tissues and blood vessels, making post-surgical recovery more challenging. In such cases, ECM-based hydrogels have been used to enhance the healing process by creating a moist environment and delivering growth factors directly to the wound site [6]. In a clinical trial involving cancer patients with radiation-impaired wounds, ECM hydrogels significantly reduced healing time and improved tissue regeneration compared to traditional dressings. This case highlighted the versatility of ECM-based materials in addressing various wound-healing challenges, particularly in complex clinical settings. ECM-based therapies have also demonstrated success in scar reduction, which is a major concern in both cosmetic and reconstructive surgeries. In one breakthrough case, an ECM-based topical treatment was applied to patients who had undergone surgery, resulting in minimal scar formation. This therapy works by modulating the inflammatory response and providing a scaffold that promotes organized tissue regeneration, reducing the risk of fibrosis and abnormal scar tissue formation. The success of this treatment in preventing hypertrophic scars and keloids has been particularly impactful for patients undergoing procedures where aesthetic outcomes are a priority, such as facial reconstruction and plastic surgery.

Lastly, ECM-based therapies have shown great promise in treating traumatic wounds sustained in accidents or combat. In one case, a soldier with a traumatic leg injury, which had resisted conventional treatments, was treated with an ECM scaffold derived from porcine bladder. The scaffold helped regenerate muscle tissue, leading to the restoration of limb function and avoiding amputation. This case demonstrated the powerful regenerative capabilities of ECM in not only promoting skin healing but also supporting the regeneration of deeper tissues such as muscle. Such breakthroughs highlight the potential of ECM-based therapies to transform the field of wound healing, offering hope to patients with even the most challenging and life-altering injuries.

9.2.2 ECM-Based Approaches in Cardiovascular Regeneration

ECM-based approaches in cardiovascular regeneration have emerged as promising solutions to address the limitations of traditional therapies for heart disease and vascular damage. One of the key challenges in cardiovascular repair is the inability of the heart to regenerate significant tissue after injury, such as in the case of a heart attack. ECM-based therapies, particularly the use of decellularized cardiac scaffolds, have shown potential in promoting heart tissue regeneration by providing a supportive structure that mimics the natural heart ECM [7]. These scaffolds not only support the survival and function of implanted cells but also help recruit the body's own cells to the injury site, facilitating the repair of damaged heart tissue. In addition to heart tissue regeneration, ECM-based materials have been instrumental in improving the outcomes of heart valve replacement and repair procedures. Traditionally, synthetic materials have been used for valve replacements, but these materials can lead to complications such as clot formation and calcification over time. ECM-derived heart valve replacements, on the other hand, are biocompatible and can integrate with the patient's own tissues, reducing the risk of immune rejection and other complications. These ECM-based valves have shown promise in both pediatric and adult patients, where the natural remodeling of the ECM materials allows the valves to grow and adapt, particularly in younger patients, reducing the need for future surgeries.

Another significant advancement in cardiovascular regeneration is the use of ECM-based therapies for vascular repair and regeneration. Synthetic grafts have long been used in vascular surgeries to replace damaged blood vessels, but these often lack the elasticity and natural integration needed for long-term success. ECM-based vascular grafts, derived from decellularized tissues, provide a more natural alternative [8]. These grafts retain the structural and biochemical components of the native ECM, promoting endothelial cell growth and the formation of new, functional blood vessels. In clinical trials, patients treated with ECM-based vascular grafts have shown better long-term outcomes, with fewer complications related to graft failure or infection. ECM-based hydrogels have also been utilized in cardiovascular regeneration, particularly in treating myocardial infarction (heart attack). These hydrogels can be injected directly into the damaged heart tissue, where they form a supportive matrix that encourages the growth of new blood vessels and the repair of damaged cardiac muscles. In preclinical studies, ECM hydrogels have been shown to improve heart function and reduce scar tissue formation after a heart attack. This approach represents a less invasive alternative to traditional treatments, such as open-heart surgery or cell transplantation, offering hope for improved recovery in patients with severe heart damage.

ECM-based therapies are being explored in tissue-engineered cardiac patches, which can be applied directly to the surface of the heart to support regeneration. These patches, composed of decellularized ECM or ECM-mimicking materials, provide mechanical support to the heart while also promoting the regeneration of cardiac tissue and blood vessels. In animal studies, these ECM-based patches have demonstrated the ability to improve heart function and prevent heart failure after myocardial injury.

Exercise 2

Fill Ups
(Stability, Biodegradable, Natural, Musculoskeletal, Growth)

1. In orthopaedics, ECM plays a crucial role in the repair and regeneration of ______ tissues, such as cartilage and bone.

2. Tissue engineering strategies often involve using ______ scaffolds that mimic the mechanical properties of natural bone.

3. The incorporation of ______ factors into ECM mimics can enhance the healing process in orthopaedic applications.

4. Decellularized ECM from bone or cartilage can be used as a ______ template to promote tissue regeneration.

5. Biomechanical properties of engineered tissues are critical for ensuring ______ in load-bearing applications in orthopaedic surgery.

9.2.3 Cancer Treatments Using ECM Mimics

ECM mimics have emerged as a cutting-edge approach in cancer treatments, offering unique advantages in targeting tumor growth and metastasis. One of the key roles of the natural extracellular matrix (ECM) in cancer is its interaction with cancer cells, influencing their behavior and progression. By replicating the structure and function of the natural ECM, researchers have developed ECM mimics that can either inhibit tumor growth or deliver targeted therapies directly to the cancerous tissues [9]. These mimics serve as scaffolds that create an environment unfavorable to cancer progression while supporting the delivery of chemotherapeutic drugs or immunotherapies in a controlled manner. The ability to influence the tumor microenvironment through ECM mimics provides new strategies to slow or even reverse cancer growth.

In particular, ECM mimics have shown great promise in developing more effective drug delivery systems. Traditional chemotherapy treatments often suffer from nonspecific targeting, which leads to significant side effects as healthy tissues are affected. ECM-based drug delivery platforms have the potential to improve this by directing treatments specifically to the tumor site. These mimics can be engineered to carry chemotherapeutic agents or nanoparticles that release the drug in response to the tumor's unique microenvironment. This targeted approach allows for higher concentrations of drugs to be delivered directly to cancer cells while minimizing damage to surrounding healthy tissues. Preclinical studies have shown that these ECM-based delivery systems can significantly reduce tumor size with fewer side effects compared to conventional chemotherapy. Another area where ECM mimics are making an impact is in the field of cancer immunotherapy. Cancer cells often evade detection by the immune system by creating a protective barrier within the tumor microenvironment, heavily influenced by the ECM. ECM mimics are being used to disrupt this barrier, making the tumor more susceptible to attack by immune cells. These mimics can be designed to release immune-activating agents that help stimulate a stronger anti-tumor immune response [10]. For example, certain ECM-derived scaffolds have been engineered to release cytokines or other immune-boosting molecules directly into the tumor site, enhancing the body's natural ability to recognize and destroy cancer cells. This approach is especially promising in combination with checkpoint inhibitors and other immunotherapies.

ECM mimics are also playing a vital role in developing cancer models that better replicate the tumor microenvironment for research and drug testing. Traditional 2D cell cultures fail to capture the complexity of tumors in vivo, often leading to poor translation of research findings to clinical applications. By creating 3D ECM-based models that mimic the architecture and biochemical cues of real tumors, researchers can study cancer progression and test new treatments in a more realistic setting. These models have been particularly valuable for studying how cancer cells invade surrounding tissues and metastasize, offering insights that can lead to more targeted and effective therapies. In addition, the use of ECM mimics in 3D cancer models enables the testing of personalized treatments, as patient-derived tumor cells can be grown in these scaffolds to evaluate how they respond to various therapies (Fig. 9.1).

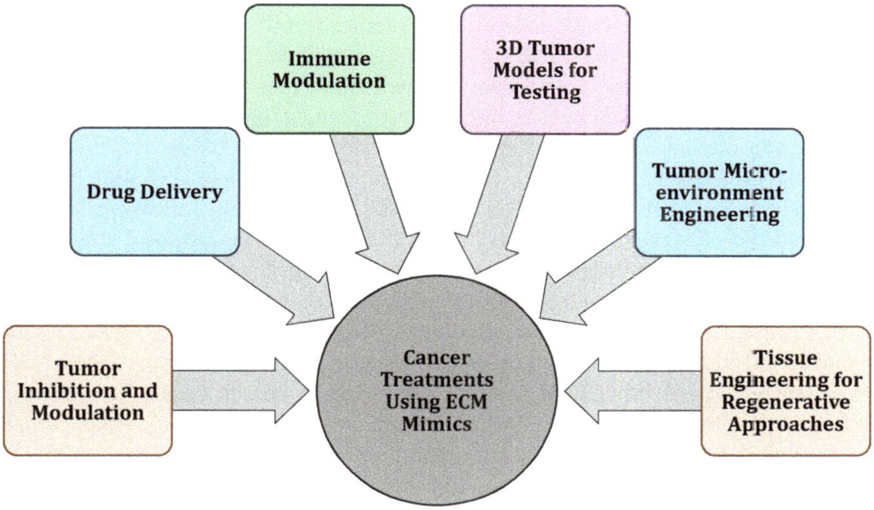

Fig. 9.1 Various cancer treatments using ECM mimics

ECM mimics are being explored in strategies to prevent cancer recurrence. One of the major challenges in oncology is that even after successful treatment, residual cancer cells can remain dormant and eventually lead to relapse. By using ECM mimics to modify the tumor microenvironment post-treatment, researchers are developing ways to prevent these dormant cells from becoming active again. For instance, ECM-based scaffolds can be designed to continuously release therapeutic agents over time or to alter the biochemical signals in the tissue to prevent cancer cells from reestablishing themselves. This approach has the potential to significantly reduce the chances of recurrence, offering long-term benefits for cancer patients who are in remission.

9.2.4 Case Studies of Regenerative Medicine Using ECM Biomaterials

Case studies in regenerative medicine using ECM biomaterials highlight the profound impact of these materials in restoring tissue function and improving patient outcomes. One notable case involves the use of decellularized ECM scaffolds to repair damaged heart tissue following a myocardial infarction (heart attack). In this case, a patient who suffered extensive cardiac damage was treated with an ECM-derived scaffold implanted into the affected area. The scaffold provided structural support, encouraging the growth of new blood vessels and cardiac muscle cells. Over time, the patient's heart function improved significantly, reducing the risk of heart failure. This case demonstrates the potential of ECM biomaterials to regenerate tissue in organs that have limited natural repair capabilities. Another significant case in regenerative medicine involved the use of ECM biomaterials to repair cartilage defects in the knee. In this study, a young athlete with a severe knee injury underwent surgery where an ECM scaffold, derived from porcine tissue, was implanted into the damaged cartilage. The scaffold facilitated the regeneration of healthy cartilage tissue by providing a matrix that mimicked the natural ECM environment of the joint [11]. Within months, the patient reported reduced pain and increased mobility, and follow-up imaging confirmed the regrowth of functional cartilage. This case underscores the role of ECM biomaterials in orthopedic regeneration, offering a promising alternative to traditional joint replacements or other invasive procedures. ECM biomaterials have also shown great success in skin regeneration, particularly in treating severe burns. In one case study, a patient with third-degree burns over a large portion of their body was treated with a dermal ECM scaffold derived from human tissue. The scaffold promoted rapid wound healing, significantly reducing the need for autografts, which are often required in burn cases. By mimicking the natural structure of skin ECM, the scaffold supported the growth of new skin layers and improved the overall healing process. The patient not only experienced faster recovery but also had less scarring and better skin function compared to traditional treatments. This case highlights the life-saving potential of ECM biomaterials in treating extensive skin damage.

A breakthrough case in regenerative medicine using ECM biomaterials involved the treatment of a bladder defect. A patient with a congenital bladder abnormality underwent a procedure where an ECM scaffold, derived from porcine bladder tissue, was used to support bladder regeneration. Over time, the scaffold was integrated into the patient's body, and new bladder tissue formed around it. The patient regained bladder function and avoided more invasive surgical options, such as bladder augmentation using synthetic materials. This case study demonstrates the ability of ECM biomaterials to promote the regeneration of complex organ structures, offering hope for patients with organ defects or diseases that require tissue reconstruction. ECM biomaterials have also been used in nerve regeneration, as illustrated by a case study involving a patient with a peripheral nerve injury. After a traumatic accident that resulted in nerve damage, the patient underwent surgery where an ECM-based nerve conduit was implanted to bridge the gap between the damaged nerve ends. The ECM conduit provided a supportive structure for the regrowth of nerve fibers, guiding them along the path to reconnect the nerve. Within several months, the patient began to regain sensation and motor function in the affected area. This case emphasizes the importance of ECM biomaterials in nerve repair, where traditional treatment options often fall short in promoting functional recovery. ECM biomaterials have been successfully used in reconstructive surgery, particularly in the repair of abdominal wall defects. In one case, a patient with a large hernia underwent surgery where an ECM scaffold was implanted to support the weakened abdominal muscles. The scaffold not only provided mechanical strength but also promoted tissue regeneration, resulting in a more natural and long-lasting repair compared to synthetic mesh alternatives. The patient experienced fewer post-surgical complications, such as infection or rejection, and had a smoother recovery. This case illustrates the versatility of

ECM biomaterials in reconstructive procedures, offering a biocompatible solution that integrates seamlessly with the body's own tissues.

These case studies demonstrate the transformative potential of ECM biomaterials across a wide range of medical conditions. From heart and cartilage regeneration to nerve repair and reconstructive surgery, ECM biomaterials provide a dynamic platform for promoting natural tissue healing and improving patient outcomes. By mimicking the body's own extracellular matrix, these materials enable more effective and personalized treatments in the field of regenerative medicine.

9.2.5 Lessons Learned from Successful Clinical Implementations

Successful clinical implementations of ECM-based therapies and biomaterials have provided valuable insights that can shape the future of regenerative medicine. One key lesson learned is the importance of biomimicry in designing effective treatments. By closely mimicking the natural extracellular matrix, these therapies are better able to support cell growth, tissue regeneration, and the repair of damaged structures. The ability of ECM biomaterials to replicate the biochemical and mechanical properties of native tissues has proven critical in enhancing treatment efficacy, improving patient outcomes, and reducing complications. This understanding has pushed researchers and clinicians to focus on creating more accurate ECM models that cater to the unique needs of different tissues and organs. Not all patients respond to treatments in the same way, especially when dealing with complex conditions like cancer or chronic wounds. Successful clinical cases have demonstrated that customizing ECM scaffolds and therapies to a patient's specific biology can lead to better outcomes. By tailoring the composition and structure of the ECM materials to match the patient's unique cellular environment, clinicians can improve tissue integration, minimize immune responses, and accelerate healing [12]. This personalized approach is becoming increasingly important in ensuring the success of ECM therapies across diverse patient populations.

The role of multidisciplinary collaboration has also emerged as a critical factor in the success of ECM-based clinical implementations. Regenerative medicine using ECM biomaterials requires the combined expertise of material scientists, biologists, engineers, and clinicians to create effective treatments. The development of these advanced biomaterials often involves intricate processes that necessitate a deep understanding of both biology and engineering. Successful clinical cases have shown that collaborative efforts between these disciplines lead to better-designed scaffolds, more accurate tissue models, and more effective treatment protocols. This lesson underscores the need for ongoing cooperation between various fields to continue advancing ECM-based therapies. Clinical trials and patient outcomes have highlighted the importance of long-term monitoring and follow-up in ECM-based treatments. Many regenerative therapies, especially those involving biomaterials, require extended periods to fully assess their efficacy and integration into the body. Some initial successes may be followed by complications or failures if proper follow-up care is not provided. Lessons from successful implementations have stressed the need for continuous monitoring of patients to detect potential issues early, ensuring that the ECM materials are functioning as intended and adapting to the patient's evolving physiological conditions. Long-term studies have also helped identify potential improvements in the design and application of ECM-based therapies.

The ethical and regulatory considerations surrounding ECM-based therapies have also provided important lessons. As these therapies are relatively new, their development has raised questions about safety, long-term effects, and accessibility. Successful clinical implementations have shown that thorough preclinical testing and rigorous regulatory approval processes are essential to ensure the safety and effectiveness of ECM biomaterials. Additionally, ethical concerns about sourcing biological materials, especially in decellularized ECMs from animal or human tissues, must be addressed to maintain public trust and acceptance. These lessons highlight the need for transparent regulatory frameworks and ethical guidelines in advancing ECM-based regenerative medicine. Successful implementations have underscored the need for scalable and cost-effective manufacturing of ECM biomaterials. While ECM-based therapies hold tremendous potential, their widespread clinical use depends on the ability to produce these materials in a consistent, scalable, and economically viable manner. Several successful cases have demonstrated that while ECM biomaterials can be highly effective, their high production costs and complex manufacturing processes can limit accessibility. Addressing these challenges through innovative production techniques and cost-reduction strategies will be critical to making ECM-based therapies more widely available to patients across the globe.

9.3 Ongoing Clinical Trials Involving ECM-Based Therapies

Ongoing clinical trials involving ECM-based therapies are expanding the possibilities for treating a wide range of medical conditions, from chronic wounds to organ regeneration. These trials are exploring how ECM biomaterials can be used to improve the effectiveness of treatments in areas such as tissue engineering, cardiovascular repair, and cancer therapy. Researchers are focusing on how these therapies can enhance healing by mimicking the body's natural extracellular matrix, promoting cell growth, and reducing complications associated with conventional treatments. With promising preliminary results, these clinical trials aim to demonstrate the safety and efficacy of ECM-based therapies, potentially revolutionizing approaches to regenerative medicine and offering new hope for patients with previously untreatable conditions.

9.3.1 Overview of Current ECM-Related Trials

Current ECM-related clinical trials are investigating a diverse range of applications, particularly in the fields of tissue regeneration, wound healing, and cancer therapy. One prominent area of research involves the use of ECM scaffolds in treating chronic and non-healing wounds, such as diabetic ulcers and severe burns. These trials are testing decellularized ECM materials that provide a natural framework for tissue regeneration, encouraging the body to repair damaged tissues more effectively. Initial findings have shown accelerated healing rates and reduced complications compared to traditional treatments, positioning ECM scaffolds as a viable solution for managing complex wounds. Researchers are focusing on optimizing these materials to ensure they are biocompatible and promote long-term healing. Another key focus of ECM-related clinical trials is in cardiovascular regeneration. Many of these trials are exploring ECM-based therapies to repair damaged heart tissue following myocardial infarction (heart attack) and to improve the performance of vascular grafts. ECM-derived scaffolds are being tested for their ability to support new tissue formation in the heart and blood vessels, offering an alternative to synthetic grafts, which can often lead to complications such as rejection or clotting. Some trials are using ECM biomaterials in the form of hydrogels that can be injected into damaged heart tissue, aiming to improve heart function and reduce scar tissue formation [13]. Early results from these studies indicate that ECM-based materials can promote the regrowth of functional heart muscle and improve overall cardiovascular health.

Cancer research is increasingly looking at how ECM mimics can be used to improve the targeting and effectiveness of treatments. Several ongoing trials are investigating how ECM-derived scaffolds can enhance the delivery of chemotherapeutic agents or immune therapies directly to tumors. By mimicking the tumor's natural microenvironment, these ECM-based materials allow for more precise drug delivery, reducing side effects and improving patient outcomes. Some studies are also focusing on using ECM scaffolds to inhibit tumor growth by disrupting the interactions between cancer cells and their surrounding matrix. These trials represent an exciting frontier in cancer treatment, offering the potential for more targeted, effective, and less invasive therapies.

9.3.2 ECM and Tissue Engineering in Orthopedics

The integration of ECM (extracellular matrix) and tissue engineering in orthopedics has revolutionized the approach to treating musculoskeletal injuries and degenerative diseases. Traditional orthopedic treatments often rely on synthetic materials or autografts, which can lead to complications such as infection, rejection, or donor site morbidity. ECM-based scaffolds offer a promising alternative by mimicking the natural composition and structure of bone, cartilage, and ligaments, providing a conducive environment for cellular activities. This biomimetic approach not only promotes the healing of damaged tissues but also enhances the body's natural regenerative capabilities, paving the way for more effective orthopedic interventions. One of the significant applications of ECM in orthopedics is the regeneration of cartilage, particularly for conditions like osteoarthritis and cartilage injuries. Researchers are developing ECM scaffolds derived from natural tissues, such as porcine or bovine cartilage, that closely resemble the native cartilage structure [14]. These scaffolds provide a supportive framework for chondrocytes to adhere, proliferate, and synthesize new cartilage matrix. Clinical studies have shown that patients receiving ECM-based treatments for cartilage defects demonstrate improved joint function and reduced pain compared to those treated with traditional methods. This advancement signifies a shift towards more natural and effective solutions for joint restoration.

In the realm of bone regeneration, ECM biomaterials are also making significant strides. Bone defects due to trauma, infection, or disease often require surgical intervention and the use of bone grafts. ECM-based scaffolds have been designed to mimic the mechanical properties and biological signals of native bone tissue, promoting osteogenesis—the formation of new bone. These scaffolds can be infused with growth factors or stem cells to enhance bone healing. Clinical trials have reported promising results, with patients exhibiting improved bone integration and regeneration when treated with ECM biomaterials, minimizing the need for extensive surgeries or long recovery periods. Ligament and tendon injuries are another area where ECM and tissue engineering are making an impact in orthopedics. The unique composition of ECM in ligaments and tendons is crucial for their mechanical strength and flexibility. Researchers are developing scaffolds that replicate this composition, allowing for better integration and healing of torn ligaments or tendons. By utilizing decellularized ECM from ligaments or tendons, scientists can create a scaffold that retains the biochemical cues necessary for cellular infiltration and tissue remodeling. Preliminary studies suggest that patients treated with ECM-based approaches for ligament reconstruction experience enhanced functional recovery and reduced re-injury rates.

The versatility of ECM-based scaffolds extends to the development of combination therapies in orthopedic applications. For instance, integrating ECM biomaterials with advanced delivery systems, such as hydrogels or nanoparticles, enables the localized and sustained release of therapeutic agents. This synergistic approach can enhance the regenerative process by providing a continuous supply of growth factors or anti-inflammatory agents at the injury site. Ongoing research is exploring these combination therapies in clinical settings, with the goal of optimizing patient outcomes and accelerating recovery times. The incorporation of ECM in tissue engineering is transforming orthopedic practice, offering innovative solutions to complex musculoskeletal challenges and improving the quality of life for patients.

9.3.3 *ECM in Dermatology: Skin Grafts and Burn Wounds*

The application of ECM (extracellular matrix) in dermatology has significantly advanced the management of skin grafts and the treatment of burn wounds. Traditional methods of treating severe burns and skin defects often involve the use of autologous skin grafts, which can lead to complications such as limited donor site availability, scarring, and prolonged recovery times. In contrast, ECM-derived scaffolds provide a promising alternative by offering a biocompatible environment that mimics the natural skin structure. These scaffolds support cellular infiltration, promote angiogenesis, and enhance the overall healing process, making them an effective solution for challenging dermatological cases. One of the major benefits of using ECM scaffolds in burn treatment is their ability to provide immediate coverage and protection to the wound site. When applied to acute burn injuries, these scaffolds create a barrier that prevents infection while facilitating the migration of skin cells into the wound [15]. The natural composition of ECM contains essential growth factors and proteins that stimulate the body's healing response, promoting the regeneration of skin layers. Studies have shown that patients treated with ECM-based grafts for burn wounds experience faster healing times and improved aesthetic outcomes compared to those receiving conventional dressings or skin grafts.

In addition to burn wounds, ECM biomaterials are also being explored for their potential in managing chronic skin conditions and improving skin graft integration. For instance, chronic wounds, such as diabetic ulcers or venous leg ulcers, often present significant challenges in terms of healing due to underlying conditions. ECM scaffolds can be used as a biological dressing, providing a conducive environment for cell proliferation and tissue regeneration. Clinical trials have demonstrated that the application of ECM in chronic wound care leads to enhanced healing rates and reduced healing times, ultimately improving the quality of life for patients suffering from these persistent conditions. The use of decellularized skin grafts derived from human or animal sources has gained traction in dermatology as well. These grafts retain the structural and biochemical properties of the natural ECM, making them highly effective for skin reconstruction following trauma or surgical interventions. Decellularized ECM grafts not only provide mechanical support but also promote cellular activity and tissue integration, reducing the risk of graft rejection and improving patient outcomes. The incorporation of these grafts into clinical practice has expanded the range of treatment options available for patients with extensive skin injuries, allowing for more successful reconstructions. The ongoing research into ECM-based therapies in dermatology continues to unveil innovative applications and improvements in wound healing. As understanding of the interactions between ECM components and skin cells deepens, researchers are exploring ways to enhance the properties of ECM scaffolds, such as incorporating bioactive molecules, stem cells, or growth factors to further optimize healing outcomes. This research holds the potential to revolutionize the treatment landscape for skin grafts and burn wounds, ultimately leading to better patient care, reduced complications, and improved recovery experiences.

9.3.4 ECM-Based Therapies in Neurology and Neurosurgery

ECM (extracellular matrix) based therapies are emerging as innovative approaches in the fields of neurology and neurosurgery, providing new avenues for treating neurological disorders and repairing nervous tissue (Fig. 9.2). The nervous system has limited intrinsic regenerative capabilities, making it challenging to address conditions such as spinal cord injuries, traumatic brain injuries, and neurodegenerative diseases. ECM-based materials can support neural regeneration by mimicking the natural microenvironment of nervous tissue, promoting cell adhesion, proliferation, and differentiation. This biomimetic approach aims to enhance recovery and functional outcomes for patients suffering from various neurological conditions. One of the key applications of ECM-based therapies in neurology is spinal cord injury repair. Current treatment options are limited, and patients often face significant long-term disabilities following such injuries. Researchers are exploring the use of ECM scaffolds to provide structural support for axonal regrowth and to facilitate the restoration of neural connections [16]. By incorporating decellularized ECM derived from peripheral nerves or other sources, these scaffolds can create a conducive environment that promotes the survival and integration of transplanted neural stem cells or progenitor cells. Early studies have shown promising results, with improved locomotor function and reduced scar formation in animal models, suggesting that ECM-based therapies may significantly enhance recovery outcomes for spinal cord injury patients. In the context of traumatic brain injuries, ECM-based strategies are being investigated for their potential to enhance recovery and repair damaged brain tissue. These injuries often result in significant inflammation, cellular death, and loss of functional neural networks. ECM scaffolds can help mitigate these effects by providing a physical structure that encourages cellular infiltration and neuroprotection. By delivering neurotrophic factors within ECM scaffolds, researchers aim to create a microenvironment that supports neuronal survival and regeneration. Preliminary clinical trials indicate that patients receiving ECM-based treatments for traumatic brain injuries exhibit improved cognitive and motor function compared to those treated with conventional methods.

ECM-based therapies are being explored in the treatment of neurodegenerative diseases such as Alzheimer's and Parkinson's disease. These conditions are characterized by the progressive degeneration of neurons, leading to cognitive decline and motor dysfunction. By utilizing ECM materials that incorporate signaling molecules or growth factors, researchers aim to enhance neurogenesis and synaptic plasticity in affected areas of the brain. Studies have demonstrated that ECM-derived scaffolds can facilitate the growth of neural cells and promote their differentiation into functional neurons, offering a potential therapeutic strategy for halting or reversing the progression of neurodegeneration. The integration of ECM-based therapies in neurosurgery is also showing promise in enhancing the outcomes of various surgical procedures. For example, ECM scaffolds can be used in cranial defect repair following tumor resections or traumatic injuries. By providing structural support and promoting tissue regeneration, these scaffolds can facilitate the integration of surrounding tissues and

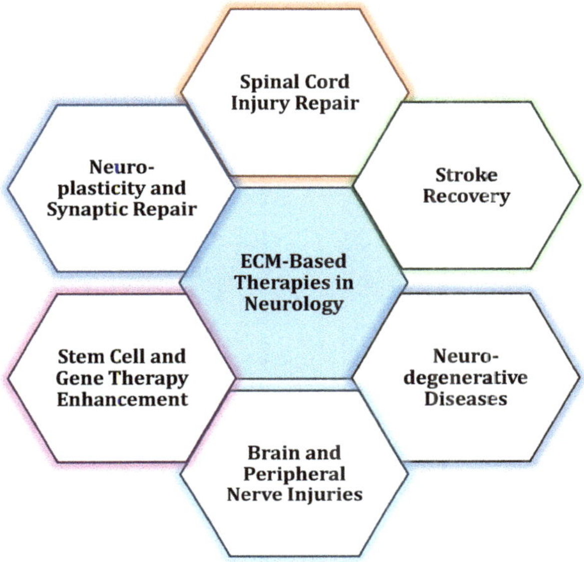

Fig. 9.2 ECM based therapies in neurology

reduce complications such as infection or graft failure. ECM materials may be used to deliver targeted therapies or stem cells during neurosurgical procedures, further enhancing healing and functional recovery.

9.3.5 Biomimetic ECM in Organ Transplants

Biomimetic extracellular matrix (ECM) is gaining significant attention in the field of organ transplantation, offering innovative solutions to enhance graft acceptance, reduce complications, and improve overall transplant outcomes (Fig. 9.3). Traditional organ transplant procedures face numerous challenges, including immune rejection, limited donor availability, and insufficient functional integration of transplanted tissues. Biomimetic ECM provides a promising approach by closely mimicking the structural and biochemical properties of native tissues, thereby creating an environment conducive to cellular adhesion, proliferation, and differentiation. This approach not only enhances the biological compatibility of the transplanted organs but also promotes the regeneration of host tissues, ultimately improving graft survival rates.

One of the most critical issues in organ transplantation is the risk of immune rejection. The body's immune system recognizes transplanted tissues as foreign, leading to an inflammatory response that can damage the graft. Biomimetic ECM can be engineered to include specific immunomodulatory molecules that help to

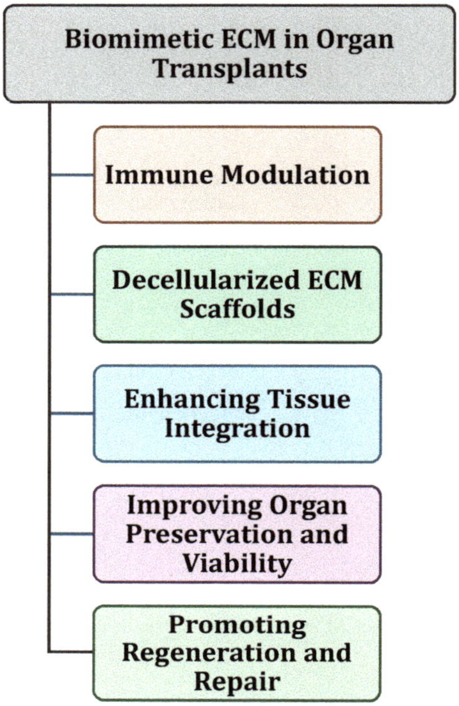

Fig. 9.3 Advantages of biomimetic ECM in organ transplants

mitigate this response. By incorporating bioactive peptides or growth factors into the ECM scaffold, researchers aim to create a microenvironment that promotes tolerance and reduces inflammation [17]. Early studies have shown that ECM scaffolds can modulate immune responses, enhancing graft acceptance and reducing the need for long-term immunosuppressive therapies, which often come with severe side effects. In addition to improving graft acceptance, biomimetic ECM can enhance the functional integration of transplanted organs. For successful transplantation, it is essential for the graft to establish a connection with the recipient's vasculature and surrounding tissues. Biomimetic scaffolds can be designed to facilitate angiogenesis—the formation of new blood vessels—which is crucial for delivering nutrients and oxygen to the transplanted tissue. By incorporating specific growth factors such as vascular endothelial growth factor (VEGF) within the ECM, researchers can promote rapid vascularization of the graft, ensuring its viability and functionality. This enhanced integration is particularly important in solid organ transplants, such as kidneys and livers, where immediate perfusion is critical to prevent tissue damage.

The application of biomimetic ECM extends beyond enhancing graft acceptance and vascularization; it also plays a role in tissue regeneration post-transplantation. After an organ transplant, the recipient's tissue often undergoes significant remodeling as it adapts to the new environment. Biomimetic ECM can provide the necessary structural support for this remodeling process, promoting the regeneration of functional tissue. The presence of specific ECM components can guide the differentiation of stem cells or progenitor cells into the desired cell types, facilitating the regeneration of damaged tissues. This capability is particularly valuable in complex transplants, where multiple tissue types need to integrate seamlessly to restore organ function. The customization of biomimetic ECM allows for the development of personalized transplant solutions. Each patient presents unique biological characteristics that can affect transplant outcomes. By tailoring the composition and architecture of biomimetic ECM scaffolds to match the recipient's specific needs, researchers can enhance the effectiveness of the transplant. Advances in 3D printing and tissue engineering technologies enable the fabrication of patient-specific ECM constructs, allowing for precise adjustments to optimize graft integration and function. This personalized approach holds great promise for improving transplant success rates and reducing complications.

Ongoing research in the field of biomimetic ECM for organ transplantation is focusing on addressing the challenges associated with donor shortages. As the demand for organ transplants continues to exceed supply, the development of bioengineered organs using biomimetic ECM presents a potential solution. By utilizing patients' own cells or stem cells to create personalized ECM scaffolds, researchers aim to develop functional organs in the laboratory that can be safely implanted into patients. This approach not only alleviates the issue of organ availability but also minimizes the risk of immune rejection, as the bioengineered organs would be autologous.

9.3.6 Success Rates and Challenges in Ongoing Trials

The success rates of ongoing clinical trials involving ECM-based therapies are showing promising results across various medical applications, from wound healing to organ transplantation. Many studies report accelerated healing times, reduced complications, and improved functional outcomes for patients treated with ECM scaffolds compared to conventional methods. For instance, trials focusing on ECM for chronic wound management have demonstrated significantly higher rates of complete wound closure and improved patient quality of life. Similarly, in orthopedic applications, ECM scaffolds have resulted in better integration and regeneration of bone and cartilage tissues. These encouraging outcomes highlight the potential of ECM-based therapies to transform clinical practice and improve patient care. Despite these successes, several challenges persist in the ongoing trials of ECM-based therapies. One of the primary obstacles is the variability in ECM composition and its sources, which can affect the consistency and reproducibility of results. Variations in the decellularization process, the type of tissues used, and the incorporation of bioactive factors can lead to differences in how scaffolds perform in vivo. Standardization of these materials and protocols is crucial for ensuring reliable outcomes across different studies. Researchers are actively working to establish guidelines and best practices to mitigate these challenges and promote uniformity in ECM-based therapeutic approaches.

Another significant challenge in ongoing trials is the immune response to ECM scaffolds. While many studies indicate that ECM materials can modulate immune reactions favorably, individual patient responses can vary widely. Factors such as genetic background, existing comorbidities, and the presence of concurrent medications can influence how the immune system interacts with ECM scaffolds. Consequently, some patients may experience adverse reactions or complications following treatment. Future research is needed to better understand these immune responses and develop strategies to optimize patient selection and minimize risks associated with ECM-based therapies. The translation of ECM-based therapies from the laboratory to clinical practice presents additional hurdles. Many promising results seen in preclinical studies have not yet been replicated in larger human trials. Issues such as funding limitations, regulatory challenges, and the complexity of human biology can impede the progress of these therapies. Ensuring that the promising results of small-scale studies are confirmed in larger, more diverse populations is vital for gaining regulatory approval and widespread clinical adoption. Researchers must navigate these challenges while maintaining a focus on the innovative potential of ECM therapies to address unmet medical needs. The cost and accessibility of ECM-based therapies pose challenges for their implementation in clinical settings. Developing and producing high-quality ECM scaffolds can be resource-intensive, leading to higher costs that may not be covered by insurance. This situation can limit access to these advanced therapies for many patients, particularly those in underserved populations. Efforts to streamline production processes, reduce costs, and establish reimbursement models will be crucial for ensuring that ECM-based therapies can be made widely available.

9.4 Future Directions and Challenges

- **Advances in Personalized ECM Therapies**: The future of ECM-based therapies lies in their personalization. Tailoring ECM scaffolds to match the specific biological and mechanical properties of individual patients can significantly enhance treatment efficacy. Advances in 3D bioprinting and tissue engineering technologies are paving the way for creating custom ECM constructs that incorporate patients' cells and specific growth factors. However, challenges remain in the standardization and validation of personalized approaches to ensure safety and effectiveness across diverse patient populations [18].
- **ECM-Based Strategies for Complex Disease Models**: Developing ECM-based therapies for complex diseases such as cancer, cardiovascular disorders, and neurodegenerative conditions is a promising direction. These strategies involve using biomimetic ECM to create in vitro models that better replicate the tumor microenvironment or the cellular composition of affected tissues. This approach can improve drug screening and personalized treatment strategies. However,

replicating the complexity of these environments poses significant scientific challenges, requiring multidisciplinary collaboration to overcome.

- **Challenges in Scaling ECM Biomaterials for Mass Production**: While ECM-based therapies show great promise, scaling up the production of ECM biomaterials for widespread clinical use remains a significant hurdle. Current production methods, which often involve labor-intensive and costly processes, may not be feasible for large-scale applications. Developing efficient, cost-effective manufacturing techniques that maintain the quality and functionality of ECM materials is crucial for translating research into practice.
- **Regulatory Hurdles and Standardization**: As ECM-based therapies advance, navigating the regulatory landscape will be a critical challenge. Regulatory agencies require robust evidence of safety and efficacy before granting approval for new therapies. Additionally, establishing standardized protocols for ECM production, characterization, and clinical testing will be essential for gaining regulatory acceptance and ensuring consistent outcomes across different trials and applications.
- **Long-term Safety and Efficacy**: Ensuring the long-term safety and efficacy of ECM-based therapies is a significant concern. While initial clinical trials may demonstrate promising results, the long-term effects of implanted ECM materials on patients must be thoroughly investigated. This includes understanding potential immune responses, the risk of degradation or calcification of the materials, and their overall impact on the host tissue over time. Ongoing surveillance and post-market studies will be essential for addressing these concerns.
- **Integration of Advanced Technologies**: Future ECM-based therapies are likely to benefit from the integration of advanced technologies, such as nanotechnology, gene editing, and smart biomaterials. These innovations can enhance the functionality of ECM scaffolds, allowing for controlled release of therapeutic agents, real-time monitoring of healing processes, and targeted delivery of cells. However, the challenge lies in ensuring that these technologies are compatible with existing ECM approaches and that they can be effectively translated into clinical applications while maintaining patient safety and regulatory compliance.

9.5 Ethical Considerations in ECM-Based Clinical Trials

- **Patient Consent and Safety in ECM Therapies**: In any clinical trial, obtaining informed consent from participants is a fundamental ethical obligation. For ECM-based therapies, it is crucial that patients are fully educated about the nature of the intervention, including potential risks and benefits, alternative treatments, and the experimental nature of the therapy. Researchers must ensure that consent forms are clear and accessible, allowing patients to make informed decisions about their participation. Ongoing monitoring of patient safety is paramount throughout the trial process. Researchers must establish robust protocols for tracking adverse events and be prepared to intervene if safety concerns arise, thereby prioritizing the well-being of trial participants.

- **Addressing Risks of ECM-Based Interventions**: While ECM-based therapies offer promising advancements in regenerative medicine, they also carry inherent risks that must be carefully managed. These risks may include immune reactions, infections, or complications related to the surgical implantation of ECM materials. Ethical considerations dictate that researchers should conduct thorough risk assessments and implement strategies to minimize potential harm to participants. Transparent communication about the risks involved, along with comprehensive follow-up care and support, is essential for fostering trust and ensuring ethical integrity in clinical trials.
- **Ethical Implications of Genetic Manipulations in ECM Engineering:** As research progresses, the use of genetic manipulation techniques in the engineering of ECM materials raises important ethical questions. For instance, the incorporation of genetically modified cells or bioactive molecules into ECM scaffolds may enhance their functionality but also raises concerns about the long-term implications of such modifications. Ethical discussions must address the potential for unintended consequences, such as unanticipated interactions with the host immune system or alterations in cellular behavior. The use of genetic modifications must be guided by strict ethical standards and regulatory frameworks to ensure that interventions do not compromise patient safety or lead to undesirable outcomes.
- **Equity and Access to ECM Therapies**: The development and implementation of ECM-based therapies should consider issues of equity and access to treatment. As these therapies may be costly and resource-intensive, there is a risk that they could exacerbate existing disparities in healthcare access. Ethical considerations necessitate that researchers and policymakers work to ensure that ECM-based treatments are accessible to all patients, regardless of socioeconomic status or geographic location. This may involve exploring alternative funding models, insurance coverage options, and public health initiatives to promote equitable access to innovative therapies.
- **Transparency and Accountability in Research**: Ethical conduct in ECM-based clinical trials also requires transparency and accountability from researchers. This includes the obligation to disclose any conflicts of interest, funding sources, and potential biases that may influence the research process. Ensuring that findings from clinical trials are shared openly with the scientific community and the public fosters trust and encourages informed dialogue about the implications of ECM therapies. Furthermore, researchers must be accountable for their actions and decisions throughout the trial process, upholding ethical standards and prioritizing the interests of participants.
- **Long-Term Ethical Monitoring**: As ECM-based therapies advance towards clinical application, the need for long-term ethical monitoring becomes increasingly important. Post-trial follow-up is essential to assess the long-term safety and effectiveness of ECM interventions. Ethical oversight should continue beyond the trial phase to ensure that any emerging concerns are addressed promptly and

transparently. Establishing independent review boards or monitoring committees to oversee ongoing research and its implications will help maintain ethical standards and protect the rights and safety of patients in the evolving landscape of ECM-based therapies.

Exercise 3

Multiple Choice Questions

1. What is the primary role of ECM in cardiovascular regeneration?
a) Energy storage
b) Providing structural support
c) Enhancing metabolic activity
d) Preventing infection

2. Which of the following ECM components is crucial for heart valve tissue engineering?
a) Hyaluronic acid
b) Collagen
c) Fibrin
d) Keratin

3. In dermatology, ECM components are essential for:
a) Hair growth
b) Skin hydration
c) Wound healing
d) All of the above

4. Which of the following biomaterials is commonly used in skin grafts to promote ECM mimicry?
a) Polyglycolic acid
b) Collagen
c) Silicone
d) Polyvinyl chloride

5. In neurology, ECM mimics are being explored for their potential in treating:
a) Alzheimer's disease
b) Spinal cord injuries
c) Stroke
d) All of the above

6. ECM-derived hydrogels are beneficial in neurosurgery for:
a) Providing mechanical support
b) Promoting neural cell growth
c) Reducing inflammation
d) Both a and b

7. Which ECM component is critical for neural tissue engineering?
a) Chitosan
b) Elastin
c) Laminin
d) Fibrinogen

8. The use of ECM scaffolds in neurological applications aims to:
a) Prevent nerve regeneration
b) Promote axonal growth
c) Increase scarring
d) Reduce blood flow

9. Which of the following is a common source of ECM for tissue engineering applications?
a) Animal tissues
b) Synthetic polymers
c) Plant materials
d) All of the above

10. ECM therapies in regenerative medicine aim to:
a) Replace damaged tissues
b) Enhance the body's natural healing processes
c) Reduce the risk of infection
d) Both a and b

9.6 Summary

This chapter delves into the significant advancements and challenges associated with extracellular matrix (ECM) based therapies across various medical applications, including wound healing, organ transplantation, and regenerative medicine. It highlights how biomimetic ECM materials are being engineered to closely mimic the native tissue microenvironment, thereby enhancing cellular interactions, promoting tissue integration, and improving patient outcomes. With success stories demonstrating accelerated healing and improved functional recovery, ECM-based therapies are emerging as a promising alternative to traditional treatment modalities. However, the chapter also addresses the various challenges that need to be overcome for successful clinical implementation. Key issues such as the standardization of ECM production, variability in patient responses, and the need for long-term safety assessments are discussed. These challenges underscore the importance of rigorous clinical trials and ongoing research to establish the efficacy and safety of ECM therapies. The ethical considerations surrounding patient consent, the risks associated with ECM interventions, and the implications of genetic manipulations in ECM engineering are examined to ensure that patient welfare remains a priority in clinical practice. The chapter explores future directions for ECM-based therapies, emphasizing the potential for personalized treatments and the development of complex disease models. The integration of advanced technologies, such as 3D bioprinting and smart biomaterials, is identified as a key area for innovation, which could lead to significant improvements in therapeutic strategies. The chapter warns that these advancements must be balanced with considerations of cost, accessibility, and equity to ensure that all patients can benefit from these innovative therapies. This chapter presents a comprehensive overview of the current landscape of ECM-based therapies, outlining both their potential and the hurdles that lie ahead. As research continues to evolve, the collaborative efforts of scientists, clinicians, and ethicists will be crucial in translating these therapies from the laboratory to clinical settings. By addressing the identified challenges and ethical considerations, the medical community can work towards realizing the full potential of ECM-based therapies, ultimately improving outcomes for patients in need of advanced treatment options.

Points to Remember

- **ECM-based therapies leverage biomimicry by replicating the natural architecture and biochemical signals found in human tissues. This design principle allows engineered materials to interact more effectively with cells, promoting tissue regeneration and integration in a way that traditional synthetic materials cannot achieve.**

- **ECM-based therapies are not limited to wound healing or organ transplantation; they are being explored for a variety of conditions, including cardiovascular diseases, neurodegenerative disorders, and even cancer treatment. This versatility underscores the potential of ECM in revolutionizing treatment strategies across multiple medical fields.**

- **Advances in 3D bioprinting technology enable the customization of ECM scaffolds tailored to individual patients. This personalized approach not only enhances the compatibility of the grafts but also optimizes therapeutic outcomes, marking a significant shift toward precision medicine in regenerative therapies.**

- **The incorporation of genetic modifications in ECM engineering raises complex ethical questions regarding the implications of altering biological materials. This chapter highlights the importance of establishing robust ethical frameworks to address concerns about safety, long-term effects, and equitable access to these advanced therapies.**

- **As ECM-based therapies gain traction in clinical settings, regulatory bodies are adapting to create guidelines that ensure patient safety and efficacy. This evolution reflects a growing recognition of the unique characteristics of ECM therapies, prompting the need for specialized frameworks that balance innovation with oversight to foster responsible development and application.**

References

1. Yue B (2014) Biology of the extracellular matrix: an overview. J Glaucoma 23(8 Suppl 1):S20–S23. https://doi.org/10.1097/IJG.0000000000000108
2. Spang MT, Christman KL (2018) Extracellular matrix hydrogel therapies: in vivo applications and development. Acta Biomater 68:1–14. https://doi.org/10.1016/j.actbio.2017.12.019
3. Hwang J, Sullivan MO, Kiick KL (2020) Targeted drug delivery via the use of ECM-mimetic materials. Front Bioeng Biotechnol 8:69. https://doi.org/10.3389/fbioe.2020.00069
4. Sharma Y, Ghatak S, Sen CK et al (2024) Emerging technologies in regenerative medicine: the future of wound care and therapy. J Mol Med. https://doi.org/10.1007/s00109-024-02493-x
5. Diller RB, Tabor AJ (2022) The role of the extracellular matrix (ECM) in wound healing: a review. Biomimetics 7(3):87. https://doi.org/10.3390/biomimetics7030087
6. Spang MT, Christman KL (2017) Extracellular matrix hydrogel therapies: in vivo applications and development. Acta Biomater 68:1–14. https://doi.org/10.1016/j.actbio.2017.12.019
7. Vu TVA, Lorizio D, Vuerich R, Lippi M, Nascimento DS, Zacchigna S (2022) Extracellular matrix-based approaches in cardiac regeneration: challenges and opportunities. Int J Mol Sci 23(24):15783. https://doi.org/10.3390/ijms232415783
8. Li H, Bao M, Nie Y (2021) Extracellular matrix-based biomaterials for cardiac regeneration and repair. Heart Fail Rev 26(5):1231–1248. https://doi.org/10.1007/s10741-020-09953-9
9. Eble JA, Niland S (2019) The extracellular matrix in tumor progression and metastasis. Clin Exp Metas 36(3):171–198. https://doi.org/10.1007/s10585-019-09966-1

10. Huang J, Zhang L, Wan D et al (2021) Extracellular matrix and its therapeutic potential for cancer treatment. Sig Transduct Target Ther 6:153. https://doi.org/10.1038/s41392-021-00544-0
11. Hinderer S, Layland SL, Schenke-Layland K (2015) ECM and ECM-like materials—biomaterials for applications in regenerative medicine and cancer therapy. Adv Drug Deliv Rev 97:260–269. https://doi.org/10.1016/j.addr.2015.11.019
12. Capella-Monsonís H, Crum RJ, Hussey GS, Badylak SF (2024) Advances, challenges, and future directions in the clinical translation of ECM biomaterials for regenerative medicine applications. Adv Drug Deliv Rev 211:115347. https://doi.org/10.1016/j.addr.2024.115347
13. Parmaksiz M, Dogan A, Odabas S, Elçin AE, Elçin YM (2016) Clinical applications of decellularized extracellular matrices for tissue engineering and regenerative medicine. Biomed Mater 11(2):022003. https://doi.org/10.1088/1748-6041/11/2/022003
14. Zhang W, Zhu Y, Li J, Guo Q, Peng J, Liu S, Yang J, Wang Y (2016) Cell-derived extracellular matrix: basic characteristics and current applications in orthopedic tissue engineering. Tissue Eng Part B Rev 22(3):193–207. https://doi.org/10.1089/ten.TEB.2015.0290
15. Hoogewerf CJ, Hop MJ, Nieuwenhuis MK, Middelkoop E, Vlies CH, Van Baar ME (2012) Early excision and grafting for burns. Cochr Datab Syst Rev 2012(3):CD009715. https://doi.org/10.1002/14651858.CD009715
16. Lau L, Cua R, Keough M et al (2013) Pathophysiology of the brain extracellular matrix: a new target for remyelination. Nat Rev Neurosci 14:722–729. https://doi.org/10.1038/nrn3550
17. Welman T, Michel S, Segaren N, Shanmugarajah K (2015) Bioengineering for organ transplantation: progress and challenges. Bioengineered 6(5):257–261. https://doi.org/10.1080/21655979.2015.1081320
18. Gambardella V, Tarazona N, Cejalvo JM, Lombardi P, Huerta M, Roselló S, Fleitas T, Roda D, Cervantes A (2020) Personalized medicine: recent progress in cancer therapy. Cancers 12(4):1009. https://doi.org/10.3390/cancers12041009

Chapter 10
Future Perspectives and Innovations

10.1 Introduction

The extracellular matrix (ECM) is a highly dynamic and complex network of macromolecules that provide structural and biochemical support to cells. It plays a critical role in tissue architecture, cellular communication, and the regulation of various physiological processes such as wound healing, cell differentiation, and tissue repair. The importance of ECM extends beyond its structural role, as it actively participates in signaling pathways that regulate cellular functions. Understanding the composition, organization, and functions of the ECM has been central to the development of regenerative medicine and tissue engineering, where it serves as a model for creating bioactive scaffolds and materials. ECM research has gained considerable attention due to its significant role in disease progression, particularly in cancer, fibrosis, and cardiovascular diseases. Changes in the composition and stiffness of the ECM can influence cellular behavior, leading to abnormal cell growth, migration, and invasion, which are hallmarks of cancer metastasis. Similarly, the disruption of ECM homeostasis in fibrotic diseases can result in excessive matrix deposition, impairing normal tissue function [1]. These insights have driven extensive research into the therapeutic potential of ECM modulation in treating a wide range of diseases.

One of the key reasons ECM related research is crucial is its involvement in tissue regeneration. The ECM serves as a scaffold for tissue repair by providing mechanical support to cells and regulating the deposition of new tissue. In wound healing, for example, the ECM provides essential cues for the migration of fibroblasts and the formation of new extracellular matrix components, such as collagen, which contribute to tissue repair. ECM degradation products can act as bioactive molecules, signaling for the recruitment of immune cells and promoting tissue regeneration. Thus, mimicking the ECM in biomaterial design has become a focal point in developing advanced therapeutic strategies. In the context of biomaterials and tissue engineering, the ECM has inspired the creation of bio-mimetic scaffolds that

D. Suhag, *Extracellular Matrix biomimicking for Theranostics*, Biomedical Materials for Multi-functional Applications, https://doi.org/10.1007/978-981-95-7900-6_10

replicate their complex structure and functions. These scaffolds, designed to simulate the natural ECM environment, are used in a variety of applications, such as wound healing, bone regeneration, and organ transplantation. By incorporating ECM components like collagen, elastin, and hyaluronic acid, researchers can design materials that promote cell adhesion, proliferation, and differentiation, closely mimicking the natural processes that occur in tissue repair and regeneration. This approach has shown great promise in improving the outcomes of regenerative therapies.

The role of the ECM is also pivotal in the development of drug delivery systems. The highly organized structure of the ECM allows for the controlled release of therapeutic agents, enabling targeted and sustained drug delivery. By embedding bioactive molecules within ECM-inspired hydrogels and scaffolds, researchers can develop systems that mimic the natural release of growth factors and other signaling molecules, thereby improving the efficacy of treatments. Such ECM-based drug delivery systems are being explored for cancer therapy, tissue repair, and chronic disease management, offering new opportunities for precision medicine [2]. ECM research is at the forefront of innovations in personalized medicine. The ability to tailor ECM-based therapies to an individual's specific biological environment holds immense potential for improving treatment outcomes. This is particularly relevant in the context of cancer, where the ECM composition varies widely between patients and plays a crucial role in tumor progression and response to treatment. By analyzing the ECM's molecular profile, researchers are developing personalized therapeutic approaches that target the tumor microenvironment, offering a new avenue for cancer treatment.

ECM research is vital for advancing our understanding of tissue biology, disease mechanisms, and the development of novel therapeutic strategies. Its diverse roles in tissue structure, cellular communication, and disease progression make it an essential focus in regenerative medicine, drug delivery, and personalized medicine.

Exercise 2

Fill Ups
(Bioink, Electrospinning, 3D printing, Cell behavior, Additive manufacturing)

1. One common biofabrication technique used to create scaffolds is ______, which allows for precise control over the structure and porosity of the material.

2. ______ is a method that uses computer-aided design to create 3D printed scaffolds that mimic the ECM architecture.

3. The process of ______ involves electrospinning polymer solutions to produce nanofibrous scaffolds that resemble natural ECM.

4. In bioprinting, living cells are often incorporated into the ______ to create functional tissue constructs.

5. The choice of scaffold material in biofabrication can significantly impact ______, which is critical for tissue regeneration.

10.2 Emerging Trends in ECM Research

Emerging trends in extracellular matrix (ECM) research are rapidly transforming the fields of regenerative medicine, biomaterials, and disease therapy. Advancements in bioengineering technologies have enabled the development of more accurate ECM-mimicking scaffolds that enhance tissue repair and regeneration. Researchers are increasingly focusing on integrating nanotechnology, smart materials, and bio fabrication techniques to create dynamic, responsive ECM models that better replicate the natural cellular environment [3]. ECM's role in disease progression, particularly in cancer and fibrosis, is leading to innovative therapeutic approaches, including ECM-targeted therapies and the use of ECM as a diagnostic tool. As ECM research evolves, its potential to drive breakthroughs in precision medicine and tissue engineering is becoming increasingly apparent.

10.2.1 Advances in ECM-Based Therapeutics

Advances in ECM-based therapeutics have significantly expanded the scope of regenerative medicine and disease treatment, offering more precise and effective approaches to tissue repair and healing. ECM-based therapies utilize biomaterials that mimic the natural matrix to promote cellular functions like adhesion, proliferation, and differentiation. These biomaterials, often derived from natural sources like collagen or synthetic polymers, are designed to interact with the body's cells, providing the necessary cues for tissue regeneration. By replicating the structure and biochemical signals of the native ECM, these therapeutics can facilitate the regeneration of damaged tissues and organs more effectively than conventional methods. The integration of ECM components into therapeutic platforms has greatly enhanced the field of drug delivery. ECM-based scaffolds and hydrogels can serve as carriers for the controlled and localized release of bioactive molecules such as growth factors, cytokines, or therapeutic drugs [4]. This targeted delivery system not only improves the efficacy of the treatment but also reduces side effects associated with systemic drug administration. In cancer therapy, for example, ECM-based materials can be engineered to deliver chemotherapeutic agents directly to tumor sites, reducing damage to healthy tissues. This precision in drug delivery, supported by the ECM's ability to influence cellular behavior, represents a major breakthrough in therapeutic design.

ECM-based therapeutics have advanced the treatment of complex conditions such as chronic wounds, cardiovascular diseases, and tissue degeneration. In chronic wound care, ECM-based dressings and grafts provide a bioactive surface that stimulates cell migration, angiogenesis, and tissue repair. Similarly, in cardiovascular applications, ECM-derived materials are being used to engineer vascular grafts and cardiac patches that closely mimic the native tissue, improving the integration and function of implants. The success of these ECM-based solutions in clinical applications underscores their transformative potential, driving further research into developing more sophisticated ECM-mimicking materials that can address a wide range of medical challenges.

10.2.2 Innovations in ECM Engineering for Regenerative Medicine

Innovations in ECM engineering has revolutionized regenerative medicine by offering novel strategies for creating biomimetic environments that closely replicate the natural extracellular matrix. One of the most significant advancements is the development of bioengineered scaffolds that mimic the composition, structure, and mechanical properties of native ECM. These scaffolds are designed to support cell adhesion, migration, and differentiation, making them ideal for tissue regeneration. Techniques such as electrospinning, 3D bioprinting, and decellularization of tissues have enabled researchers to create highly specialized scaffolds that promote specific cellular behaviors. These innovations have enhanced the ability to repair complex tissues, such as skin, bone, and cartilage, by providing a more supportive and functional framework for cell growth. The integration of smart materials into ECM engineering has further expanded its potential in regenerative medicine [5]. These materials are responsive to environmental stimuli, such as temperature, pH, or mechanical stress, and can change their properties to match the dynamic nature of the body's tissues. For instance, shape-memory polymers and hydrogels that mimic ECM properties can be designed to swell, degrade, or release bioactive molecules in response to physiological conditions. This adaptability enables a more controlled and sustained healing process, particularly in tissues that experience mechanical strain, such as the heart and musculoskeletal system. By mimicking the ECM's ability to adapt to changes in the cellular environment, these materials offer new possibilities for improving the longevity and functionality of tissue-engineered constructs.

In addition to structural and smart materials, ECM engineering has seen major innovations in the incorporation of bioactive molecules, such as growth factors and signaling proteins, into biomaterials. These molecules play a critical role in regulating cellular functions and tissue regeneration. By embedding bioactive cues within ECM-inspired scaffolds, researchers can create materials that actively participate in tissue healing by promoting angiogenesis, cell proliferation, and differentiation. This approach has proven particularly effective in accelerating wound healing and regenerating vascularized tissues. The ability to control the release of bioactive molecules from ECM-engineered materials allows for more precise therapeutic interventions, enhancing the effectiveness of treatments in regenerative medicine. These ongoing innovations are positioning ECM engineering as a cornerstone of future regenerative therapies.

10.2.3 Integration of ECM with Nanotechnology and Smart Materials

The integration of ECM with nanotechnology and smart materials has opened up new frontiers in biomedical research, particularly in the fields of regenerative medicine and tissue engineering. Nanotechnology allows the precise manipulation of materials at the nanoscale, which closely mimics the fibrous structure and nanoscale features of the natural extracellular matrix. By incorporating nanoparticles, nanofibers, and nanocomposites into ECM-based scaffolds, researchers can create highly detailed, biocompatible structures that promote cell attachment, migration, and differentiation. These nanostructured ECM models can be tailored to mimic specific tissue environments, enhancing tissue regeneration and repair processes. Moreover, the increased surface area provided by nanomaterials allows for better interaction with cells and the integration of therapeutic agents. Smart materials, which are designed to respond to external stimuli like temperature, pH, and mechanical stress, are increasingly being incorporated into ECM-based constructs to mimic the dynamic nature of tissues. These materials can change their physical or chemical properties in response to changes in the body's environment, allowing them to deliver drugs, release growth factors, or alter their structure to promote tissue regeneration [6]. For example, smart hydrogels can swell or degrade in response to physiological conditions, providing a controlled release of bioactive molecules or adjusting to changes in tissue stress. This adaptability makes smart materials highly useful for applications in wound healing, bone regeneration, and cardiac tissue engineering, where the mechanical and biological environment is constantly changing.

The combination of ECM with nanotechnology and smart materials not only enhances the functionality of tissue scaffolds but also enables more precise therapeutic interventions. Nanoparticles and nanofibers can be functionalized with bioactive molecules, enabling targeted drug delivery or localized treatment of diseases, such as cancer or chronic wounds. The smart behavior of these materials allows them to release therapeutic agents only when needed, reducing side effects and improving the efficacy of treatments. In cancer therapy, for instance, smart ECM-based scaffolds can deliver chemotherapy drugs directly to tumor cells in response to specific environmental triggers, such as pH changes in the tumor microenvironment. This integration of ECM with nanotechnology and smart materials represents a promising avenue for developing more effective, patient-specific treatments in regenerative medicine and beyond.

Exercise 2

Fill Ups
(Bioink, Electrospinning, 3D printing, Cell behavior, Additive manufacturing)

1. One common biofabrication technique used to create scaffolds is ______, which allows for precise control over the structure and porosity of the material.

2. ______ is a method that uses computer-aided design to create 3D printed scaffolds that mimic the ECM architecture.

3. The process of ______ involves electrospinning polymer solutions to produce nanofibrous scaffolds that resemble natural ECM.

4. In bioprinting, living cells are often incorporated into the ______ to create functional tissue constructs.

5. The choice of scaffold material in biofabrication can significantly impact ______, which is critical for tissue regeneration.

10.2.4 Role of ECM in Immunomodulation and Cancer Therapy

The extracellular matrix (ECM) plays a crucial role in immunomodulation, influencing immune cell behavior and regulating the immune response (Fig. 10.1). The composition and structural properties of the ECM can affect the infiltration, activation, and differentiation of immune cells, such as macrophages, T-cells, and dendritic cells, within tissues. In inflammatory conditions, for instance, ECM undergoes remodeling, releasing bioactive molecules like cytokines and growth factors that signal immune cells to migrate to the site of injury or infection. These interactions between the ECM and immune cells are essential for maintaining tissue homeostasis and facilitating wound healing. By modulating the immune response, ECM can either promote or inhibit inflammation, making it a key target in therapies aimed at controlling excessive immune activity or enhancing immune defenses. In the context of cancer therapy, ECM plays a dual role. On one hand, it can support tumor growth and metastasis by creating a protective microenvironment that shields cancer cells from immune surveillance. Tumors often modify the surrounding ECM, increasing its stiffness and altering its composition to promote cancer cell survival, invasion, and resistance to therapy. This remodeled ECM can suppress the activity of immune cells, such as cytotoxic T-cells and natural killer cells, that are crucial for recognizing and eliminating cancer cells. Understanding how ECM contributes to immune evasion in tumors has become a focal point in cancer research, leading to the development of therapies aimed at targeting the tumor ECM to improve immune cell infiltration and activity [4]. Targeting ECM for cancer therapy has opened up new avenues for immunomodulation in cancer treatment. Strategies such as ECM-targeted drugs, enzymes that degrade ECM components, and the use of ECM-mimicking scaffolds to deliver immunotherapies are being explored to disrupt the tumor microenvironment and enhance the effectiveness of immune-based treatments. For example, degrading specific ECM proteins, like collagen, can soften the tumor stroma, making it easier for immune cells to penetrate the tumor and attack cancer cells. ECM-based biomaterials are being developed to deliver immunomodulatory agents directly to the tumor site, boosting the immune system's ability to recognize and destroy cancer cells. These approaches demonstrate the pivotal role of ECM in cancer immunotherapy and highlight its potential as a therapeutic target in the fight against cancer.

10.3 Potential Breakthroughs in ECM Biomimicry

Potential breakthroughs in ECM biomimicry are poised to revolutionize the fields of regenerative medicine and tissue engineering by creating more accurate, functional replicas of the natural extracellular matrix. ECM biomimicry aims to replicate the complex architecture, mechanical properties, and biochemical signals of native tissues, providing an optimal environment for cell growth, differentiation, and tissue repair. Advances in bio fabrication techniques, such as 3D bioprinting and electrospinning, along with the integration of bioactive molecules, have enabled the creation of highly sophisticated scaffolds that closely mimic the dynamic nature of the ECM. These innovations have the potential to accelerate tissue regeneration, improve the efficacy of drug delivery systems, and even advance personalized medicine by designing ECM-mimetic materials tailored to individual patient needs. As research in this area progresses, ECM biomimicry is expected to play a key role in overcoming current challenges in tissue engineering and developing more effective therapeutic solutions.

10.3.1 Bio Fabrication Techniques for ECM-Like Scaffolds

Bio fabrication techniques have transformed the development of ECM-like scaffolds, providing highly structured and functional materials for tissue engineering and regenerative medicine (Fig. 10.2). One of the most prominent bio fabrication methods is 3D bioprinting, which allows for the precise deposition of cells and biomaterials layer by layer to recreate the complex architecture of the extracellular matrix. With 3D bioprinting, researchers can design scaffolds with controlled porosity, stiffness, and biochemical composition, which are crucial for mimicking the natural environment of different tissues. This technique also enables the incorporation of bioactive molecules, such as growth factors, that support cellular processes like proliferation and differentiation, making it an ideal tool for creating personalized, tissue-specific scaffolds [7].

Electrospinning is another widely used technique in fabricating ECM-like scaffolds. It involves the creation of nanofibrous structures by applying a high voltage to a polymer solution, resulting in fibers that closely resemble the fibrous nature of the ECM. The nanofibers produced by electrospinning provide a large surface area for cell attachment and proliferation, while also allowing for the incorporation of bioactive agents for enhanced tissue regeneration. The versatility of electrospinning allows for the use of various natural and synthetic polymers, enabling the fabrication

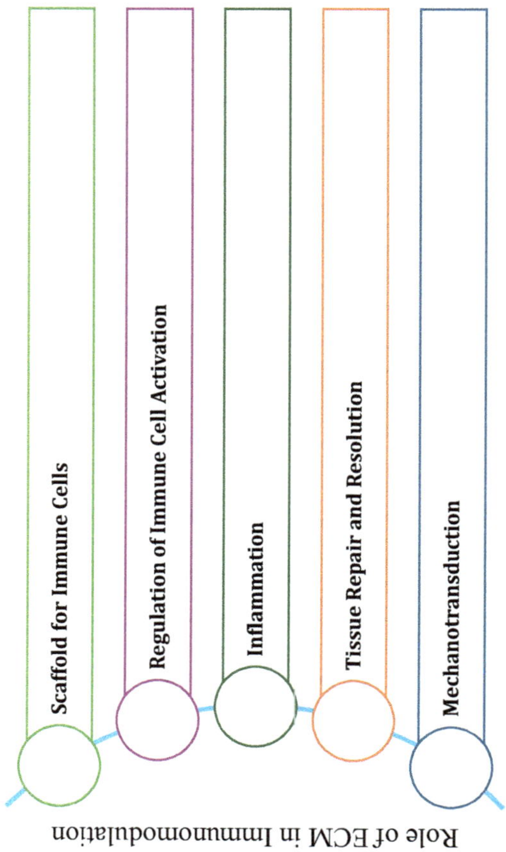

Fig. 10.1 Role of ECM in immunomodulation

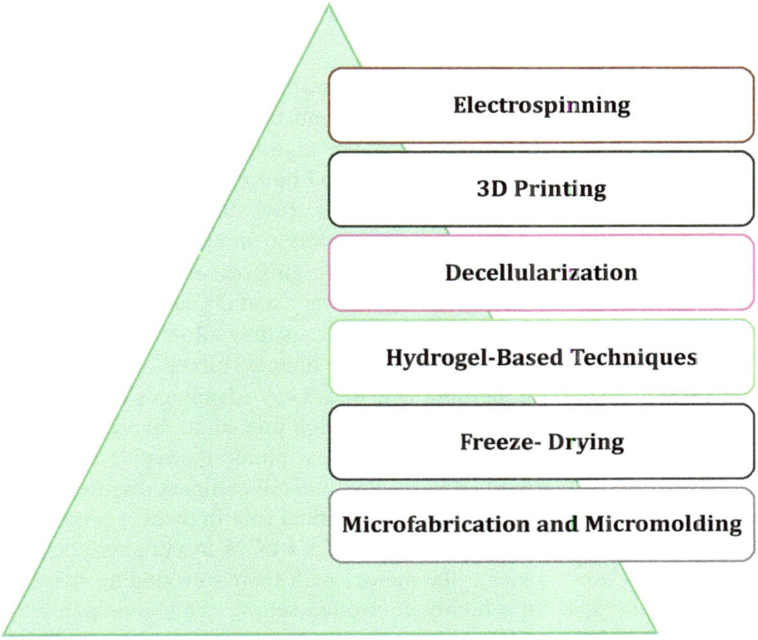

Fig. 10.2 Bio fabrication techniques for ECM-like scaffolds

of scaffolds with tailored mechanical and biological properties. The fibrous structure facilitates the delivery of cells and nutrients, making it highly effective in applications such as wound healing, skin regeneration, and bone tissue engineering.

Another cutting-edge bio fabrication method is decellularization, which involves removing the cellular components from native tissues, leaving behind an ECM scaffold that retains the original tissue architecture and composition. This technique is especially useful for developing scaffolds for organ regeneration, as it preserves the intricate vascular networks and biochemical signals needed for tissue function. Decellularized ECM scaffolds can be repopulated with patient-derived cells, reducing the risk of immune rejection and improving the integration of the scaffold with the host tissue. Bio fabrication techniques like decellularization are advancing the field by providing more biologically relevant scaffolds that can better mimic the structure and function of native tissues, leading to improved outcomes in regenerative medicine.

10.3.2 Synthetic ECM Models for Precision Medicine

Synthetic ECM models are emerging as a powerful tool in precision medicine, enabling more customized and effective treatments by mimicking the intricate structures and functions of the natural extracellular matrix. These synthetic models are designed to replicate the physical, chemical, and biological properties of the ECM, providing a supportive environment for cells to grow, differentiate, and respond to therapeutic interventions. By tailoring these models to the specific needs of individual patients or disease types, researchers can create highly personalized treatment strategies that enhance tissue regeneration, drug delivery, and disease modeling. Precision medicine benefits greatly from synthetic ECMs, as they allow for the development of therapies that can be fine-tuned to target the unique cellular and molecular characteristics of each patient's condition. One of the key advantages of synthetic ECM models is their ability to be engineered with high precision, incorporating specific mechanical properties and biochemical cues that mimic the native tissue environment. This is particularly important in the treatment of complex diseases like cancer, where the tumor microenvironment plays a critical role in disease progression and response to therapy [8]. By recreating the tumor's ECM in vitro, synthetic models can be used to test how cancer cells interact with their surrounding matrix and to screen potential therapies in a highly controlled setting. This approach allows for the identification of the most effective treatments based on how cancer cells respond to changes in the synthetic ECM, enabling more targeted and personalized cancer therapies.

Synthetic ECM models also hold great potential for advancing tissue engineering and regenerative medicine. By mimicking the unique properties of specific tissues, these models can be used to create scaffolds that promote the regeneration of damaged tissues in a patient-specific manner. For example, synthetic ECMs designed to mimic the architecture of cartilage or bone can be implanted to aid in the repair of orthopedic injuries, while models that replicate the ECM of the heart or lungs can support the regeneration of cardiovascular or respiratory tissues. This precision in ECM design ensures that the synthetic scaffolds provide the appropriate signals and mechanical support needed for optimal tissue repair, leading to improved therapeutic outcomes. Synthetic ECM models can be functionalized with bioactive molecules, such as growth factors or cytokines, to enhance their therapeutic potential. These bioactive agents can be incorporated into the scaffold in a controlled manner, allowing for the sustained release of signaling molecules that guide cellular behavior and tissue regeneration. This ability to modulate the release of therapeutic agents based on the needs of the patient makes synthetic ECM models highly adaptable for precision medicine. The use of synthetic materials reduces the risk of immunogenicity or pathogen transmission, which can be a concern with naturally derived ECMs. As a result, synthetic ECM models offer a safer and more customizable platform for developing patient-specific therapies, driving the future of personalized medicine.

10.3.3 Biomimetic Hydrogels for Controlled Drug Delivery

Biomimetic hydrogels have gained significant attention as advanced materials for controlled drug delivery, offering a platform that closely mimics the structure and functionality of the natural extracellular matrix (ECM). These hydrogels are designed to replicate the physical and biochemical properties of the ECM, providing an ideal environment for the sustained and targeted release of therapeutic agents. The highly porous structure of hydrogels allows for the encapsulation of drugs, proteins, and other bioactive molecules, while their tunable properties enable precise control over the release kinetics. This ability to regulate the release of drugs over time reduces the frequency of dosing, enhances the therapeutic effect, and minimizes side effects, making biomimetic hydrogels a promising tool in precision medicine.

Advantages of biomimetic hydrogels includes their responsiveness to environmental stimuli, such as pH, temperature, and enzyme activity, which can trigger the release of the encapsulated drug in a controlled manner. For example, in cancer therapy, hydrogels can be engineered to release chemotherapeutic agents in response to the acidic environment of a tumor, ensuring that the drug is delivered directly to the cancer cells while sparing healthy tissues. Similarly, hydrogels used in wound healing can release growth factors or antimicrobial agents in response to changes in the wound environment, promoting faster and more efficient healing. This smart drug delivery capability not only improves treatment efficacy but also reduces the risk of systemic toxicity [9]. Biomimetic hydrogels also offer versatility in their design, allowing for the incorporation of various natural and synthetic materials to achieve specific therapeutic outcomes. For instance, hydrogels can be functionalized with ECM proteins like collagen or hyaluronic acid to enhance cell adhesion and proliferation, making them particularly useful for tissue engineering applications. In controlled drug delivery, these ECM-mimicking materials can interact with the body's cells, promoting localized healing while delivering drugs over extended periods. The ability to modify the mechanical properties, degradation rate, and drug release profile of these hydrogels makes them a highly customizable and adaptable platform for addressing a wide range of medical conditions, from cancer to chronic wounds and beyond.

10.3.4 Application of AI in ECM-Related Diagnostics

The application of artificial intelligence (AI) in ECM-related diagnostics is revolutionizing the way diseases are detected and monitored, particularly in complex conditions like cancer, fibrosis, and cardiovascular diseases (Fig. 10.3). AI, through machine learning algorithms and data analytics, has the ability to process vast amounts of biomedical data and identify subtle changes in the ECM that may indicate early stages of disease.

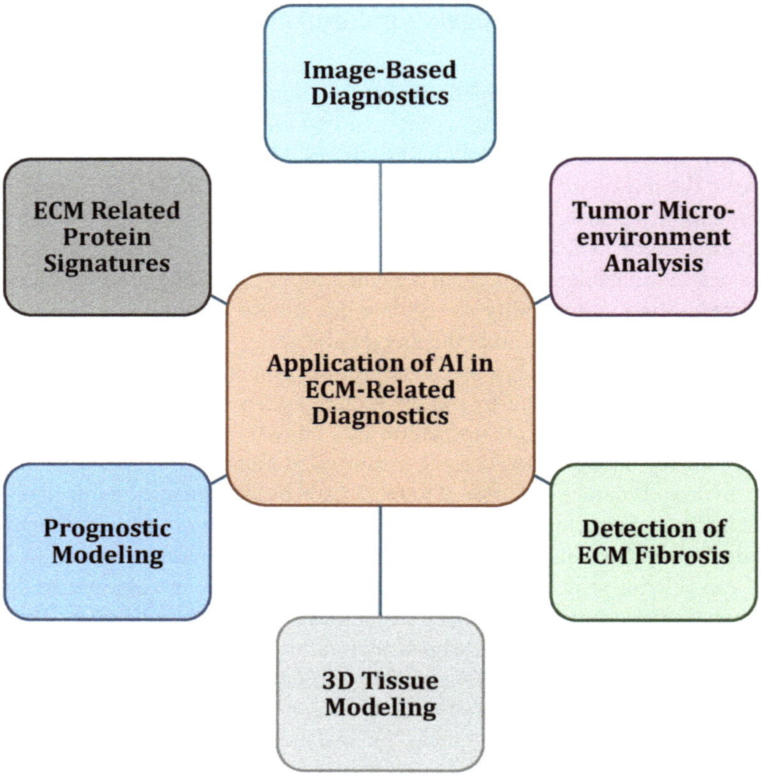

Fig. 10.3 Applications of AI in ECM-related diagnostics

- **Early Detection of Diseases**: AI algorithms can analyze imaging data from various modalities, such as MRI, CT scans, and histopathological slides, to identify subtle changes in the extracellular matrix (ECM) associated with early stages of diseases like cancer and fibrosis. By detecting alterations in ECM structure or composition before clinical symptoms appear, AI enhances the potential for timely interventions, leading to improved patient outcomes [10].
- **Tumor Microenvironment Analysis**: In oncology, AI plays a crucial role in evaluating the tumor microenvironment, where the ECM significantly influences tumor behavior and treatment response. AI systems can be trained to recognize patterns of ECM remodeling linked to specific cancer types, stages, or aggressiveness [11]. This capability allows for non-invasive assessments of tumor characteristics and can help predict a patient's likelihood of metastasis or response to therapies, facilitating personalized treatment strategies.
- **Quantitative Assessment of Fibrosis**: AI can effectively quantify ECM changes in diseases characterized by excessive fibrosis, such as liver cirrhosis or pulmonary fibrosis. Machine learning models can analyze histological images to measure

as the U.S. Food and Drug Administration (FDA) and the European Medicines Agency (EMA), have established frameworks to ensure that ECM-based products are safe, effective, and manufactured in compliance with Good Manufacturing Practices (GMP). These regulations often require extensive preclinical and clinical testing to evaluate the biological responses and long-term effects of ECM materials on human health. Navigating the regulatory landscape can be complex, as the classification of ECM products—whether they are considered medical devices, biologics, or drugs—affects the type of regulatory pathways and data requirements needed for approval.

Ethical considerations are equally important in the context of ECM applications, particularly when it involves human tissues or cells. The use of human-derived ECM materials raises concerns about donor consent, tissue sourcing, and the potential for exploitation or commodification of biological materials. Ethical guidelines must be established to ensure that all donor materials are obtained with informed consent and that donors are adequately informed about the potential uses of their tissues. Additionally, considerations surrounding the privacy and confidentiality of donor information must be prioritized. As ECM technologies continue to evolve, ongoing ethical discussions are essential to address the implications of using human tissues in research and clinical applications, ensuring that the rights and welfare of donors are upheld.

The integration of new technologies, such as AI and biomaterials, into ECM applications raises additional ethical questions regarding data security, patient privacy, and the potential for bias in AI-driven diagnostics or therapies. Ensuring that AI algorithms are trained on diverse datasets is critical to avoid disparities in healthcare outcomes. As ECM research progresses, it is essential for stakeholders, including researchers, clinicians, and policymakers, to engage in dialogues about the ethical implications of these technologies. Developing comprehensive ethical frameworks that encompass regulatory compliance, donor rights, and responsible innovation will be crucial for fostering public trust and promoting the responsible advancement of ECM applications in medicine.

10.4.3 Overcoming Barriers in ECM-Integrated Therapeutics

Overcoming barriers in ECM-integrated therapeutics is essential for translating laboratory innovations into effective clinical applications. One of the primary challenges is the complexity of designing ECM materials that not only mimic the natural matrix but also effectively deliver therapeutic agents. ECM's diverse composition and dynamic nature necessitate advanced materials that can respond to biological signals and provide a suitable environment for cellular interactions. To address this, researchers are increasingly exploring biomimetic materials and smart drug delivery systems that can release therapeutics in a controlled manner, tailored to the specific needs of the tissue or disease state. By incorporating responsive elements, such as hydrogels that change properties in response to pH or temperature, scientists can create ECM-integrated therapeutics that enhance the efficacy of treatment while minimizing side effects. Another significant barrier is the limited understanding of the complex interactions between ECM components, cells, and therapeutic agents [13]. The effectiveness of ECM-integrated therapeutics relies heavily on how these elements interact at the cellular and molecular levels. Inadequate knowledge of these interactions can lead to suboptimal therapeutic outcomes. To overcome this challenge, interdisciplinary approaches that combine insights from materials science, molecular biology, and pharmacology are crucial. Utilizing advanced in vitro and in vivo models that closely mimic the physiological conditions can provide valuable insights into how therapeutic agents interact with the ECM and the surrounding cells. Incorporating technologies such as artificial intelligence and machine learning can facilitate the identification of optimal ECM compositions and formulations that enhance therapeutic efficacy.

Regulatory and manufacturing challenges also present obstacles to the successful implementation of ECM-integrated therapeutics. Navigating the regulatory landscape requires comprehensive testing to ensure safety, efficacy, and quality, which can be resource-intensive and time-consuming. To streamline this process, researchers must work closely with regulatory bodies to clarify requirements early in the development stage. Scaling up production while maintaining the quality and functional properties of ECM materials is essential for commercial viability. Collaborations between academic institutions and industry partners can help overcome these hurdles by sharing expertise in manufacturing processes and regulatory compliance. By addressing these barriers collectively, the full potential of ECM-integrated therapeutics can be realized, leading to innovative treatments that improve patient outcomes across various medical fields.

10.4.4 Future Directions in ECM-Driven Tissue Engineering

Future directions in ECM-driven tissue engineering hold immense promise for advancing regenerative medicine and addressing unmet clinical needs. One of the most exciting prospects is the integration of bioprinting technologies with ECM-based materials to create complex, multi-layered tissue constructs. 3D bioprinting allows for precise control over the spatial arrangement of cells and ECM components, enabling the recreation of the intricate architecture of natural tissues. This approach not only facilitates the development of more realistic tissue models for drug testing and disease modeling but also enhances the potential for successful tissue grafts that can integrate seamlessly with the host tissue. As bioprinting techniques continue to evolve, the ability to incorporate multiple cell types and bioactive factors into ECM constructs will further enhance their functionality and relevance in clinical applications. Another significant direction for ECM-driven tissue engineering is the incorporation of smart biomaterials that can respond dynamically to environmental cues [14]. These materials can be engineered to change their properties in response to specific stimuli, such as changes in pH, temperature, or the presence of certain biomolecules. For example, smart hydrogels that release therapeutic agents in response to inflammatory signals can provide localized treatment and promote tissue healing more effectively. The development of such responsive materials will enable more tailored therapeutic approaches, allowing for the modulation of the ECM in real-time to better support tissue regeneration and repair. Combining these smart materials with advanced imaging techniques can facilitate the monitoring of tissue development and healing processes, providing valuable insights into the efficacy of the engineered constructs.

The incorporation of nanotechnology into ECM-driven tissue engineering is another promising avenue for future research. Nanomaterials can enhance the mechanical properties of ECM constructs, improve drug loading and release profiles, and facilitate cellular interactions at the nanoscale level. By integrating nanoparticles or nanofibers into ECM scaffolds, researchers can create structures that mimic the native ECM more closely while providing improved functionality. For example, nanomaterials can be used to deliver growth factors or other bioactive molecules in a controlled manner, enhancing cellular responses and promoting tissue regeneration. As our understanding of nanoscale interactions between materials and biological systems improves, the potential for developing innovative ECM-based therapeutics will continue to expand.

Lastly, the future of ECM-driven tissue engineering will likely involve a more personalized approach to patient care, leveraging advances in genomics, proteomics, and bioinformatics. By understanding individual patients' unique ECM profiles and their specific tissue microenvironments, tailored ECM-based therapies can be developed to meet their needs. This precision medicine approach will not only enhance the effectiveness of tissue engineering solutions but also improve patient outcomes by minimizing the risks of rejection and complications. As interdisciplinary collaborations between engineers, biologists, and clinicians become more common, the integration of ECM-driven strategies into personalized treatment plans will be crucial for advancing the field of regenerative medicine and addressing the complex challenges faced in tissue repair and regeneration.

10.5 Summary

This chapter explores the dynamic landscape of extracellular matrix (ECM) research and its transformative impact on tissue engineering and regenerative medicine. As a critical component of the tissue microenvironment, ECM plays a pivotal role in cell behavior, influencing processes such as cell adhesion, proliferation, and differentiation. Recent advances in biomimetic materials and fabrication techniques, such as 3D bioprinting and smart hydrogels, have enhanced the ability to create ECM-inspired constructs that replicate the natural tissue architecture and biochemical cues essential for effective tissue regeneration. These innovations pave the way for improved therapeutic strategies and personalized medicine approaches that address a variety of clinical challenges. The integration of artificial intelligence (AI) and nanotechnology into ECM research further amplifies its potential. AI-driven analytics enable the identification of key ECM biomarkers and facilitate the development of predictive models for disease progression and treatment response. In parallel, the incorporation of nanomaterials into ECM constructs enhances mechanical properties and allows for targeted drug delivery, optimizing therapeutic outcomes. However, the path forward is not without its challenges, as researchers must navigate regulatory landscapes and ethical considerations, ensuring that ECM-based innovations are safe, effective, and ethically sourced. Collaborative efforts between academia and industry are essential for overcoming these barriers and fostering the responsible advancement of ECM technologies.

Looking ahead, the future of ECM-driven tissue engineering promises significant advancements in personalized medicine, with an emphasis on tailored therapies that consider individual patient profiles and specific tissue microenvironments. Continued exploration of smart biomaterials that respond to biological signals, along with the application of cutting-edge technologies such as bioprinting and nanotechnology, will drive innovation in the field. As researchers deepen their understanding of ECM dynamics and interactions, the potential for developing effective ECM-integrated therapies expands, ultimately aiming to improve patient care and outcomes in regenerative medicine. This chapter underscores the critical importance of ECM research as a cornerstone of modern therapeutic strategies, highlighting both the opportunities and challenges that lie ahead.

Exercise 3

Multiple Choice Questions

1. What is a primary goal of ECM-based therapeutics?
a) To enhance mechanical properties of tissues
b) To restore or improve tissue function
c) To reduce manufacturing costs
d) To eliminate all cells

2. Which of the following ECM components is commonly targeted for therapeutic interventions in tissue engineering?
a) Collagen
b) Cellulose
c) Polyethylene
d) DNA

3. A significant breakthrough in ECM biomimicry involves the use of:
a) Completely synthetic materials
b) Decellularized tissues
c) Non-biodegradable polymers
d) Heavy metals

4. Which of the following techniques is being explored to enhance ECM biomimicry?
a) Nanotechnology
b) Thermal treatment
c) Mechanical grinding
d) Electrolysis

5. In ECM biomimicry, the replication of biochemical cues is essential for:
a) Enhancing structural integrity
b) Promoting specific cellular responses
c) Increasing weight
d) Reducing costs

6. Biomimetic hydrogels are designed to:
a) Mimic the properties of natural ECM
b) Increase toxicity
c) Prevent drug absorption
d) Inhibit cell growth

7. Which property of biomimetic hydrogels makes them suitable for controlled drug delivery?
a) Low permeability
b) High elasticity
c) Tunable degradation rates
d) Heavy weight

8. The incorporation of ______ into biomimetic hydrogels can enhance their drug release profiles.
a) Enzymes
b) Synthetic dyes
c) Non-biodegradable polymers
d) Heavy metals

9. What is a common application of ECM-based therapeutics in medicine?
a) Blood transfusions
b) Organ transplantation
c) Cancer treatment
d) All of the above

10. Advances in ECM research may lead to improved therapies for conditions such as:
a) Osteoarthritis
b) Cardiovascular diseases
c) Neurodegenerative disorders
d) All of the above

Points to Remember

- **The extracellular matrix (ECM) is not just a structural support system; it actively regulates cellular behavior, influencing key processes like adhesion, proliferation, and differentiation, which are essential for tissue repair and regeneration.**

- **Advances in 3D bioprinting technologies enable researchers to create complex, multi-layered tissue constructs that closely mimic the architecture of natural tissues, enhancing the potential for successful grafts in clinical applications.**

- **Artificial intelligence (AI) is revolutionizing ECM research by enabling the analysis of large datasets to identify ECM biomarkers, facilitating early disease detection, and predicting patient responses to therapies, thereby enhancing precision medicine.**

- **The development of smart biomaterials that respond dynamically to environmental stimuli (like pH or temperature changes) is paving the way for more effective ECM-integrated therapeutics that can adapt in real-time to support tissue regeneration.**

- **Future directions in ECM-driven tissue engineering emphasize personalized approaches, tailoring ECM-based therapies to individual patients' unique ECM profiles and specific tissue environments, ultimately aiming to improve treatment outcomes.**

References

1. Papanicolaou M, Cox TR (2021) Extracellular matrix (ECM). In: Offermanns S, Rosenthal W (eds) Encyclopedia of molecular pharmacology. Springer, Cham. https://doi.org/10.1007/978-3-030-57401-7_5691
2. Huang J, Zhang L, Wan D et al (2021) Extracellular matrix and its therapeutic potential for cancer treatment. Sig Transd Target Ther 6:153. https://doi.org/10.1038/s41392-021-00544-0
3. Naba A (2024) Mechanisms of assembly and remodelling of the extracellular matrix. Nat Rev Mol Cell Biol 25:865–885. https://doi.org/10.1038/s41580-024-00767-3
4. Zhang Y, Dong X, Zhang Y, Chen Z, Zhou G, Chen N, Shen W, Yang K, Pei P (2024) Biomaterials to regulate tumor extracellular matrix in immunotherapy. J Control Release 376:149–166. https://doi.org/10.1016/j.jconrel.2024.10.010
5. Vilaça-Faria H, Noro J, Reis RL, Pirraco RP (2024) Extracellular matrix-derived materials for tissue engineering and regenerative medicine: a journey from isolation to characterization and application. Bioactive Mater 34:494–519. https://doi.org/10.1016/j.bioactmat.2024.01.004
6. Narkar AR, Tong Z, Soman P, Henderson JH (2022) Smart biomaterial platforms: controlling and being controlled by cells. Biomaterials 283:121450. https://doi.org/10.1016/j.biomaterials.2022.121450
7. Ananth KP, Jayram ND (2023) A comprehensive review of 3D printing techniques for biomaterial-based scaffold fabrication in bone tissue engineering. Ann 3D Printed Med 13:100141. https://doi.org/10.1016/j.stlm.2023.100141
8. Aazmi A, Zhang D, Mazzaglia C, Yu M, Wang Z, Yang H, Huang YYS, Ma L (2023) Biofabrication methods for reconstructing extracellular matrix mimetics. Bioactive Mater 31:475–496. https://doi.org/10.1016/j.bioactmat.2023.08.018
9. Lienemann PS, Lutolf MP, Ehrbar M (2012) Biomimetic hydrogels for controlled biomolecule delivery to augment bone regeneration. Adv Drug Deliv Rev 64(12):1078–1089. https://doi.org/10.1016/j.addr.2012.03.010
10. Khalifa M, Albadawy M (2024) AI in diagnostic imaging: Revolutionising accuracy and efficiency. Comput Methods Prog Biomed Update 5:100146. https://doi.org/10.1016/j.cmpbup.2024.100146
11. Tan P, Chen X, Zhang H, Wei Q, Luo K (2023) Artificial intelligence aids in development of nanomedicines for cancer management. Semin Cancer Biol 89:61–75. https://doi.org/10.1016/j.semcancer.2023.01.005
12. Scaife AD (2023) Improve predictive maintenance through the application of artificial intelligence: a systematic review. Results Eng 21:101645. https://doi.org/10.1016/j.rineng.2023.101645
13. Ojaimi YA, Blin T, Lamamy J, Gracia M, Pitiot A, Denevault-Sabourin C, Joubert N, Pouget J, Gouilleux-Gruart V, Heuzé-Vourc'h N, Lanznaster D, Poty S, Sécher T (2021) Therapeutic antibodies—natural and pathological barriers and strategies to overcome them. Pharmacol Therap 233:108022. https://doi.org/10.1016/j.pharmthera.2021.108022
14. Chandra P, Soker S, Atala A (2020) Tissue engineering: current status and future perspectives. In: Elsevier eBooks, pp 1–35. https://doi.org/10.1016/b978-0-12-818422-6.00004-6

Chapter 11
Conclusion

11.1 Summary of Key Findings

The extracellular matrix (ECM) serves as a critical framework for cellular processes, providing structural support and biochemical cues that regulate cell behavior. Its complex composition, which includes proteins like collagen and elastin, along with glycosaminoglycans and proteoglycans, is tailored to the specific needs of tissues. The understanding of ECM as a dynamic entity rather than a static scaffold has transformed perspectives in cell biology, emphasizing its role in maintaining tissue integrity and facilitating communication between cells [1]. This realization has highlighted its importance in both normal physiological processes and pathological conditions. In functioning of human body, ECM orchestrates essential cellular functions such as adhesion, migration, proliferation, and differentiation. By acting as a reservoir for growth factors and providing mechanical signals, the ECM ensures a balance between tissue homeostasis and repair. The adaptability of ECM components to varying physiological demands underscores its significance in sustaining life. These dynamics are particularly evident in developmental biology, where ECM remodeling guides organogenesis and tissue specialization. Conversely, alterations in ECM composition and architecture have been identified as hallmarks of various diseases, including cancer, cardiovascular disorders, and chronic inflammatory conditions [2]. In cancer, for example, ECM stiffening and abnormal remodeling foster tumor progression and metastasis. Similarly, in cardiovascular diseases, ECM dysregulation contributes to atherosclerosis and heart failure. These insights have paved the way for ECM-based diagnostic innovations, such as biomarker identification and advanced imaging techniques, which allow earlier detection and more precise disease characterization.

The concept of ECM biomimicry has emerged as a transformative approach in medicine, aiming to replicate the native ECM's properties to develop therapeutic solutions [3]. Synthetic and natural ECM-based scaffolds have revolutionized tissue

D. Suhag, *Extracellular Matrix biomimicking for Theranostics*, Biomedical Materials for Multi-functional Applications, https://doi.org/10.1007/978-981-95-7900-6_11

engineering and regenerative medicine by providing a conducive microenvironment for cell growth and tissue repair. These advancements have enabled significant progress in treating skin injuries, cartilage defects, and even complex organs. Wound healing and drug delivery are additional fields where ECM research has made notable contributions. Mimicking ECM's unique capability to regulate growth factor release and cell migration has led to the development of biomaterials that enhance healing in challenging environments, such as radiation-impaired wounds. ECM-inspired drug delivery systems ensure targeted, sustained release, reducing side effects and improving therapeutic efficacy. Technological advancements, including 3D printing, nanotechnology, and bio fabrication, have further propelled ECM research [4]. The ability to create ECM-like structures with precision has expanded possibilities in tissue engineering, offering hope for personalized medicine and organ transplantation. Nanotechnology has enhanced understanding at a molecular level, enabling the design of nanoscale ECM materials that interact seamlessly with cells. Despite these achievements, challenges persist in translating ECM-based innovations to clinical settings. Issues such as immune responses, scalability, and regulatory approvals require attention to ensure safe and effective therapies. However, case studies and ongoing clinical trials have provided encouraging evidence of ECM's potential to address unmet medical needs, particularly in cancer therapy and chronic wound management.

The exploration of ECM dynamics has provided profound insights into its multifaceted roles in health and disease. The progress made in mimicking its properties and functions has transformed both research and clinical applications, offering a promising foundation for future innovations in healthcare. This journey underscores the ECM's vital role as a cornerstone of life sciences, bridging fundamental research with translational medicine.

Exercise 1
Fill Ups
(Regeneration, Compatibility, Structural, Bioavailability, Biochemical)

1. The extracellular matrix (ECM) plays a crucial role in regulating cellular behavior by providing _______ cues for cell adhesion, migration, and differentiation.

2. ECM mimicry in healthcare involves designing biomaterials that replicate the _______ and mechanical properties of natural ECM.

3. ECM-based biomaterials are often used in tissue engineering to promote _______ and functional integration of implants with host tissues.

4. The integration of ECM mimicry in drug delivery systems improves therapeutic outcomes by enhancing drug _______ and targeting specific tissues.

5. Mimicking the ECM environment can reduce the risk of immune rejection and improve the _______ of transplanted cells and tissues.

11.2 Impact of ECM Mimicry on Healthcare

Extracellular matrix (ECM) mimicry has emerged as a revolutionary concept in healthcare, bridging the gap between basic biology and clinical applications. By replicating the structural, mechanical, and biochemical properties of native ECM,

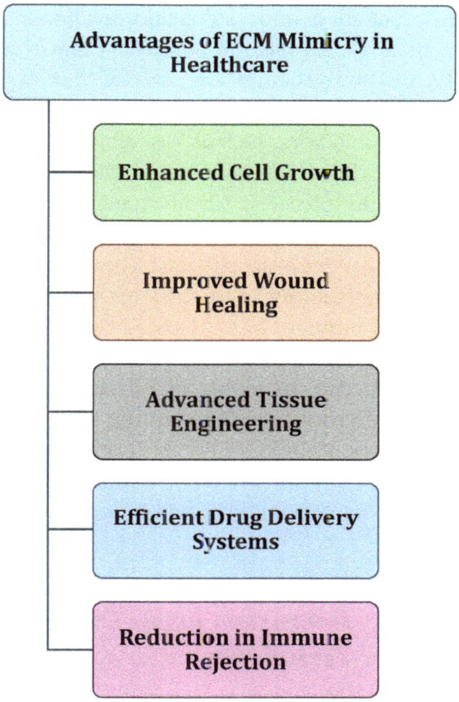

Fig. 11.1 Major advantages of ECM mimicry in healthcare

researchers and clinicians have unlocked innovative pathways to improve patient care. From tissue engineering to diagnostic advancements and personalized medicine, ECM mimicry provides solutions for some of the most challenging medical problems (Fig. 11.1) [5]. Its capacity to replicate the dynamic microenvironment of human tissues has enabled significant progress in wound healing, organ repair, disease detection, and drug delivery.

11.2.1 Contributions to Tissue Engineering and Regenerative Medicine

ECM mimicry has profoundly influenced tissue engineering and regenerative medicine, offering solutions to regenerate damaged or lost tissues. Engineered ECM scaffolds, both synthetic and natural, mimic the physical structure and biochemical signals of native ECM, creating a supportive environment for cell adhesion, migration, and proliferation. These scaffolds have shown remarkable success in applications like bone and cartilage regeneration, where traditional approaches often fall short. For example, biomimetic ECM scaffolds have been used to guide stem cell differentiation into osteogenic or chondrogenic lineages, leading to the development of bioengineered constructs that restore structural and functional properties [6]. Beyond structural support, ECM mimicry has addressed challenges in vascularization and integration of engineered tissues with host systems. The incorporation of growth factors, such as vascular endothelial growth factor (VEGF), within ECM-based scaffolds has enhanced angiogenesis, ensuring nutrient and oxygen supply to regenerating tissues. Advancements in 3D bioprinting allow precise fabrication of ECM-like structures, facilitating the creation of complex, functional tissues like skin, liver, and even heart patches. These innovations have accelerated the translation of tissue engineering solutions from the lab to clinical settings.

ECM mimicry has expanded the horizons of regenerative medicine by enabling the repair of complex tissues and organs. Biomaterials that simulate the mechanical properties of native ECM have been used in nerve regeneration and corneal repair, offering hope for restoring sensory and functional capabilities. With ongoing advancements in ECM bio fabrication, the potential for creating entire bioengineered organs for transplantation is within reach, presenting groundbreaking opportunities for treating organ failure and reducing reliance on donor tissues.

11.2.2 Improvements in Diagnostic Tools and Imaging Techniques

The application of ECM mimicry in diagnostics has revolutionized how diseases are detected and monitored. ECM's intricate involvement in disease progression has led to the identification of ECM-derived biomarkers, enabling early and precise disease detection [7]. For instance, alterations in ECM proteins, such as collagen and fibronectin, are now utilized as biomarkers for cancer and fibrosis. By replicating ECM properties in diagnostic platforms, researchers have developed tools that enhance the sensitivity and specificity of disease detection, offering clinicians actionable insights at earlier stages. Imaging techniques have also benefited significantly from ECM mimicry. Techniques like matrix-assisted laser desorption/ionization imaging mass spectrometry (MALDI-IMS) leverage ECM properties to map tissue architecture and molecular changes. Furthermore, ECM-based hydrogels have been employed as contrast agents in magnetic resonance imaging (MRI) to visualize tumor microenvironments or fibrotic tissues. These advancements not only improve diagnostic accuracy but also allow for real-time monitoring of therapeutic interventions, enabling personalized treatment approaches.

Microfluidic devices integrated with ECM-mimicking materials have become essential tools in disease modeling and drug screening. These platforms replicate the physical and biochemical cues of the ECM, providing a realistic environment to study cell behavior and interactions. By using such models, researchers can analyze disease progression and therapeutic responses at a cellular level, paving the way for innovative diagnostic strategies.

11.2.3 Innovations in Drug Delivery and Personalized Medicine

ECM mimicry has transformed drug delivery systems, offering controlled, targeted, and sustained release mechanisms. Inspired by the ECM's natural ability to store and release growth factors, researchers have developed biomaterials that encapsulate therapeutic agents and deliver them directly to affected tissues. Hydrogels, for example, are widely used in ECM-inspired drug delivery, ensuring prolonged and localized release of drugs in cancer, wound healing, and cardiovascular therapies. This reduces systemic side effects and enhances treatment efficacy, making therapies safer and more effective. In cancer treatment, ECM-mimicking materials are being utilized to overcome the challenges of drug resistance and tumor heterogeneity. By engineering biomaterials that mimic the tumor ECM, researchers can better understand drug-tumor interactions and design delivery systems that penetrate dense tumor matrices [8]. Moreover, ECM-based nanoparticles are being developed to target specific receptors in diseased tissues, ensuring precision medicine approaches that adapt to the unique molecular profiles of individual patients. ECM mimicry has also contributed to the advancement of combination therapies, where multiple drugs or bioactive agents are delivered through a single platform. For example, ECM-inspired scaffolds can release both growth factors and anti-inflammatory agents to promote healing in chronic wounds. These innovations highlight the potential of ECM mimicry to revolutionize personalized medicine, allowing treatments to be tailored to the specific needs of each patient based on their genetic and molecular profiles.

11.3 Future Directions in ECM Research

The future of extracellular matrix (ECM) research holds immense promise, driven by the need to address current limitations and explore untapped opportunities (Fig. 11.2). Innovations in biomaterials, advanced technologies like artificial intelligence (AI) and machine learning, and enhanced regulatory frameworks are paving the way for groundbreaking applications. By building on existing knowledge, researchers aim to develop ECM-based solutions that tackle unmet medical needs, improve global healthcare access, and ensure safe clinical translation. However, these advancements come with ethical and logistical challenges that require careful consideration.

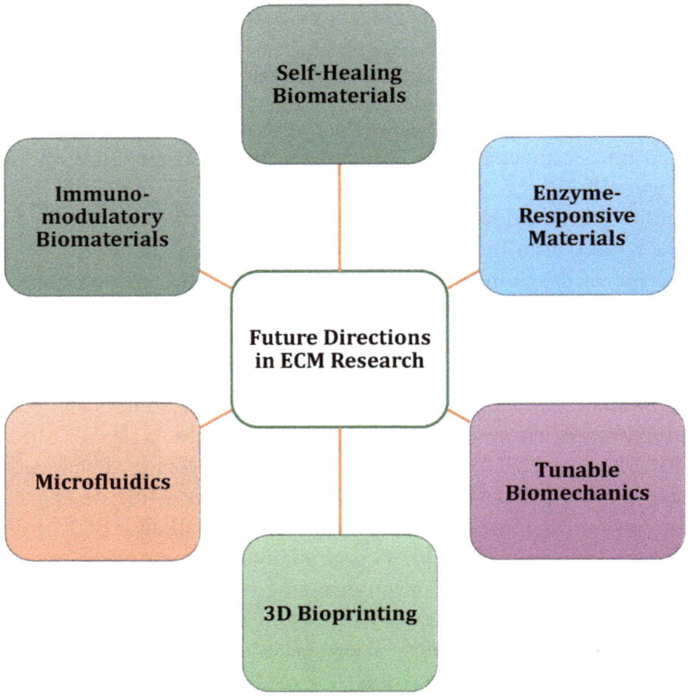

Fig. 11.2 Future Directions in ECM Research

11.3.1 Exploring Novel Biomaterials for ECM Mimicry

Developing novel biomaterials that mimic the properties of native ECM remains a cornerstone of future research. Current materials, though effective, often face challenges like limited biocompatibility, suboptimal mechanical properties, and inconsistent biodegradation rates. Researchers are now exploring hybrid biomaterials that combine synthetic polymers with natural ECM components, such as collagen, elastin, and hyaluronic acid, to enhance functionality and adaptability. These advanced materials aim to better replicate the complex architecture and biochemical cues of the ECM, providing more effective platforms for tissue engineering and regenerative medicine. The integration of nanotechnology into biomaterial development has opened new avenues for ECM mimicry. Nanostructured materials, such as electrospun nanofibers and self-assembling peptides, are designed to emulate the nanoscale organization of the ECM. These materials not only improve cell adhesion and proliferation but also allow for precise control over drug delivery and mechanical properties [9]. Furthermore, bioactive coatings and stimuli-responsive biomaterials are being investigated to create dynamic ECM environments that adapt to changing biological conditions, enhancing their therapeutic potential.

Another promising area is the use of decellularized ECM from allogeneic or xenogeneic sources, which provides a native-like scaffold for tissue regeneration. Advances in decellularization techniques aim to retain ECM's structural and functional integrity while minimizing immune responses. Coupled with gene-editing tools like CRISPR, researchers are also exploring ways to engineer ECM biomaterials with enhanced regenerative capabilities, offering personalized solutions for complex medical conditions.

11.3.2 Integration of Advanced Technologies Like AI and Machine Learning in ECM Studies

The integration of AI and machine learning into ECM research has the potential to revolutionize the field by accelerating data analysis, material design, and clinical applications. AI algorithms can analyze large datasets from genomics, proteomics, and imaging studies to uncover patterns and relationships within the ECM that were previously difficult to discern. For instance, machine learning models can predict how specific ECM alterations influence disease progression, aiding in the development of targeted therapies. In biomaterial development, AI is being utilized to design ECM-mimicking materials with optimized properties. Algorithms can simulate and predict the performance of various materials under different biological conditions, significantly reducing the time and cost of experimental trials [10]. This computational approach enables the creation of highly customized ECM scaffolds tailored to specific tissue types or diseases. Additionally, AI-driven techniques like generative design are being explored to fabricate complex ECM structures for 3D bioprinting, improving precision and reproducibility.

AI and machine learning are also transforming clinical applications of ECM-based therapies. From personalized drug delivery systems to predictive models for wound healing, these technologies offer tools to monitor patient outcomes in real-time and refine treatments accordingly. By integrating AI with ECM research, healthcare systems can achieve more efficient, data-driven solutions that improve patient care while addressing challenges in scalability and resource allocation.

11.3.3 Ethical and Regulatory Challenges in Clinical Translation

The clinical translation of ECM-based innovations faces significant ethical and regulatory challenges that must be addressed to ensure safe and equitable use. One major concern is the potential for immune reactions and long-term effects of biomaterials used in ECM mimicry. Regulatory bodies require extensive preclinical and clinical trials to assess the safety, efficacy, and biocompatibility of these materials, which can delay the approval process and increase costs. Simplifying and standardizing these procedures while maintaining rigorous safety standards is a key area of focus for future research. Ethical issues surrounding the sourcing of ECM components also demand attention. The use of animal-derived ECM materials raises concerns about sustainability, animal welfare, and potential zoonotic risks. Researchers are exploring alternatives, such as recombinant proteins and plant-based ECM analogs, to address these issues while maintaining functionality [11]. Additionally, ethical considerations related to the use of gene-editing tools and synthetic biology in ECM engineering highlight the need for transparent governance and public engagement. Another challenge lies in ensuring equitable access to ECM-based therapies across diverse populations. High costs associated with ECM biomaterials and advanced fabrication technologies can limit their availability in low-resource settings. Collaborative efforts between researchers, policymakers, and industry stakeholders are crucial to developing cost-effective solutions and implementing fair pricing models that make ECM-based healthcare innovations accessible to all.

11.3.4 Potential for Addressing Unmet Medical Needs and Global Health Challenges

ECM research offers transformative potential for addressing unmet medical needs and tackling global health challenges. In oncology, for instance, ECM-based drug delivery systems and tumor models are paving the way for more effective cancer treatments, especially for aggressive forms like triple-negative breast cancer. These innovations enable targeted therapies that minimize side effects and improve patient outcomes, addressing a critical gap in cancer care. In the field of wound healing, ECM biomaterials are being designed to treat chronic and radiation-impaired wounds, which often fail to heal with conventional methods [12]. By mimicking the native ECM's regenerative properties, these materials accelerate healing and reduce the risk of infection, offering hope to millions of patients worldwide. Additionally, ECM-inspired solutions are being explored for regenerative treatments in resource-limited settings, where access to advanced medical care is restricted. Beyond individual treatments, ECM research has implications for global health, particularly in addressing diseases linked to aging and lifestyle changes, such as cardiovascular disorders and diabetes. ECM-based therapies that restore tissue function and prevent disease progression could reduce the burden on healthcare systems and improve quality of life for aging populations. By prioritizing affordability, scalability, and sustainability, ECM innovations can play a pivotal role in achieving universal health coverage and improving global health equity.

11.4 Summary

This chapter explores the transformative potential of extracellular matrix (ECM) research, highlighting its role in advancing healthcare and addressing global medical challenges. The development of novel biomaterials, such as hybrid polymers, nanostructured materials, and decellularized ECM scaffolds, aims to replicate the intricate properties of native ECM, paving the way for breakthroughs in tissue engineering and regenerative medicine. These innovations are expected to enhance biocompatibility, functionality, and adaptability in clinical applications.

The integration of advanced technologies like artificial intelligence (AI) and machine learning has further expanded the possibilities in ECM research. AI-driven tools enable faster material design, personalized treatments, and predictive modeling of ECM-related disease progression. By leveraging these technologies, researchers can accelerate innovation and improve the precision and efficiency of ECM-based therapies. However, ethical and regulatory challenges remain a significant concern in the clinical translation of ECM-based solutions. Addressing issues such as immune reactions, sustainable sourcing of materials, and equitable access to therapies is critical to ensuring the safe and ethical application of ECM innovations. Collaborative efforts are necessary to streamline regulatory frameworks and make these technologies accessible across diverse populations. The chapter also underscores the potential of ECM research to address unmet medical needs and global health challenges. From developing advanced cancer treatments to enhancing wound healing and regenerative therapies, ECM mimicry offers hope for tackling complex medical conditions. By focusing on affordability, sustainability, and scalability, ECM innovations can significantly contribute to improving healthcare equity and quality worldwide. The future of ECM research lies in its ability to combine cutting-edge science with ethical and practical considerations, delivering transformative solutions for modern medicine and global health.

Exercise 2
Multiple Choice Questions

1. Which diagnostic imaging technique is commonly used to study the structural properties of ECM in tissues?
a) MRI
b) X-ray
c) PET
d) Ultrasound

2. What is the primary goal of developing biomaterials for ECM mimicry?
a) To increase the durability of medical implants
b) To replicate the structure and function of natural ECM
c) To enhance imaging resolution
d) To improve drug solubility

3. How does AI contribute to ECM-related healthcare advancements?
a) By automating surgical procedures
b) By predicting ECM behavior under various conditions
c) By designing patient-specific ECM scaffolds
d) Both b and c

4. Machine learning models in ECM research are primarily used to:
a) Analyze complex datasets
b) Perform minimally invasive surgeries
c) Develop new biomaterials
d) Replace manual imaging techniques

5. A major ethical challenge in ECM mimicry research is:
a) Limited availability of imaging tools
b) High cost of biomaterials
c) Patient consent for experimental procedures
d) Lack of AI integration

6. Which of the following is a regulatory challenge in the clinical translation of ECM-mimicking technologies?
a) Standardizing material composition
b) Obtaining FDA approval for new biomaterials
c) Ensuring reproducibility of results
d) All of the above

7. What novel material is commonly explored for ECM mimicry due to its biocompatibility?
a) Titanium
b) Polycaprolactone (PCL)
c) Silicone
d) Gold nanoparticles

8. Which of the following advancements can improve imaging for ECM studies?
a) Combining MRI with AI for better analysis
b) Increasing the resolution of X-rays
c) Using low-energy ultrasound
d) Limiting imaging to 2D planes

Points to Remember

- **ECM biomimicry refers to replicating the complex architecture and biochemical signals of the native extracellular matrix to support cellular functions, tissue repair, and regeneration. This approach is crucial in bridging the gap between biological complexity and artificial scaffolds.**

- **Natural materials like collagen, fibrin, and hyaluronic acid serve as excellent ECM mimics due to their innate bioactivity and compatibility. However, challenges such as immunogenicity and inconsistent mechanical properties limit their widespread use.**

- **Synthetic polymers like polylactic acid (PLA) and polyethylene glycol (PEG) offer tunable mechanical and degradation properties. These scaffolds can be engineered for specific applications but may lack the bioactivity of natural materials.**

- **Hybrid scaffolds combine natural and synthetic materials to merge bioactivity with mechanical robustness. For example, hydrogels enriched with growth factors and nanoparticles provide both structural support and therapeutic functionality.**

- **Hydrogels are widely used due to their high-water content, biocompatibility, and ability to encapsulate cells or bioactive molecules. They mimic the ECM environment and are particularly valuable in wound healing and drug delivery.**

References

1. Frantz C, Stewart KM, Weaver VM (2010) The extracellular matrix at a glance. J Cell Sci 123(Pt 24):4195–4200. https://doi.org/10.1242/jcs.023820
2. Yue B (2014) Biology of the extracellular matrix: an overview. J Glaucoma 23(8 Suppl 1):S20–S23. https://doi.org/10.1097/IJG.0000000000000108
3. Pellowe AS, Gonzalez AL (2016) Extracellular matrix biomimicry for the creation of investigational and therapeutic devices. Wiley Interdisc Rev. Nanomed Nanobiotechnol 8(1):5–22. https://doi.org/10.1002/wnan.1349
4. Hu M, Ling Z, Ren X (2022) Extracellular matrix dynamics: tracking in biological systems and their implications. J Biol Eng 16(1):13. https://doi.org/10.1186/s13036-022-00292-x
5. Kim Y, Ko H, Kwon IK, Shin K (2016) Extracellular matrix revisited: roles in tissue engineering. Int Neurourol J 20(Suppl 1):S23–S29. https://doi.org/10.5213/inj.1632600.318
6. Chen G, Kawazoe N (2018) Biomimetic extracellular matrices and scaffolds prepared from cultured cells. Adv Exp Med Biol 1078:465–474. https://doi.org/10.1007/978-981-13-0950-2_24
7. Mohammed Butt A, Rupareliya V, Hariharan A, Kumar H (2023) Building a pathway to recovery: targeting ECM remodeling in CNS injuries. Brain Res 1819:148533. https://doi.org/10.1016/j.brainres.2023.148533
8. Hwang J, Sullivan MO, Kiick KL (2020) Targeted drug delivery via the use of ECM-mimetic materials. Front Bioeng Biotechnol 8:69. https://doi.org/10.3389/fbioe.2020.00069
9. Setiawati A, Nguyen HT, Jung Y, Shin K (2018) Future research directions in the design of versatile extracellular matrix in tissue engineering. Int Neurourol J 22(Suppl 2):S66–S75. https://doi.org/10.5213/inj.1836154.077

10. Greenberg ZF, Graim KS, He M (2023) Towards artificial intelligence-enabled extracellular vesicle precision drug delivery. Adv Drug Deliv Rev 199:114974. https://doi.org/10.1016/j.addr.2023.114974
11. Capella-Monsonís H, Crum RJ, Hussey GS, Badylak SF (2024) Advances, challenges, and future directions in the clinical translation of ECM biomaterials for regenerative medicine applications. Adv Drug Deliv Rev 211:115347. https://doi.org/10.1016/j.addr.2024.115347
12. Huang R, Zhu J, Fan R, Tang Y, Hu L, Lee H, Chen S (2024) Extracellular vesicle-based drug delivery systems in cancer. Extracell Vesicle 4:100053. https://doi.org/10.1016/j.vesic.2024.100053

Resources and Further Reading

Chapter 1: Introduction to Extracellular Matrix (ECM) Dynamics

This chapter provides a foundational overview of the extracellular matrix (ECM), highlighting its structural components, physical properties, and biological significance. It sets the stage for understanding ECM's role in cellular communication, tissue organization, and overall microenvironment behavior.

Textbooks

1. "Extracellular Matrix Biology" by Richard O. Hynes and Kenneth M. Yamada
   - A detailed resource that explores the molecular structure and function of ECM in tissue biology and development.
2. "Tissue Engineering" by Bernhard O. Palsson and Sangeeta N. Bhatia
   - Covers ECM architecture and its role in scaffolding, cell adhesion, and regenerative medicine.
3. "Molecular Biology of the Cell" by Alberts et al.
   - Includes comprehensive sections on ECM structure, integrins, and cell–matrix interactions in health and disease.

Journals

1. Matrix Biology
   - A leading journal focusing on ECM structure, function, and its implications in physiology and pathology.
2. Journal of Cell Science
   - Regularly publishes articles on ECM biology and cell–matrix interactions.

D. Suhag, *Extracellular Matrix biomimicking for Theranostics*, Biomedical Materials for Multi-functional Applications, https://doi.org/10.1007/978-981-95-7900-6

3. Nature Reviews Molecular Cell Biology
    - Offers high-impact reviews on ECM remodeling, signaling, and tissue homeostasis.

Online Resources

1. Matrisome Project (matrisome.org)
    - An online database cataloging ECM-related genes and proteins across different species.
2. Nature Collection: Extracellular Matrix
    - Curated collection of ECM research, available through Nature's website.

Research Papers

1. "The extracellular matrix: not just pretty fibrils" by Mina J. Bissell and Derrick J. Stetler-Stevenson
    - A classic review on how ECM governs cell function beyond structure.
2. "Dynamic reciprocity in the wound microenvironment" by D. W. Gilkes et al.
    - Discusses bidirectional communication between cells and ECM during physiological changes.

Professional Organizations and Conferences

1. International Society for Matrix Biology (ISMB)
    - Promotes ECM-related research and hosts biennial international matrix biology meetings.
2. American Society for Matrix Biology (ASMB)
    - Provides training, workshops, and funding opportunities for ECM researchers.

Further Readings

1. "Extracellular Matrix and the Cell Surface" by Klaus Kühn
    - Explores molecular mechanisms of ECM-cell surface interactions.
2. "Cell-Extracellular Matrix Interactions in Cancer" edited by Peter P. Lee
    - Focuses on ECM's role in cancer progression and microenvironment remodeling.
3. "Biomimetic Approaches to ECM Research" by Thomas Seufferlein
    - Discusses experimental models and biomaterials that mimic ECM behavior.

Online Courses

1. Coursera—"Cell Biology: Mitochondria and Signaling" (University of California, San Diego)
    - Includes modules on ECM-cell communication and signaling.
2. edX—"Tissue Engineering" (Columbia University)
    - Covers ECM fundamentals as part of its tissue regeneration focus.

Chapter 2: Cellular Interactions with the ECM

This chapter delves into how cells engage with the extracellular matrix (ECM) through integrins, cadherins, and other adhesion molecules. It explores signaling cascades, mechanical transduction, and dynamic feedback between the cell cytoskeleton and the ECM, which regulate cell fate, motility, and tissue integrity.

Textbooks

1. "Molecular Biology of the Cell" by Alberts et al.
    - Offers comprehensive explanations of cell-ECM interactions, integrins, focal adhesions, and signal transduction pathways.
2. "Cell Biology" by Thomas D. Pollard, William C. Earnshaw, and Jennifer Lippincott-Schwartz
    - Focuses on cytoskeletal dynamics and cellular adhesion mechanisms with detailed ECM interaction insights.
3. "Cell-Extracellular Matrix Interactions in Cancer" edited by Peter P. Lee
    - Provides a disease-focused perspective on cellular communication with the ECM.

Journals

1. Journal of Cell Biology
    - Features in-depth studies on cell adhesion, migration, and ECM-related signaling.
2. Developmental Cell
    - Publishes high-impact research on developmental processes involving cell-ECM crosstalk.
3. FASEB Journal
    - Covers cell physiology and ECM interactions in development, aging, and disease.

Online Resources

1. The Cell Image Library (cellimagelibrary.org)
   - A curated repository of annotated cellular structures including ECM interactions.
2. Nature Education—Scitable: "Cell-ECM Adhesions"
   - Educational articles explaining ECM contact points and associated cell signaling.

Research Papers

1. "Integrins and cell signaling" by Michael H. Ginsberg, A. Partridge, and J. E. Hemler (Current Opinion in Cell Biology, 2005)
   - An influential paper on integrin-mediated cell-ECM interactions.
2. "Mechanical control of tissue morphogenesis through cytoskeletal tension" by E. Hannezo and B. L. Goode
   - Discusses the biomechanical aspect of cell-ECM interactions.

Professional Organizations and Conferences

1. American Society for Cell Biology (ASCB)
   - Offers workshops and symposia on cell-ECM interactions.
2. European Cell Biology Organization (ECBO)
   - Provides grants and publishes articles on ECM-cell interface studies.

Further Readings

1. "Mechanobiology Handbook" edited by Jiro Nagatomi
   - Explores the mechanotransduction mechanisms in cell-ECM interaction.
2. "Integrins in Development" edited by Erik H. Danen
   - Reviews how integrin-ECM binding governs embryogenesis and differentiation.
3. "Cellular Biomechanics and Mechanobiology" by Amit Gefen
   - Covers the forces and cellular responses in ECM-mediated environments.

Online Courses

1. Coursera—"Cell Biology: Signaling and Cytoskeleton" (University of Geneva)
   - Explains cytoskeletal rearrangements and ECM signal reception.
2. edX—"Mechanobiology of Cells" (EPFL)
   - A specialized course on cellular response to ECM-derived mechanical cues.

Chapter 3: Mechanisms of ECM Remodelling

This chapter focuses on the dynamic remodeling of the extracellular matrix (ECM) by enzymes, mechanical forces, and cellular activity. It discusses the roles of matrix metalloproteinases (MMPs), tissue inhibitors of metalloproteinases (TIMPs), and the balance between ECM synthesis and degradation. The chapter also explores the implications of ECM remodeling in development, tissue repair, fibrosis, and cancer progression.

Textbooks

1. "Extracellular Matrix Biology" edited by Richard O. Hynes and Kenneth M. Yamada
   - A foundational text detailing ECM composition, function, and remodeling mechanisms.
2. "Matrix Metalloproteinase Biology" by Rafael Fridman and Bonnie F. Sloane
   - Offers comprehensive coverage of MMP structure, function, and role in ECM turnover.
3. "Tissue Engineering" by Clemens Van Blitterswijk
   - Explain how ECM remodeling supports engineered tissue development and integration.

Journals

1. Matrix Biology
   - Specializes in ECM composition, structure, signaling, and remodeling processes.
2. The Journal of Cell Science
   - Publishes research on ECM degradation, integrin signaling, and cell–matrix interactions.
3. American Journal of Pathology
   - Contains studies on ECM remodeling in disease pathology and regeneration.

Online Resources

1. Nature Scitable—"The Extracellular Matrix and Cell Signaling"
   - Educational overview of how remodeling influences cell behavior.
2. EMBL-EBI Matrix Database
   - Annotated datasets of ECM proteins and their remodeling enzymes.

Research Papers

1. "Matrix metalloproteinases: regulators of the tumor microenvironment" by Egeblad and Werb (Cell, 2002)
   - Landmark paper detailing MMPs' role in tumor-associated ECM remodeling.
2. "Dynamic remodeling of the extracellular matrix underlies morphogenesis" by Rozario and DeSimone (Nature Reviews Molecular Cell Biology, 2010)
   - Reviews ECM remodeling during development and tissue architecture formation.

Professional Organizations and Conferences

1. International Society for Matrix Biology (ISMB)
   - Promotes research on ECM composition and remodeling with biennial congresses.
2. American Association for Cancer Research (AACR)
   - Offers sessions and workshops on ECM remodeling in cancer.

Further Readings

1. "ECM Proteolysis in Tissue Remodelling and Disease" by Roy Zent and Ambra Pozzi
   - Covers proteolytic enzymes and their regulation during physiological and pathological remodeling.
2. "Cell–Matrix Junctions" edited by Carole A. Parent
   - Explores junctional complexes and remodeling-related mechanisms.
3. "Fibrosis: Causes, Consequences and Therapeutic Strategies" edited by T. Wynn and D. Sheppard
   - Discusses how ECM remodeling drives fibrotic disease.

Online Courses

1. Coursera—"The Biology of Tissue Repair" (University of Manchester)
   - Includes modules on ECM degradation, remodeling, and tissue regeneration.
2. Future Learn—"Understanding Human Cell and Tissue Biology"
   - Covers basic and applied ECM remodeling concepts.

Chapter 4: ECM in Health and Disease

This chapter explores how the extracellular matrix (ECM) contributes to normal tissue homeostasis and how its dysregulation can lead to various diseases. It discusses the role of ECM in tissue integrity, wound healing, immune responses, and its alterations in conditions like fibrosis, cancer, and cardiovascular diseases. Understanding these dynamics is crucial for developing ECM-targeted diagnostics and therapies.

Textbooks

1. "The Extracellular Matrix: An Overview" by Francesco DeMayo and John McLachlan
   - Provides insight into the physiological roles of ECM and its implications in disease.
2. "Fibrosis: Causes, Consequences and Therapeutic Strategies" edited by T. Wynn and D. Sheppard
   - Covers ECM remodeling in chronic diseases such as liver cirrhosis, pulmonary fibrosis, and kidney disease.
3. "Cancer and the Extracellular Matrix" edited by Roy Zent and Ambra Pozzi
   - Offers detailed perspectives on how ECM influences cancer progression and metastasis.

Journals

1. The Journal of Pathology
   - Publishes articles on ECM's involvement in both health and disease pathophysiology.
2. Nature Reviews Disease Primers
   - Contains primer reviews on diseases with strong ECM-related etiologies.
3. Trends in Molecular Medicine
   - Features research on ECM signaling in inflammation, cancer, and regenerative biology.

Online Resources

1. National Center for Biotechnology Information (NCBI)—ECM and Disease Topic Collection
   - A curated set of articles on the roles of ECM in disease mechanisms.
2. ECM Atlas (https://ecmatlas.org)
   - A resource on ECM proteins in different tissues and pathological states.

Research Papers

1. "The extracellular matrix: not just pretty fibrils" by Hynes RO (Science, 2009)
   - A foundational paper discussing ECM's roles in development and disease.
2. "ECM remodeling in health and disease: implications for tissue engineering" by Theocharis et al. (Advanced Drug Delivery Reviews, 2016)
   - Reviews how ECM remodeling is both essential and pathological.

Professional Organizations and Conferences

1. European Matrix Biology Organization (EMBO Matrix Club)
   - Organizes symposia and resources focusing on ECM in health and disease.
2. American Society for Matrix Biology (ASMB)
   - Offers webinars and meetings dedicated to ECM dysfunction in disease.

Further Readings

1. "Matrix Biology in Regenerative Medicine and Disease" by Paulette Spencer and Jeffrey W. Holmes
   - Discusses ECM's structural and signaling roles in organ-specific diseases.
2. "Tumor Microenvironment and Cellular Stress" edited by Rebecca G. Bagley
   - Highlights ECM as a central component of the tumor microenvironment.
3. "The Biology of the Extracellular Matrix: Regulation of Homeostasis and Remodeling" edited by Monica Neagu
   - Reviews regulatory mechanisms governing ECM balance and its disruption.

Online Courses

1. edX—"Cell Biology: Mitochondria, Cytoskeleton and ECM" by Kyoto University
   - Includes a module on ECM roles in tissue physiology and pathology.
2. Coursera—"Mechanisms of Human Disease" by University of Zurich
   - Features a segment on ECM remodeling in degenerative and inflammatory disorders.

Chapter 5: Biomimicry of ECM in Medical Applications

This chapter explores how the principles of ECM structure and function are replicated in synthetic or natural materials for therapeutic use. From hydrogels and scaffolds to engineered matrices, ECM biomimicry has revolutionized tissue engineering, wound healing, and regenerative medicine. This chapter provides insight into material selection, design considerations, and translational applications of ECM-inspired systems.

Textbooks

1. "Biomimetic Biomaterials: Structure and Applications" edited by Andrew S. H. Wong and Jun Ma
   - A detailed look at designing biomaterials based on ECM architecture and function.
2. "Tissue Engineering" by Bernhard O. Palsson and Sangeeta N. Bhatia
   - Discusses ECM mimics for regenerative medicine and cellular therapy.
3. "Principles of Tissue Engineering" by Robert Lanza, Robert Langer, and Joseph Vacanti
   - Covers biomaterials and scaffold designs that mimic ECM for clinical applications.

Journals

1. Biomaterials
   - Publishes extensive research on ECM-inspired materials and their biomedical applications.
2. Advanced Healthcare Materials
   - Focuses on the intersection of material science and biology, including biomimetic ECM scaffolds.
3. Biofabrication
   - Features articles on bioprinting and fabrication of ECM-mimicking structures.

Online Resources

1. Nature Materials—ECM Biomimicry Collection
   - A curated selection of high-impact articles on ECM mimetic materials.
2. Biomaterials Network (Biomat.net)
   - Offers resources and news related to ECM-based innovations in biomedicine.

Research Papers

1. "ECM-mimetic hydrogels for tissue engineering" by Tibbitt MW and Anseth KS (Biotechnology and Bioengineering, 2009)
   - A foundational article describing synthetic hydrogels that replicate ECM properties.

2. "Designing bioactive ECM-mimetic materials for regenerative medicine" by Salinas CN and Anseth KS (Progress in Materials Science, 2010)
   - Discusses principles for creating ECM-mimetic materials with biological cues.

Professional Organizations and Conferences

1. Tissue Engineering and Regenerative Medicine International Society (TERMIS)
   - Hosts conferences and publishes findings on ECM-inspired biomaterials.
2. Society For Biomaterials (SFB)
   - Includes focus groups on ECM mimicry and scaffold technologies.

Further Readings

1. "Bioinspired Materials for Medical Applications" edited by Lijie Grace Zhang and Ali Khademhosseini
   - A compilation of biomimetic technologies for various biomedical purposes.
2. "Engineering Biomaterials for Regenerative Medicine" by Gordana Vunjak-Novakovic and R. Langer
   - Focus on design strategies for ECM-mimetic biomaterials.
3. "Extracellular Matrix-derived Biomaterials in Tissue Engineering" by Yao Yao and Rui L. Reis (Trends in Biotechnology, 2021)
   - Reviews recent advancements in decellularized ECM-based scaffolds.

Online Courses

1. Coursera—"Biomaterials for Health" by École Polytechnique Fédérale de Lausanne
   - Includes modules on ECM-based scaffold engineering.
2. MIT OpenCourseWare—"Tissue Engineering"
   - Offers downloadable lectures and readings on biomimetic scaffold design.

Chapter 6: ECM in Diagnostic Innovations

This chapter focuses on how the extracellular matrix (ECM) contributes to emerging diagnostic technologies. By leveraging ECM components and their interactions with cells, scientists are developing biosensors, imaging agents, and lab-on-chip systems that can detect diseases at early stages. The chapter discusses innovations in ECM-based biomarkers, microfluidics, and imaging techniques that enhance diagnostic accuracy.

Textbooks

1. "Extracellular Matrix in Development and Disease" edited by Jeffrey H. Miner and Douglas S. Darling
   - Offers in-depth insight into the diagnostic implications of ECM changes during disease progression.
2. "Biomedical Applications of Hydrogels Handbook" edited by Raphael M. Ottenbrite, Kinam Park, and Teruo Okano
   - Discusses the use of ECM-mimicking hydrogels in biosensing and diagnostics.
3. "Principles of Bionanotechnology" by David Goodsell
   - Covers nanotechnological tools for exploiting ECM components in diagnostics.

Journals

1. ACS Sensors
   - Publishes innovative studies on ECM-based biosensors and diagnostic systems.
2. Lab on a Chip (RSC Publishing)
   - Features articles on microfluidic platforms and ECM-integrated diagnostics.
3. Biosensors and Bioelectronics
   - Covers advances in ECM-inspired sensing platforms for disease detection.

Online Resources

1. Nature Reviews Molecular Cell Biology—ECM and Diagnostics Section
   - Curated articles exploring ECM as a diagnostic target.
2. NIH ExRNA Atlas & ECM Biomarker Programs
   - U.S. government initiative profiling ECM-associated extracellular RNAs and biomarkers.

Research Papers

1. "The ECM as a biomarker reservoir" by Theocharis AD et al., *Nature Reviews Cancer, 2016*
   - A comprehensive review on how ECM molecules serve as biomarkers for cancer and other diseases.
2. "Label-free detection of ECM changes in cancer" by Wang J et al., *Biosensors and Bioelectronics, 2021*

- Demonstrates how sensor platforms detect ECM changes for diagnostic purposes.

3. "Integration of ECM cues into diagnostic microfluidic chips" by Nguyen T et al., *Lab Chip, 2020*
   - Details use of ECM in developing personalized lab-on-chip diagnostics.

Professional Organizations and Conferences

1. IEEE Engineering in Medicine and Biology Society (EMBS)
   - Sponsors sessions and publishes work on biosensors and ECM-influenced diagnostics.
2. Society for Laboratory Automation and Screening (SLAS)
   - Focus on high-throughput ECM-based diagnostic platforms.

Further Readings

1. "Biomarkers and Biosensors: Detection and Molecular Approaches" by Ajit Sadana and Neeti Sadana
   - A technical overview of biosensor mechanisms, including ECM-based applications.
2. "Smart Biosensor Technology" by George K. Knopf and Amarjeet S. Bhalla
   - Discusses biosensors that respond to ECM-related biochemical changes.
3. "Molecular and Cellular Probes of the ECM" by Patel VN and Kessenbrock K, *Trends in Biotechnology, 2019*
   - Explores ECM visualization and probing methods used in diagnostics.

Online Courses

1. edX—"Biosensors for Health Monitoring" by EPFL
   - Introduces biosensors including those interacting with ECM molecules.
2. Coursera—"Biomedical Imaging" by the University of Queensland
   - Offers insights into ECM-targeted imaging techniques used for diagnostics.

Chapter 7: Therapeutic Applications of ECM Mimicry

This chapter explores how mimicking the extracellular matrix (ECM) enables advanced therapeutic strategies across regenerative medicine, oncology, and tissue repair. By designing scaffolds and materials that replicate the structural and biochemical cues of natural ECM, researchers can enhance drug delivery, promote cell regeneration, and target pathological microenvironments more effectively.

Textbooks

1. "Tissue Engineering" by Bernhard O. Palsson and Sangeeta N. Bhatia
   - Offers detailed concepts of tissue engineering and therapeutic scaffolds mimicking ECM.
2. "Biomimetic Materials for Tissue Regenerations" edited by Pankaj Vadgama
   - Focuses on material science for ECM-mimetic therapies in various tissues.
3. "Regenerative Engineering: Translating Tissue Engineering Principles to Biomedical Applications" by Yusuf Khan and Cato T. Laurencin
   - Discusses clinical translation of ECM-inspired regenerative platforms.

Journals

1. Tissue Engineering Part A, B, and C
   - Leading source of articles on engineered ECM scaffolds and regenerative therapies.
2. Advanced Healthcare Materials
   - Publishes cutting-edge studies on ECM-mimicking biomaterials for therapeutic use.
3. Journal of Controlled Release
   - Features research on ECM-based drug delivery strategies.

Online Resources

1. NIH Regenerative Medicine Program
   - Resource hub for ECM-related therapeutic applications and funding initiatives.
2. Nature Portfolio: ECM and Tissue Engineering
   - Collection of papers and commentaries on ECM therapeutic biomaterials.

Research Papers

1. "Mimicking the ECM for regenerative medicine and cancer therapy" by Chaudhuri O. et al., *Nature Materials, 2020*
   - Examines ECM biomimicry in mechanobiology and disease-targeted therapies.
2. "Engineering ECM-mimetic materials for enhanced tissue repair" by Wang X. et al., *Advanced Functional Materials, 2021*
   - Demonstrates how ECM-inspired materials improve healing outcomes.

3. "Biomimetic matrices for therapeutic cell delivery" by Place ES et al., *Nature Materials, 2009*
   - Classic paper discussing the therapeutic potential of ECM scaffolds.

Professional Organizations and Conferences

1. TERMIS (Tissue Engineering and Regenerative Medicine International Society)
   - Premier organization supporting ECM-based therapeutic development.
2. Regenerative Medicine Foundation (RMF)
   - Promotes translation of ECM-based therapies into clinical practice.

Further Readings

1. "Biofabrication and 3D Tissue Modeling" by Gilson Khang
   - Provides insight into ECM-inspired platforms for therapeutic engineering.
2. "Smart Scaffolds for Tissue Engineering and Regenerative Medicine" by Masoud Mozafari
   - Discusses multifunctional ECM-like scaffolds in clinical applications.
3. "Matrix Biology in Regenerative Medicine and Drug Delivery" edited by Lanza R. and Vacanti J.
   - Covers therapeutic implications of ECM biology and its mimicry.

Online Courses

1. FutureLearn—"Biomaterials and Stem Cells in Regenerative Medicine" by University of Sheffield
   - Covers ECM scaffolds and therapeutic materials in depth.
2. edX—"Tissue Engineering for Regenerative Medicine" by Columbia University
   - Offers practical and theoretical approaches to ECM-mimicking therapies.

Chapter 8: Advanced Technologies in ECM Mimicry

This chapter highlights state-of-the-art technologies used to design and fabricate extracellular matrix (ECM) mimetic platforms. These technologies include 3D bioprinting, electrospinning, microfluidics, and nanotechnology, which are revolutionizing the way ECM architecture and function are replicated for both in vitro and in vivo applications.

Textbooks

1. "3D Bioprinting and Nanotechnology in Tissue Engineering and Regenerative Medicine" by Lijie Grace Zhang, John P. Fisher, and Kam Leong
   - A definitive guide to combining advanced manufacturing with ECM mimicry.
2. "Electrospinning: Principles, Practice and Possibilities" by Shilpa Sant and Murugan Ramalingam
   - Explores how electrospun scaffolds replicate ECM nanostructures.
3. "Micro and Nanotechnologies in Engineering Stem Cells and Tissues" edited by Murugan Ramalingam
   - Discusses microfabrication technologies applied to biomimetic ECM.

Journals

1. Biofabrication
   - Dedicated to publishing research on advanced technologies like bioprinting and ECM fabrication.
2. Advanced Materials
   - Features high-impact papers on ECM-inspired materials and technologies.
3. ACS Nano
   - Regularly covers nanotech-based strategies for ECM mimicry.

Online Resources

1. NIH 3D Bioprinting Initiative
   - Offers protocols, research updates, and training resources on ECM-focused bioprinting.
2. Electrospinning Co. UK
   - A hub for electrospinning techniques and scaffold development for ECM research.

Research Papers

1. "3D bioprinting of tissue-engineered constructs for in vitro modeling and regenerative medicine" by Murphy SV, Atala A., *Nature Biotechnology, 2014*
   - A foundational paper on how bioprinting replicates complex ECM architectures.
2. "Microfluidic platforms for mechanobiology and ECM interaction studies" by Polacheck WJ et al., *Lab on a Chip, 2013*

- Describes the role of microfluidics in mimicking ECM microenvironments.

3. "Electrospun nanofibers for regenerative medicine and ECM mimicry" by Bhardwaj N, Kundu SC, *Biotechnology Advances, 2010*
   - Demonstrates how nanofibrous scaffolds act as ECM surrogates.

Professional Organizations and Conferences

1. Biofabrication (International Society for Biofabrication - ISBF)
   - Supports interdisciplinary work in ECM mimicry via advanced fabrication methods.
2. IEEE Engineering in Medicine and Biology Society (EMBS)
   - Hosts symposia on biofabrication and ECM technology development.

Further Readings

1. "Nanostructured Biomaterials for Regenerative Medicine" edited by Vincenzo Guarino
   - Covers fabrication strategies mimicking ECM at the nanoscale.
2. "Bioprinting in Regenerative Medicine" by Kursad Turksen
   - Focus on ECM replication using 3D printing for medical applications.
3. "Smart Polymers and Their Applications" by Maria Rosa Aguilar and Julio San Román
   - Discusses stimulus-responsive materials relevant for ECM mimicry.

Online Courses

1. Coursera—"3D Printing Applications" by University of Illinois
   - Includes modules on bioprinting and its role in tissue engineering.
2. edX—"Nanotechnology and Nanosensors" by Technion—Israel Institute of Technology
   - Provides insights into nano-enabled ECM scaffold development.

Chapter 9: Clinical Case Studies and Trials

This chapter presents real-world applications of ECM biomimicking technologies in clinical settings. It explores translational research outcomes, patient-centered studies, preclinical models, and ongoing clinical trials that assess the safety, efficacy, and regenerative potential of ECM-mimetic therapies in wound healing, cancer, cardiovascular diseases, and orthopedic applications.

Textbooks

1. "Translational Regenerative Medicine" by Anthony Atala and Julie Allickson
   - Offers case studies of ECM-based therapies transitioning from bench to bedside.
2. "Tissue Engineering and Regenerative Medicine: A Nano Approach" by Murugan Ramalingam and Seeram Ramakrishna
   - Includes sections on clinical testing and regulatory aspects of ECM scaffolds.
3. "Principles of Regenerative Medicine" by Anthony Atala, Robert Lanza, James A. Thomson, and Robert Nerem
   - Comprehensive overview including clinical translation of ECM biomaterials.

Journals

1. Regenerative Medicine
   - Publishes studies and trials focused on clinical use of ECM-mimetic technologies.
2. Journal of Translational Medicine
   - Features case-based studies of ECM-inspired therapies in patient populations.
3. Stem Cells Translational Medicine
   - Reports on clinical outcomes involving ECM scaffolds and stem cell integration.

Online Resources

1. ClinicalTrials.gov
   - A searchable database of global clinical trials involving ECM and related biomaterials.
2. NIH National Center for Advancing Translational Sciences (NCATS)
   - Offers information on translational studies including ECM applications.

Research Papers

1. "Clinical translation of stem cell therapy for cardiovascular disease: a focus on ECM scaffolds" by Madonna R et al., *European Heart Journal, 2016*
   - Evaluates ECM-scaffold-integrated therapies in cardiac tissue repair.
2. "Decellularized matrix scaffolds in regenerative medicine: Clinical applications and challenges" by Crapo PM et al., *Biomaterials, 2011*
   - A comprehensive review of ECM-derived scaffolds in clinical practice.

3. "Use of ECM-mimetic scaffolds in wound care and chronic ulcer treatment: A clinical perspective" by Greaves NS et al., *Advances in Wound Care, 2015*
   - Real-world clinical data on ECM-based dressings and templates.

Professional Organizations and Conferences

1. International Society for Stem Cell Research (ISSCR)
   - Hosts sessions on clinical trials involving ECM mimics and regenerative therapy.
2. Tissue Engineering and Regenerative Medicine International Society (TERMIS)
   - Focused on clinical implementation and trial data related to ECM biomaterials.
3. World Regenerative Medicine Congress
   - A premier platform for presenting case studies and clinical breakthroughs.

Further Readings

1. "Clinical Translation of Tissue Engineering Products" by Mona Elsayed and Moustapha Kassem
   - Discusses regulatory, clinical, and ethical issues in ECM-based product deployment.
2. "Decellularized Scaffolds and Organogenesis" by Laura Niklason
   - Explores clinical implementation of ECM-mimicking decellularized scaffolds.
3. "Clinical Trials in Regenerative Medicine" edited by David Warburton
   - Offers curated studies and outcome analyses of ECM applications in human trials.

Online Courses

1. edX—"Clinical Trials Management and Advanced Operations" by University of Rochester
   - Covers the design and execution of clinical studies involving biomimetic systems.
2. Coursera—"Translational Research: Bench to Bedside" by University of Michigan
   - Learn the full pipeline of ECM-based product development and clinical trials.

Chapter 10: Future Perspectives and Innovations

This chapter explores the future landscape of ECM biomimicking, focusing on emerging trends like biofabrication, 4D bioprinting, synthetic biology, AI-integrated design, and next-generation theranostics. It highlights innovations that will redefine regenerative medicine, personalized therapy, and smart biosensing platforms driven by ECM-inspired strategies.

Textbooks

1. "4D Printing: Fundamentals, Materials, and Applications" by Md. Ibrahim H. Mondal
    - Introduces shape-morphing smart biomaterials including ECM-mimetic applications.
2. "Synthetic Biology: A Primer" by Geoff Baldwin, Darren N. Nesbeth, and Isak S. Pretorius
    - Describes engineered biological systems that can mimic or regulate ECM environments.
3. "Biofabrication: Micro- and Nanofabrication, Printing, Patterning and Assemblies for Bioengineering, and Cell and Tissue Applications" by Gabor Forgacs and Thomas Boland
    - A deep dive into future bioprinting techniques relevant to ECM mimicry.

Journals

1. Advanced Healthcare Materials
    - Covers next-gen ECM-inspired materials and smart platforms.
2. Nature Biomedical Engineering
    - Publishes cutting-edge innovations and predictive technologies for ECM applications.
3. Trends in Biotechnology
    - Features foresight articles on bioengineered ECM systems and future diagnostics.

Online Resources

1. Future Science Group—Regenerative Medicine Gateway
    - Offers news and forecasts on ECM-focused research and commercial platforms.
2. Singularity Hub—Biotech Futures

- Tracks developments in synthetic ECM systems, AI, and tissue interface technologies.

Research Papers

1. "Smart biomaterials—synthetic biology meets bioengineering" by Appiah C et al., *Nature Reviews Materials, 2021*
   - Discusses integration of smart sensing with ECM mimicry.
2. "4D bioprinting for dynamic ECM-based scaffolds" by Kirillova A et al., *Advanced Materials, 2017*
   - Breakthrough in programmable ECM-mimetic hydrogels.
3. "Artificial extracellular matrices for the next generation of personalized medicine" by Zhang D et al., *Trends in Pharmacological Sciences, 2020*
   - Focus on patient-specific ECM innovation.

Professional Organizations and Conferences

1. Biofabrication (IUPESM affiliated group)
   - Highlights innovations in ECM-based structures through biofabrication.
2. Society for Biomaterials—Emerging Technologies Section
   - Showcases future-forward biomimetic research.
3. IEEE EMBS Conference on Biomedical and Health Informatics
   - Tracks convergence of ECM mimicry with AI, sensors, and smart health systems.

Further Readings

1. "Engineering in Translational Medicine" by Weibo Cai and Younan Xia
   - Future-facing perspectives on ECM technologies in clinical innovation.
2. "Next-Generation Biomaterials for Bone and Cartilage Regeneration" by D. Katti and A. Katti
   - Includes chapters on ECM-mimetic scaffolds driven by nanotech and AI.
3. "Artificial Intelligence in Precision Medicine" by Debmalya Barh
   - Explores AI-based predictive modeling in ECM therapeutic design.

Online Courses

1. FutureLearn—"Biofabrication and Bioink Design" by Utrecht University
   - Learn to design next-gen ECM-based materials using bioinks and 3D printing.

2. Coursera—"Artificial Intelligence for Medicine" by DeepLearning.AI
   - Applies to future AI-integrated ECM therapeutic planning and delivery.

Chapter 11: Conclusion

This final chapter consolidates insights from the previous chapters, emphasizing the transformative role of ECM biomimicry in diagnostics, therapeutics, and personalized medicine. It reflects on interdisciplinary innovations, discusses translational challenges, and underscores the necessity for collaborative approaches in future ECM-based theranostics.

Textbooks

1. "Principles of Regenerative Medicine" by Anthony Atala, Robert Lanza, James A. Thomson, and Robert Nerem
   - Summarizes regenerative approaches including ECM-based strategies for clinical translation.
2. "Biomimetic Biomaterials: Structure and Function" by Andrew Atkinson
   - Highlights the structure–function relationship of ECM-mimetic materials with concluding insights into clinical applicability.
3. "Biomedical Engineering and Design Handbook" by Myer Kutz
   - Contains practical summaries on biomaterial applications, technology design, and interdisciplinary impact.

Journals

1. Regenerative Medicine
   - Includes perspectives and closing commentaries on ECM's future in clinical settings.
2. BioDesign and Manufacturing
   - Offers insight into the end-to-end pathway of biomimetic design toward real-world applications.
3. Frontiers in Bioengineering and Biotechnology—Translational Bioengineering Section
   - Features concluding case reports and opinion articles on ECM translational research.

Online Resources

1. NIH 3D Bioprinting and ECM Resources
    - Provides conclusive reports on federally supported ECM translational initiatives.
2. Nature Outlook: Regenerative Medicine
    - Offers expert commentary and outlooks consolidating the ECM and biomimetic innovations landscape.

Research Papers

1. "Extracellular matrix-based therapeutics: Translational challenges and opportunities" by Ferreira SA et al., *NPJ Regenerative Medicine, 2022*
    - Discusses final-stage implementation issues for ECM-based products.
2. "Towards clinical translation of ECM-mimetic scaffolds: Lessons learned and future directions" by Yu C et al., *Bioactive Materials, 2021*
    - A comprehensive review reflecting on ECM translation and concluding lessons.
3. "Bridging design and application: ECM mimics in the clinic" by Schmidt CE et al., *Nature Reviews Materials, 2020*
    - End-stage review connecting material development with clinical practice.

Professional Organizations and Conferences

1. TERMIS (Tissue Engineering and Regenerative Medicine International Society)
    - Concludes the biomimetic innovation pipeline from lab to clinical translation.
2. World Biomaterials Congress
    - Offers summarizing reports on global ECM-related advancements.
3. American Institute for Medical and Biological Engineering (AIMBE)
    - Bridges policy, innovation, and future of biomimetic materials.

Further Readings

1. "Biodesign: The Process of Innovating Medical Technologies" by Zenios, Makower, Yock
    - Closing insights into productization and real-world ECM-inspired innovations.
2. "Translational Regenerative Medicine" by Anthony Atala and Julie Allickson

- Emphasizes the translation process of ECM research into therapies.

3. "Tissue Engineering Using Ceramics and Polymers" by Aldo R. Boccaccini and Peter X. Ma
    - Final chapters discuss biomimicry impact on future medical engineering.

Online Courses

1. edX—"The Science of Stem Cells and Regenerative Medicine" by Harvard University
    - Offers a concluding synthesis of regenerative strategies including ECM focus.
2. Coursera—"Translational Science" by the University of Michigan
    - End-to-end training from bench research to bedside implementation.

Answers

Chapter 1

Introduction to Extracellular Matrix (ECM) Dynamics

Exercise 1

1. Extracellular matrix
2. Transmembrane receptors
3. Fibronectin
4. Cytoskeleton
5. Matrix metalloproteinases

Exercise 2

1. Matrix metalloproteinases
2. Extracellular matrix
3. Tissue inhibitors of metalloproteinases
4. Fibroblasts
5. Collagen

Exercise 3

1. C) Structural support and regulation of cell behavior
2. C) Cytoplasm
3. B) Through signaling pathways initiated by ECM Components
4. C) Matrix metalloproteinases
5. B) Collagen
6. B) It serves as a scaffold for tissue repair
7. B) Elastin

D. Suhag, *Extracellular Matrix biomimicking for Theranostics*, Biomedical Materials
for Multi-functional Applications, https://doi.org/10.1007/978-981-95-7900-6

Chapter 2

Cellular Interactions with the ECM

Exercise 1

1. Integrins
2. Fibronectin
3. MAPK/ERK
4. Elastin
5. Matrix metalloproteinases (MMPs)

Exercise 2

1. Matrix metalloproteinases (MMPs)
2. Fibrosis
3. Homeostasis
4. Fibronectin
5. ECM turnover

Exercise 3

1. C) Structural support and signaling
2. B) Integrins
3. D) Osteoblasts (bone cells)
4. B) Collagen
5. A) They degrade ECM components, facilitating tissue remodeling.
6. B) Facilitating cell adhesion and migration
7. B) MAPK/ERK pathway
8. B) It can be remodeled to facilitate tumor invasion and metastasis.
9. B) Laminin

Chapter 3

Mechanisms of ECM Remodeling

Exercise 1

1. Matrix metalloproteinases
2. Collagen
3. Cell migration
4. Tissue Inhibitors of Metalloproteinases (TIMPs)
5. Myofibroblasts

Exercise 2

1. Mechanotransduction
2. Collagen
3. Proteolytic enzymes
4. Traction forces

5. Mechanical

Exercise 3

1. b) Matrix metalloproteinases (MMPs)
2. c) Wound healing
3. b) Fibroblasts
4. b) Regulate cell-ECM adhesion
5. b) MMPs and TIMPs (Tissue Inhibitors of Metalloproteinases)
6. b) Mechanotransduction
7. c) Actin
8. c) Regulating both ECM synthesis and degradation
9. a) Metastasis
10. c) Stiffness

Chapter 4

ECM in Health and Disease

Exercise 1

1. Adhesion
2. Matrix metalloproteinases (MMPs)
3. Laminins
4. Cell Signaling
5. Glycosaminoglycans (GAGs)

Exercise 2

1. Cell adhesion, migration, and differentiation
2. valvular heart diseases
3. cell signaling, proliferation, and survival
4. contractility
5. Reducing fibrosis and improving tissue regeneration

Exercise 3

1. Collagen and fibrosis
2. ECM degradation and remodeling
3. Deposition
4. Tissue destruction
5. Inflammation and disease progression

Exercise 4

1. Structural support
2. Cell adhesion and migration
3. Neuronal degeneration
4. Hyaluronic acid
5. Microglia

Chapter 5

Biomimicry of ECM in Medical Applications

Exercise 1

1. Biochemical, mechanical
2. 3D
3. Porosity
4. Growth
5. Stiffness, flexibility

Exercise 2

1. Stiffness
2. Growth factors
3. Elasticity
4. Protein
5. Biochemical

Exercise 3

1. a) Structural support
2. a) Collagen
3. b) By replicating natural healing environments
4. b) By mimicking biochemical cues of natural ECM
5. b) Porosity and mechanical strength
6. b) Achieving the right biomechanical and biochemical properties
7. b) By providing a 3D network for cell growth
8. b) Fibroblasts
9. a) It regulates cell differentiation and proliferation

Chapter 6

ECM in Diagnostic Innovations

Exercise 1

1. Collagen, elastin
2. Cancer, fibrosis
3. Fibronectin
4. Matrix metalloproteinases (MMPs)
5. Hyaluronic acid

Exercise 2

1. Fluids, molecules
2. 3D
3. Biochemical
4. Fast, cost-effective

5. Sensors

Exercise 3

1. d) All of the above
2. a) Collagen
3. d) Both b and c
4. d) All of the above
5. c) Mimic the tissue microenvironment
6. c) Collagen
7. b) By altering mechanical properties
8. d) All of the above
9. b) They provide information on tissue health
10. d) All of the above

Chapter 7

Therapeutic Applications of ECM Mimicry

Exercise 1

1. Structure, composition
2. Adhesion, proliferation
3. Natural
4. Fibronectin
5. Migration, proliferation

Exercise 2

1. Microenvironment
2. Collagen
3. Cell
4. Immune
5. Cosmetic

Exercise 3

1. b) To replicate biological functions
2. b) Collagen
3. b) Support cell adhesion and proliferation
4. b) Dermatology
5. b) Promote cell migration and differentiation
6. b) Texture
7. d) All of the above
8. b) Ensuring biocompatibility
9. d) All of the above
10. c) Stem cells

Chapter 8

Advanced Technologies in ECM Mimicry

Exercise 1

1. Chitosan, alginate
2. Polylactic acid (PLA)
3. Growth factors
4. Hydrated
5. Mechanical, biological

Exercise 2

1. Water
2. Water
3. 3D
4. Composition
5. Extracellular matrix (ECM)

Exercise 3

1. b) 3D Bioprinting
2. b) Modifying the surface properties at the nanoscale
3. b) Rheology
4. b) Removing cellular components from tissue
5. b) Promote cellular responses
6. b) It creates fibrous structures for scaffolds
7. b) Confocal microscopy
8. b) Simulate the mechanical and biochemical properties of the ECM
9. d) All of the above
10. b) Gene editing

Chapter 9

Clinical Case Studies & Trials

Exercise 1

1. Supportive
2. Chemotherapy
3. Scaffolds
4. Invasion
5. Cytokines

Exercise 2

1. Musculoskeletal
2. Biodegradable
3. Growth

4. Natural
5. Stability

Exercise 3

1. b) Providing structural support
2. b) Collagen
3. d) All of the above
4. b) Collagen
5. d) All of the above
6. d) Both a and b
7. c) Laminin
8. b) Promote axonal growth
9. d) All of the above
10. d) Both a and b

Chapter 10

Future Perspectives & Innovations

Exercise 1

1. Synthetic
2. Bioprinting
3. Big data
4. Imaging
5. Immune

Exercise 2

1. 3D printing
2. Additive manufacturing
3. Electrospinning
4. Bioink
5. Cell behavior

Exercise 3

1. b) To restore or improve tissue function
2. a) Collagen
3. b) Decellularized tissues
4. a) Nanotechnology
5. b) Promoting specific cellular responses
6. a) Mimic the properties of natural ECM
7. c) Tunable degradation rates
8. a) Enzymes
9. d) All of the above
10. d) All of the above

Chapter 11

Conclusion

Exercise 1

1. Biochemical
2. Structural
3. Regeneration
4. Bioavailability
5. Compatibility

Exercise 2

1. a) MRI
2. b) To replicate the structure and function of natural ECM
3. d) Both b and c
4. a) Analyze complex datasets
5. c) Patient consent for experimental procedures
6. d) All of the above
7. b) Polycaprolactone (PCL)
8. a) Combining MRI with AI for better analysis

Glossary

Apoptosis Programmed form of cell death that occurs in a controlled manner, allowing the body to eliminate damaged, unwanted, or dysfunctional cells without causing inflammation. It involves a series of biochemical events leading to characteristic cell changes, such as cell shrinkage, chromatin condensation, and fragmentation into apoptotic bodies. Apoptosis is essential for normal development, tissue homeostasis, and the immune response, playing a crucial role in processes like embryogenesis and the removal of infected or cancerous cells.

Atherosclerosis Cardiovascular disease characterized by the buildup of fatty deposits, cholesterol, and other substances in the walls of arteries, leading to the formation of plaques. In the context of the extracellular matrix (ECM).

Autophosphorylation Process where a kinase enzyme phosphorylates itself, usually on specific tyrosine, serine, or threonine residues. This self-phosphorylation activates the kinase, which in turn can regulate various cellular processes related to the ECM, such as cell adhesion, migration, and signal transduction. For example, receptors like integrins, which interact with ECM components, can trigger autophosphorylation events to modulate cell-ECM interactions and tissue remodeling.

Cardiomyocytes Specialized muscle cells of the heart responsible for its contraction and overall function. They make up the myocardium (the heart muscle) and are essential for pumping blood throughout the body. In the context of the extracellular matrix (ECM), cardiomyocytes interact with the ECM to maintain heart structure and function. The ECM provides mechanical support, facilitates communication between cells, and plays a role in regulating cardiomyocyte behavior, such as growth, repair, and response to stress. Changes in the ECM can influence heart health and contribute to conditions like heart failure and fibrosis.

Decorin A small leucine-rich proteoglycan found in the extracellular matrix (ECM) that plays a key role in collagen organization and regulation. It consists of a core protein with glycosaminoglycan (GAG) side chains and is involved in various functions.

D. Suhag, *Extracellular Matrix biomimicking for Theranostics*, Biomedical Materials for Multi-functional Applications, https://doi.org/10.1007/978-981-95-7900-6

Desmoplasia The formation of an excessive fibrous or connective tissue response in an organ or tissue, often associated with the presence of tumors. In the context of the extracellular matrix (ECM), desmoplasia involves the deposition of collagen and other ECM components, leading to increased stiffness and density of the tissue. This process can be a response to tumor growth, inflammation, or injury, and is commonly observed in certain types of cancer, where it can contribute to tumor progression and create a barrier to effective treatment.

Echogenicity The ability of a tissue or substance to reflect ultrasound waves, which is used in imaging techniques like ultrasound. In the context of the extracellular matrix (ECM), echogenicity can provide valuable information about the composition, structure, and pathology of tissues.

Fibrogenic The ability to stimulate or promote the formation of fibrous tissue, particularly collagen, in the body. This term is often used in the context of wound healing, tissue repair, and fibrosis, where certain cells, cytokines, or growth factors trigger the proliferation and activation of fibroblasts, leading to increased production of extracellular matrix components. Fibrogenic processes can be beneficial in normal healing but may also contribute to pathological conditions characterized by excessive fibrosis, such as in liver cirrhosis or pulmonary fibrosis.

Glycosaminoglycans Glycosaminoglycans (GAGs) are long, unbranched polysaccharides composed of repeating disaccharide units. They are key components of the extracellular matrix (ECM) and play crucial roles in maintaining tissue hydration, providing structural support, and facilitating cell signaling. GAGs, such as hyaluronic acid, chondroitin sulfate, and heparan sulfate, help regulate the ECM's physical properties, influence cell adhesion and migration, and participate in processes like wound healing and tissue repair.

Hemidesmosome Specialized structures that anchor epithelial cells to the extracellular matrix (ECM), specifically to the underlying basement membrane. They connect the cell's internal cytoskeleton to ECM proteins, like laminin, ensuring tissue stability and mechanical strength, especially in skin and other epithelial layers.

Hyaluronidase Enzyme that breaks down hyaluronic acid, a major component of the extracellular matrix (ECM). By hydrolyzing hyaluronic acid, hyaluronidase facilitates the degradation of the ECM, promoting tissue remodeling and influencing cell migration and proliferation. This enzyme plays a significant role in various physiological processes, such as wound healing and inflammation, and is also used clinically to enhance the dispersion of drugs and improve the absorption of fluids in tissues.

Immunohistochemistrya Technique used to detect specific ECM proteins or molecules within tissue samples. By using antibodies that bind to ECM components, IHC helps visualize the distribution, localization, and expression levels of proteins like collagen, fibronectin, or laminin, providing insights into tissue structure, disease progression, and ECM-related pathologies.

Microencapsulation A process in which tiny particles or droplets are coated with a protective layer to form microcapsules. In the context of the extracellular matrix (ECM), microencapsulation can be used to deliver cells, growth factors, or therapeutic agents in a controlled manner to enhance tissue engineering and regenerative medicine applications.

Myofibroblasts Play a key role in wound healing and tissue repair. They have properties of both fibroblasts and smooth muscle cells, allowing them to produce ECM components (like collagen) and contract the matrix to close wounds. Myofibroblasts are essential for tissue remodeling but can also contribute to fibrosis when they accumulate excessively, leading to scarring and tissue stiffness.

Orchestrating The coordinated regulation and interaction of various components, such as proteins, enzymes, and cells, to maintain tissue structure and function. This involves controlling processes like cell signaling, adhesion, migration, and matrix remodeling, ensuring that the ECM supports normal tissue development, repair, and homeostasis.

Parenchymal cells Functional cells of an organ that perform its specific physiological tasks. They are distinct from stromal cells, which provide structural support and form the connective tissue framework. For example, in the liver, hepatocytes are the parenchymal cells responsible for metabolic functions, while in the lungs, alveolar cells perform gas exchange. Parenchymal cells are essential for the organ's overall function and are typically involved in processes like secretion, absorption, and metabolism.

Pathophysiological The abnormal processes and changes in the ECM that contribute to disease development and progression. The ECM is a dynamic network of proteins and other molecules that provides structural support to tissues and regulates cellular behavior, such as growth, migration, and differentiation. When the ECM is disrupted or altered, it can contribute to various pathologies.

Pemphigoid A group of autoimmune blistering disorders where the immune system attacks components of the basement membrane in the skin and mucous membranes, part of the extracellular matrix (ECM). This leads to separation between the epidermis and dermis, causing blisters and inflammation. The most common form is bullous pemphigoid.

Spatiotemporal The combination of spatial (space-related) and temporal (time-related) aspects in the study of biological processes. In the context of the extracellular matrix (ECM), spatiotemporal analysis focuses on how the composition, organization, and functions of the ECM change over time and in different locations within tissues.

Syndecans Transmembrane proteoglycans that play a key role in the extracellular matrix (ECM) by mediating cell-ECM interactions. They bind to ECM components like fibronectin, collagen, and growth factors, regulating cell adhesion, migration, and signaling pathways. Syndecans help modulate wound healing, tissue repair, and inflammation within the ECM.

Thrombospondins Thrombospondins (TSPs) are a family of multifunctional glycoproteins that play important roles in the extracellular matrix (ECM). They are involved in regulating cell-to-cell and cell-to-matrix interactions. Thrombospondins contribute to processes such as tissue remodeling, wound healing, angiogenesis, and the maintenance of tissue integrity. They act as mediators of cellular adhesion, influence the behavior of endothelial cells and fibroblasts, and modulate the activity of growth factors.

Transcriptomics Study of the complete set of RNA transcripts produced by the genome of a cell or organism at a specific time under specific conditions. In the context of the extracellular matrix (ECM), transcriptomics focuses on understanding how gene expression related to ECM components (like collagens, proteoglycans, and glycoproteins) is regulated in different physiological and pathological states.

Tropoelastin The soluble precursor to elastin, a key protein in the extracellular matrix (ECM) that provides elasticity to tissues such as skin, lungs, and blood vessels. Tropoelastin is secreted by cells and assembles into elastin fibers, allowing tissues to stretch and return to their original shape. It's essential for maintaining the flexibility and resilience of tissues.

Vasculogenesis Process by which new blood vessels form from endothelial precursor cells during embryonic development. In the context of the extracellular matrix (ECM), vasculogenesis involves the interaction of these precursor cells with the ECM components, which provide the necessary signals and structural support for blood vessel formation.

Versican Large extracellular matrix (ECM) proteoglycan that plays a crucial role in cell adhesion, migration, and proliferation. It consists of a core protein with glycosaminoglycan (GAG) side chains, which contribute to its ability to bind water and create a gel-like matrix. Versican is involved in tissue organization, development, and repair, and it influences various biological processes such as wound healing, inflammation, and tumor progression. Its presence in the ECM helps regulate the behavior of cells in the surrounding environment.